AF2S2281

Operator Theory: Advances and
Applications
Vol. 126

Editor:
I. Gohberg

Editorial Office:
School of Mathematical
Sciences
Tel Aviv University
Ramat Aviv, Israel

Editorial Board:
J. Arazy (Haifa)
A. Atzmon (Tel Aviv)
J. A. Ball (Blacksburg)
A. Ben-Artzi (Tel Aviv)
H. Bercovici (Bloomington)
A. Böttcher (Chemnitz)
K. Clancey (Athens, USA)
L. A. Coburn (Buffalo)
K. R. Davidson (Waterloo, Ontario)
R. G. Douglas (Stony Brook)
H. Dym (Rehovot)
A. Dynin (Columbus)
P. A. Fillmore (Halifax)
P. A. Fuhrmann (Beer Sheva)
S. Goldberg (College Park)
B. Gramsch (Mainz)
G. Heinig (Chemnitz)
J. A. Helton (La Jolla)
M.A. Kaashoek (Amsterdam)

H.G. Kaper (Argonne)
S.T. Kuroda (Tokyo)
P. Lancaster (Calgary)
L.E. Lerer (Haifa)
E. Meister (Darmstadt)
B. Mityagin (Columbus)
V. V. Peller (Manhattan, Kansas)
J. D. Pincus (Stony Brook)
M. Rosenblum (Charlottesville)
J. Rovnyak (Charlottesville)
D. E. Sarason (Berkeley)
H. Upmeier (Marburg)
S. M. Verduyn-Lunel (Amsterdam)
D. Voiculescu (Berkeley)
H. Widom (Santa Cruz)
D. Xia (Nashville)
D. Yafaev (Rennes)

Honorary and Advisory
Editorial Board:
C. Foias (Bloomington)
P. R. Halmos (Santa Clara)
T. Kailath (Stanford)
P. D. Lax (New York)
M. S. Livsic (Beer Sheva)

Partial Differential Equations and Spectral Theory

PDE2000 Conference in Clausthal, Germany

Michael Demuth
Bert-Wolfgang Schulze
Editors

Springer Basel AG

Editors:

Michael Demuth
Institut für Mathematik
TU Clausthal
Erzstr. 1
38678 Clausthal-Zellerfeld
Germany
e-mail: demuth@math.tu-clausthal.de

Bert-Wolfgang Schulze
Institut für Mathematik
Universität Potsdam
Am Neuen Palais 10
14469 Potsdam
Germany
e-mail: schulze@math.uni-potsdam.de

2000 Mathematics Subject Classification 35-99, 46-99, 47-99, 81-99

A CIP catalogue record for this book is available from the Library of Congress, Washington D.C., USA

Deutsche Bibliothek Cataloging-in-Publication Data

Partial differential equations and spectral theory : PDE2000 Conference in Clausthal,
Germany / Michael Demuth ; Bert-Wolfgang Schulze, ed.. - Basel ; Boston ; Berlin :
Birkhäuser, 2001
   (Operator theory ; Vol. 126)
   ISBN 978-3-0348-9483-8   ISBN 978-3-0348-8231-6 (eBook)
   DOI 10.1007/978-3-0348-8231-6

This work is subject to copyright. All rights are reserved, whether the whole or part of the material is concerned,
specifically the rights of translation, reprinting, re-use of illustrations, recitation, broadcasting, reproduction on microfilm
or in other ways, and storage in data banks. For any kind of use permission of the copyright owner must be obtained.

© 2001 Springer Basel AG
Originally published by Birkhäuser Verlag in 2001
Softcover reprint of the hardcover 1st edition 2001
Printed on acid-free paper produced from chlorine-free pulp. TCF ∞
Cover design: Heinz Hiltbrunner, Basel

ISBN 978-3-0348-9483-8

Preface

This volume contains the proceedings of "PDE 2000", the international conference on partial differential equations held July 24 - 28, 2000, in Clausthal. The conference took place during the EXPO 2000 and was sponsored by the Land Niedersachsen, the Deutsche Forschungsgemeinschaft, the Bergstadt Clausthal-Zellerfeld and the Kreissparkasse Clausthal-Zellerfeld.

This conference continues a series: Ludwigfelde 1976, Reinhardsbrunn 1985, Holzhau 1988, Breitenbrunn 1990, Lambrecht 1991 (proceedings in Operator Theory: Advances and Applications, Vol. 57, Birkhäuser Verlag 1992), Potsdam 1992 and 1993, Holzhau 1994 (proceedings in Operator Theory: Advances and Applications, Vol. 78, Birkhäuser Verlag 1995), Caputh 1995 and Potsdam 1996 (proceedings in Mathematical Research, Vol. 100, Akademie Verlag 1997).

The intention of the organizers was to bring together specialists from different areas of modern analysis, mathematical physics and geometry, to discuss not only the recent progress in their own fields but also the interaction between these fields. The special topics of the conference were spectral and scattering theory, semiclassical and asymptotic analysis, pseudodifferential operators and their relation to geometry, as well as partial differential operators and their connection to stochastic analysis and to the theory of semigroups.

The scientific advisory board of the conference in Clausthal consisted of M. Ben-Artzi (Jerusalem), Chen Hua (Peking), M. Demuth (Clausthal), T. Ichinose (Kanazawa), L. Rodino (Turin), B.-W. Schulze (Potsdam) and J. Sjöstrand (Paris).

The organizers would like to thank Mrs. S. Freiberg and Dr. A. Noll for coordinating many technical details of the conference together with Dr. W. Renger, Dr. E. Giere, Mr. M. Baro, Mr. S. Djawadi and the staff of the Institute of Mathematics in Clausthal. Moreover, we thank Mr. M. Baro for copy-editing the present proceedings.

Michael Demuth
Clausthal

Bert-Wolfgang Schulze
Potsdam

Contents

Operator Theory:
Advances and Applications, Vol. 126
© 2001 Birkhäuser Verlag Basel/Switzerland

Instability in the Spectral and the Fredholm Properties of an Infinite Dimensional Dirac Operator on the Abstract Boson-Fermion Fock Space

ASAO ARAI *

Abstract. A perturbed Dirac operator $Q(\alpha)$ on the abstract Boson-Fermion Fock space is considered, where $\alpha \in \mathbb{C}$ is a perturbation (coupling) parameter and the unperturbed operator $Q(0)$ is taken to be a free infinite dimensional Dirac operator introduced by the author (A. Arai, J. Funct. Anal. **105** (1992), 342–408). The following results are reported: (i) Under some conditions, the kernel of $Q(\alpha)$ is one dimensional for all $\alpha \neq \alpha_0$ with some $\alpha_0 \neq 0$ and degenerate at $\alpha = \alpha_0$, while, under another condition, the kernel of $Q(\alpha)$ is one dimensional for all $\alpha \in \mathbb{C}$. (ii) There are cases where, for all sufficiently large $|\alpha|$ with $\alpha < 0$, $Q(\alpha)$ has infinitely many non-zero eigenvalues even if $Q(0)$ has no non-zero eigenvalues. This is a strong coupling effect. (iii) Fredholm property of $Q(\alpha)$ also depends on the coupling parameter α.

1   Introduction

Let $\mathcal{H}$ and $\mathcal{K}$ be separable complex Hilbert spaces. Then the abstract Boson-Fermion Fock space $\mathcal{F}(\mathcal{H}, \mathcal{K})$ associated with the pair $\langle \mathcal{H}, \mathcal{K} \rangle$ is defined by

$$\mathcal{F}(\mathcal{H}, \mathcal{K}) := \mathcal{F}_{\mathrm{b}}(\mathcal{H}) \otimes \mathcal{F}_{\mathrm{f}}(\mathcal{K}), \tag{1.1}$$

where $\mathcal{F}_{\mathrm{b}}(\mathcal{H}) := \oplus_{n=0}^{\infty} \otimes_{\mathrm{s}}^{n} \mathcal{H}$ is the Boson Fock space over $\mathcal{H}$ ($\otimes_{\mathrm{s}}^{n}\mathcal{H}$ denotes the n-fold symmetric tensor product Hilbert space of $\mathcal{H}$; $\otimes_{\mathrm{s}}^{0}\mathcal{H} := \mathbb{C}$) and $\mathcal{F}_{\mathrm{f}}(\mathcal{K}) := \oplus_{p=0}^{\infty} \wedge^{p}\mathcal{K}$ is the Fermion Fock space over $\mathcal{K}$ ($\wedge^{p}\mathcal{K}$ denotes the p-fold anti-symmetric tensor product Hilbert space of $\mathcal{K}$; $\wedge^{0}\mathcal{K} := \mathbb{C}$).

Let $\mathsf{C}(\mathcal{H}, \mathcal{K})$ be the set of densely defined closed linear operators from $\mathcal{H}$ to $\mathcal{K}$. Then, for each $A \in \mathsf{C}(\mathcal{H}, \mathcal{K})$, one can define a Dirac-type operator Q_A on $\mathcal{F}(\mathcal{H}, \mathcal{K})$ [Ar92] (for the definition of Q_A, Section 2 below). The operator Q_A is an infinite dimensional version of free Dirac operators on finite dimensional spaces.

In [Ar92] some fundamental properties of Q_A were established. Moreover, a perturbed Dirac operator of the form $Q_A(V) := Q_A + V$ was considered in view of

* Supported by the Grant-In-Aid No.11440036 for Scientific Research from the Ministry of Education, Science, Sports and Culture, Japan.

index theory, where V is a symmetric operator on $\mathcal{F}(\mathcal{H}, \mathcal{K})$, and a functional integral representation for the index of $Q_A(V)$ restricted to a subspace, called the "bosonic subspace", was derived (for related aspects and further developments, see [Ar89], [Ar91], [Ar93a], [Ar93b], [Ar94], [Ar96], [Ar97], [AM91], [AM93]). It still remains, however, as an important problem, to investigate spectral properties of $Q_A(V)$. In this paper, as a first step towards this direction, we present a perturbation of Q_A which is not of the form V considered in [Ar92] and rather simple, but gives rise to interesting *nonperturbative instability phenomena* in spectral and Fredholm properties. For proofs on the results reported in this paper, see [Ar00].

2   A Class of Perturbed Dirac Operators

We denote by $a(f)$ ($f \in \mathcal{H}$) and $b(u)$ ($u \in \mathcal{K}$) the annihilation operators on $\mathcal{F}_{\mathrm{b}}(\mathcal{H})$ and on $\mathcal{F}_{\mathrm{f}}(\mathcal{K})$ respectively (e.g., [BR97, §5.2]). Let $\Omega_{\mathrm{b}} := \{1, 0, 0, \cdots\} \in \mathcal{F}_{\mathrm{b}}(\mathcal{H})$ (resp. $\Omega_{\mathrm{f}} := \{1, 0, 0, \cdots\} \in \mathcal{F}_{\mathrm{f}}(\mathcal{K})$) be the Fock vacuum in $\mathcal{F}_{\mathrm{b}}(\mathcal{H})$ (resp. $\mathcal{F}_{\mathrm{f}}(\mathcal{K})$). Let $A \in \mathsf{C}(\mathcal{H}, \mathcal{K})$ and

$$\mathcal{D}_A^\infty := \mathcal{L}\left\{ a(f_1)^* \cdots a(f_n)^* \Omega_{\mathrm{b}} \otimes b(u_1)^* \cdots b(u_p)^* \Omega_{\mathrm{f}} \,\middle|\, n, p \geq 0, f_j \in C^\infty(A^*A), \right.$$

$$\left. j = 1, \cdots, n, \ u_k \in C^\infty(AA^*), k = 1, \cdots, p \right\}, \quad (2.1)$$

where $\mathcal{L}\{\cdots\}$ means the subspace algebraically spanned by the vectors in the set $\{\cdots\}$ and $C^\infty(T) := \cap_{n=1}^\infty D(T^n)$ for a linear operator T on a Hilbert space ($D(T^n)$ denotes the domain of T^n) . It follows that $\mathcal{D}_A^\infty$ is dense in $\mathcal{F}(\mathcal{H}, \mathcal{K})$.

Let $\{e_n\}_{n=1}^\infty$ be a complete orthonormal system of $\mathcal{K}$ such that $e_n \in D(A^*), n \in \mathbb{N}$. Then it is shown that there is a unique densely defined closed linear operator d_A on $\mathcal{F}(\mathcal{H}, \mathcal{K})$ such that (i) for all $\Psi \in \mathcal{D}_A^\infty$

$$d_A \Psi = \sum_{n=1}^\infty a(A^* e_n) b(e_n)^* \Psi \quad (2.2)$$

independently of the choice of $\{e_n\}_{n=1}^\infty$ and (ii) $\mathcal{D}_A^\infty$ is a core of d_A. It follows that $d_A^2 = 0$. The operator may be regarded as an infinite dimensional version of finite dimensional exterior differential operators. *A free Dirac operator* on $\mathcal{F}(\mathcal{H}, \mathcal{K})$ is defined by

$$Q_A = d_A + d_A^*. \quad (2.3)$$

We have an orthogonal decomposition

$$\mathcal{F}(\mathcal{H}, \mathcal{K}) = \mathcal{F}_+(\mathcal{H}, \mathcal{K}) \oplus \mathcal{F}_-(\mathcal{H}, \mathcal{K}) \quad (2.4)$$

with

$$\mathcal{F}_+(\mathcal{H},\mathcal{K}) := \oplus_{p=0}^{\infty} \mathcal{F}_{\mathrm{b}}(\mathcal{H}) \otimes \wedge^{2p}(\mathcal{K}), \quad \mathcal{F}_-(\mathcal{H},\mathcal{K}) := \oplus_{p=0}^{\infty} \mathcal{F}_{\mathrm{b}}(\mathcal{H}) \otimes \wedge^{2p+1}(\mathcal{K}). \quad (2.5)$$

Let P_+ and P_- be the orthogonal projections onto $\mathcal{F}_+(\mathcal{H},\mathcal{K})$ and $\mathcal{F}_-(\mathcal{H},\mathcal{K})$ respectively and define

$$\Gamma := P_+ - P_-. \quad (2.6)$$

For a self-adjoint operator S on $\mathcal{H}$ (resp. $\mathcal{K}$), $d\Gamma_{\mathrm{b}}(S)$ (resp. $d\Gamma_{\mathrm{f}}(S)$) denotes the second quantization of S in the Boson Fock space $\mathcal{F}_{\mathrm{b}}(\mathcal{H})$ (resp. the Fermion Fock space $\mathcal{F}_{\mathrm{f}}(\mathcal{H})$) [BR97, §5.2].

Basic properties of Q_A are as follows [Ar92]:

(i) The operator Q_A is self-adjoint and essentially self-adjoint on $\mathcal{D}_A^{\infty}$.
(ii) The operator Γ leaves $D(Q_A)$ invariant and $\Gamma Q_A + Q_A \Gamma = 0$ on $D(Q_A)$.
(iii) The following operator equations hold :

$$Q_A^2 = d_A^* d_A + d_A d_A^* = d\Gamma_{\mathrm{b}}(A^*A) \otimes I + I \otimes d\Gamma_{\mathrm{f}}(AA^*). \quad (2.7)$$

where I denotes the identity.

We perturb Q_A through a perturbation of d_A. Let $g \in D(A)$, $g \neq 0$ and $v \in D(A^*), v \neq 0$ and define

$$d(\alpha) := d_A + \alpha a(g) \otimes b(v)^*. \quad (2.8)$$

where $\alpha \in \mathbb{C}$ is a coupling constant. It is easy to see that $D(d(\alpha)) \supset \mathcal{D}_A^{\infty}$ and $d(\alpha)|\mathcal{D}_A^{\infty}$ is closable. Let

$$\bar{d}(\alpha) := \overline{d(\alpha)|\mathcal{D}_A^{\infty}}, \quad (2.9)$$

the closure of $d(\alpha)|\mathcal{D}_A^{\infty}$, and

$$Q(\alpha) := \bar{d}(\alpha) + \bar{d}(\alpha)^*. \quad (2.10)$$

This is the perturbed Dirac operator considered in this paper.

3 Results

Theorem 3.1. (i) *For all $\alpha \in \mathbb{C}$, $Q(\alpha)$ is self-adjoint, and essentially self-adjoint on $\mathcal{D}_A^{\infty}$ with $Q(\alpha) = \overline{Q_A + V_{g,v}}$, where $V_{g,v} := \alpha a(g) \otimes b(v)^* + \alpha^* a(g)^* \otimes b(v)$. Moreover, Γ leaves $D(Q(\alpha))$ invariant and $\Gamma Q(\alpha) + Q(\alpha)\Gamma = 0$ on $D(Q(\alpha))$.*
(ii) *Suppose that A is injective and $g \in D(|A|^{-1})$. Then, for all $|\alpha| < (\|v\| \, \| |A|^{-1}g\|)^{-1}$, $D(Q(\alpha)) = D(Q_A)$.*
(iii) *For all $z \in \mathbb{C} \setminus \mathbb{R}$, $(Q(\alpha) - z)^{-1}$ is strongly continuous in $\alpha \in \mathbb{C}$.*

To describe the spectral properties of $Q(\alpha)$, we introduce a bounded linear operator $T_{g,v}$ from $\mathcal{H}$ to $\mathcal{K}$ by

$$T_{g,v}f := (g, f)v, \quad f \in \mathcal{H}, \tag{3.1}$$

where $(\cdot, \cdot)$ denotes inner product, and define

$$A(\alpha) := A + \alpha T_{g,v}. \tag{3.2}$$

For a linear operator T on a Hilbert space, we denote by $\sigma(T)$ (resp. $\sigma_{\mathrm{p}}(T)$) the spectrum (resp. point spectrum) of T. A general feature of the spectra of $Q(\alpha)$ is given in the following theorem:

Theorem 3.2. *For all $\alpha \in \mathbb{C}$, $\sigma(Q(\alpha))$ and $\sigma_{\mathrm{p}}(Q(\alpha))$ are symmetric with respect to the origin and*

$$\sigma(Q(\alpha)) = \{0\} \bigcup \overline{\left(\bigcup_{n=1}^{\infty} \left\{ \pm\sqrt{\sum_{j=1}^{n} \lambda_j} \,\middle|\, \lambda_j \in \sigma(A(\alpha)^*A(\alpha)), j = 1, \cdots, n \right\} \right)},$$
$$\tag{3.3}$$

$$\sigma_{\mathrm{p}}(Q(\alpha)) = \{0\} \bigcup \left(\bigcup_{n=1}^{\infty} \left\{ \pm\sqrt{\sum_{j=1}^{n} \lambda_j} \,\middle|\, \lambda_j \in \sigma_{\mathrm{p}}(A(\alpha)^*A(\alpha)), j = 1, \cdots, n \right\} \right),$$
$$\tag{3.4}$$

with

$$\dim \ker(Q(\alpha) - \lambda) = \dim \ker(Q(\alpha) + \lambda), \quad \lambda \in \sigma_{\mathrm{p}}(Q(\alpha)). \tag{3.5}$$

This theorem shows that the spectrum and the point spectrum of $Q(\alpha)$ are completely determined from those of $A(\alpha)^*A(\alpha)$.

To state properties of the kernel of $Q(\alpha)$, we introduce the following conditions on $\{A, g, v\}$:

(C.1) A is injective and $v \in D(A^{-1})$ with $(g, A^{-1}v) \neq 0$. In this case we introduce a constant

$$\alpha_0 := -\frac{1}{(g, A^{-1}v)}. \tag{3.6}$$

(C.2) A^* is injective and $g \in D(A^{*-1})$ with $(A^{*-1}g, v) \neq 0$. In this case we introduce a constant

$$\beta_0 := -\frac{1}{(A^{*-1}g, v)}. \tag{3.7}$$

(C.3) A is injective, and $v \notin D(A^{-1})$ or $v \in D(A^{-1})$ with $(g, A^{-1}v) = 0$.

(C.4) A^* is injective, and $g \notin D(A^{*-1})$ or $g \in D(A^{*-1})$ with $(A^{*-1}g, v) = 0$.

For a linear operator T on a Hilbert space, we set $\operatorname{nul} T := \dim \ker T$.

Theorem 3.3. (i) *Suppose that (C.1) and (C.2) hold. Let*

$$\Psi_{n,j} := a(A^{-1}v)^{*\,n}\Omega_{\mathrm{b}} \otimes b(A^{*\,-1}g)^{*\,j}\Omega_{\mathrm{f}}, \quad n = 0, 1, 2, \cdots, \quad j = 0, 1. \quad (3.8)$$

Then $\operatorname{nul} Q(\alpha_0) = \infty$ *with* $\ker Q(\alpha_0) = \overline{\mathcal{L}\{\Psi_{n,j} | n \geq 0, j = 0, 1\}}$. *Moreover, for all* $\alpha \neq \alpha_0$,

$$\operatorname{nul} Q(\alpha) = 1, \quad \ker Q(\alpha) = \{c\Omega_{\mathrm{b}} \otimes \Omega_{\mathrm{f}} | c \in \mathbb{C}\}. \quad (3.9)$$

(ii) *Suppose that (C.1) and (C.4) hold. Then* $\operatorname{nul} Q(\alpha_0) = \infty$ *with* $\ker Q(\alpha_0) = \mathcal{L}\{\Psi_{n,0} | n \geq 0\}$. *Moreover, for all* $\alpha \neq \alpha_0$, *(3.9) holds.*
(iii) *Suppose that (C.2) and (C.3) hold. Then* $\operatorname{nul} Q(\beta_0) = 2$ *with* $\ker Q(\beta_0) = \mathcal{L}\{\Psi_{0,j} | j = 0, 1\}$.
(iv) *Suppose that (C.3) and (C.4) hold. Then, for all* $\alpha \in \mathbb{C}$, *(3.9) holds.*

As for non-zero eigenvalues of $Q(\alpha)$, we have the following result:

Theorem 3.4. *Consider the case where* $\mathcal{H} = \mathcal{K}$, A *is a nonnegative self-adjoint operator with* $\ker A = \{0\}$ *and* $g = v \in D(A^{-1})$ *(then* $\alpha_0 = -1/(v, A^{-1}v) < 0$*). Let* $\alpha < \alpha_0$. *Then there exists a constant* $x_0(\alpha) < 0$ *such that* $\alpha(v, (x_0(\alpha) - A)^{-1}v) = 1$, *and for all* $n \in \{0\} \cup \mathbb{N}$,

$$\pm\sqrt{n}|x_0(\alpha)| \in \sigma_{\mathrm{p}}(Q(\alpha)). \quad (3.10)$$

Note that this theorem holds even if Q_A has no non-zero eigenvalues. As the condition that $\alpha < \alpha_0$ shows, this is a *strong coupling effect*.

By Theorem 3.1(i)-(ii), the operator $Q_+(\alpha)$ defined by

$$D(Q_+(\alpha)) := D(Q(\alpha)) \cap \mathcal{F}_+(\mathcal{H}, \mathcal{K}), \quad Q_+(\alpha)\Psi := Q(\alpha)\Psi, \ \Psi \in D(Q_+(\alpha)) \quad (3.11)$$

is a densely defined closed linear operator from $\mathcal{F}_+(\mathcal{H}, \mathcal{K})$ to $\mathcal{F}_-(\mathcal{H}, \mathcal{K})$. We define an index of $Q(\alpha)$ by

$$\operatorname{ind}_\Gamma(Q(\alpha)) := \operatorname{nul} Q_+(\alpha) - \operatorname{nul} Q_+(\alpha)^*, \quad (3.12)$$

the index of $Q_+(\alpha)$, provided that $\operatorname{nul} Q_+(\alpha) < \infty$ or $\operatorname{nul} Q_+(\alpha)^* < \infty$.

Results on the (semi-) Fredholm property and the index $\operatorname{ind}_\Gamma Q(\alpha)$ are as follows:

Theorem 3.5. (i) *Suppose that (C.1) and (C.2) hold. Then* $Q(\alpha_0)$ *is not semi-Fredholm. Moreover, for all* $\alpha \neq \alpha_0$, $Q(\alpha)$ *is Fredholm with* $\operatorname{ind}_\Gamma Q(\alpha) = 1$.
(ii) *Suppose that (C.1) and (C.4) hold. Then* $Q(\alpha_0)$ *is semi-Fredholm with* $\operatorname{ind}_\Gamma Q(\alpha) = \infty$. *Moreover, for all* $\alpha \neq \alpha_0$, $Q(\alpha)$ *is Fredholm with* $\operatorname{ind}_\Gamma Q(\alpha) = 1$.
(iii) *Suppose that (C.2) and (C.3) hold. Then* $Q(\beta_0)$ *is Fredholm with* $\operatorname{ind}_\Gamma Q(\beta_0) = 0$. *Moreover, for all* $\alpha \neq \beta_0$, $Q(\alpha)$ *is Fredholm with* $\operatorname{ind}_\Gamma Q(\alpha) = 1$.
(iv) *Suppose that (C.3) and (C.4) hold. Then, for all* $\alpha \in \mathbb{C}$, $Q(\alpha)$ *is Fredholm with* $\operatorname{ind}_\Gamma Q(\alpha) = 1$.

References

[Ar89] A. Arai, Path integral representation of the index of Kähler-Dirac operators on an infinite dimensional manifold, J. Funct. Anal. **82**(1989), 330-369.

[Ar91] A. Arai, A general class of infinite dimensional Dirac operators and related aspects, *Functional Analysis & Related Topics* (Ed. S. Koshi), pages 85–98, World Scientific, Singapore, 1991.

[Ar92] A. Arai, A general class of infinite dimensional Dirac operators and path integral representation of their index, J. Funct. Anal. **105**(1992), 342–408.

[Ar93a] A. Arai, Dirac operators in Boson-Fermion Fock spaces and supersymmetric quantum field theory, J. Geom. Phys. **11**(1993), 465–490.

[Ar93b] A. Arai, Supersymmetric extension of quantum scalar field theories, *Quantum and Non-Commutative Analysis* (Ed. H.Araki et al), pages 73–90, Kluwer Academic Publishers, Dordrecht 1993.

[Ar94] A. Arai, On self-adjointness of Dirac operators in boson-fermion Fock spaces, Hokkaido Math. Jour. **23**(1994), 319-353.

[Ar96] A. Arai, Supersymmetric quantum field theory and infinite dimensional analysis, Sugaku Expositions **9**(1996), 87-98.

[Ar97] A. Arai, Strong anticommutativity of Dirac operators on Boson-Fermion Fock spaces and representations of a supersymmetry algebra, Math. Nachr. **207** (1999), 61–77.

[Ar00] A. Arai, Spectral properties of a Dirac operator on the abstract Boson-Fermion Fock space, in preparation.

[AM91] A. Arai and I. Mitoma, De Rham-Hodge-Kodaira decomposition in ∞-dimensions, Math. Ann. **291**(1991), 51–73.

[AM93] Arai, A. and Mitoma, I.: Comparison and nuclearity of spaces of differential forms on topological vector spaces, J. Funct. Anal. **111**(1993), 278–294.

[BR97] O. Bratteli and D. W. Robinson, Operator Algebras and Quantum Statistical Mechanics 2, Second Edition, Springer, Berlin, Heidelberg, 1997.

Address

ASAO ARAI, Department of Mathematics, Hokkaido University, Sapporo 060-0810, Japan

E-MAIL: arai@math.sci.hokudai.ac.jp

2000 Mathematics Subject Classification. Primary 47A53 ; Secondary 81Q10

Operator Theory:
Advances and Applications, Vol. 126
© 2001 Birkhäuser Verlag Basel/Switzerland

Well-Posedness of Nonlinear Parabolic Equations of Viscous Hamilton-Jacobi Type

M. Ben-Artzi

Abstract. In this paper we review some recent results concerning the class of nonlinear equations of evolution given by,

$$u_t - \Delta u = \mu |\nabla u|^p, \ \mu \in \mathbb{R}, \ p \geq 1.$$

The equation is studied as a Cauchy problem in $\mathbb{R}^n$. While it is well-posed in spaces of functions with two bounded derivatives (for all p), it is well-posed in L^q only for supercritical q (depending on p). It is shown that for initial-data measures, there are in general no solutions if the measure is singular. Finally, decay results for long-time are given, when $\mu \leq 0$.

1 Introduction

In this paper we review some recent results concerning the class of nonlinear equations of evolution given by,

$$u_t - \Delta u = \mu |\nabla u|^p, \ \mu \in \mathbb{R}, \ p \geq 1, \tag{1.1}$$

$$u(x,0) = u_0(x), \ x \in \mathbb{R}^n. \tag{1.2}$$

We denote $\nabla = \nabla_x = \left(\frac{\partial}{\partial x_1}, \ldots, \frac{\partial}{\partial x_n}\right), \Delta = \sum_{i=1}^{n} \left(\frac{\partial}{\partial x_i}\right)^2$. While equations of the type $u_t - \Delta u = u^p$ have been extensively studied (see e.g. [BF83,B⊃T86,We80] and references there), the same is not true for (1.1). We note that most of the results mentioned in this paper apply to the more general case where the right-hand side of (1.1) is replaced by $F(\nabla u)$, with suitable growth conditions on F. Thus, (1.1) can be viewed as a model for a "viscous Hamilton-Jacobi" equation. Indeed, this equation appears naturally in a variety of studies. Some examples include:

(a) The one-dimensional case $n = 1$. In this case the equation appears in the study of growth of surfaces and is labeled as the "generalized KPZ equation" [GGK98,GKV98,KS88,KS91].

(b) Still in the one-dimensional case, we take $\mu = -1$ and $p = 2$, thus obtaining the equation $u_t + u_x^2 = u_{xx}$. Differentiating with respect to x and setting $v = u_x$ we get for $v(x,t)$ the equation $v_t + (v^2)_x = v_{xx}$, which is the well-known Burgers equation.

(c) Consider the Navier-Stokes equations in the plane ($n = 2$), which in vorticity form can be written as

$$\xi_t + (\underline{u} \cdot \underline{\nabla})\xi = \nu \Delta \xi$$

(ξ is the vorticity $\partial_2 u^1 - \partial_1 u^2$ of the velocity field $\underline{u} = (u^1, u^2)$). Suppose we know in advance that $|\underline{u}|$ is bounded. Then ξ satisfies the inequality $\xi_t - \nu \Delta \xi \leq C|\nabla \xi|$. Thus, the methods used in the study of (1.1) are also applicable in the case of the inequality.

In the following sections we shall discuss the global well-posedness of (1.1) in various spaces and the decay properties of solutions as $t \to +\infty$.

2   Existence of Global Solutions

Let $C_b^2(\mathbb{R}^n) := C^2(\mathbb{R}^n) \cap W^{2,\infty}(\mathbb{R}^n)$, namely, the space of twice continuously differentiable functions with bounded derivatives. It was proved in [AB98] that $C_b^2(\mathbb{R}^n)$ is a "persistence space" to classical solutions of (1.1). Namely, we have the theorem.

Theorem 2.1. *[AB98] Let $u_0 \in C_b^2(\mathbb{R}^n)$. Then for any $\mu \in \mathbb{R}$, $p \geq 1$, there exists a unique classical solution to (1.1) - (1.2), such that $u(\cdot, t) \in C_b^2(\mathbb{R}^n)$ for all $t \geq 0$ and the mapping*

$$u_0 \in C_b^2(\mathbb{R}^n) \to u \in C(\overline{\mathbb{R}_+}, C_b^2(\mathbb{R}^n))$$

is continuous.

Furthermore, the solution satisfies the following maximum-minimum principles.

$$\sup_{\substack{x \in \mathbb{R}^n \\ t \in (0,T]}} u(x,t) = \sup_{x \in \mathbb{R}^n} u_0(x), \quad \inf_{\substack{x \in \mathbb{R}^n \\ t \in (0,T]}} u(x,t) = \inf_{x \in \mathbb{R}^n} u_0(x), \ \forall T > 0, \tag{2.1}$$

$$\|\nabla u(\cdot, t)\|_{L^\infty(\mathbb{R}^n)} \leq \|\nabla u_0\|_{L^\infty(\mathbb{R}^n)}, \ \forall t \geq 0. \tag{2.2}$$

In the proof, one shows that the solution exists in a time interval $(0, T]$, where T depends only on $\|\nabla u_0\|_{L^\infty(\mathbb{R}^n)}$. The inequality (2.2) then allows the continuation of the solution to $[T, 2T], \ldots$. We remark that to prove (2.2) the equation (1.1) is differentiated with respect to x_j. Denoting $u_j = \frac{\partial}{\partial x_j} u$, we get

$$\frac{\partial}{\partial t} u_j - \Delta u_j = \sum_{i=1}^{n} \Psi_i(x,t) \frac{\partial u_j}{\partial x_i}, \tag{2.3}$$

where $\Psi_i(x,t) = \mu p |\nabla u|^{p-2} \frac{\partial u}{\partial x_i} \in L^\infty(\mathbb{R}^n \times (0,T])$. However, the solution u_j to the linear parabolic equation (2.3) is not twice continuously differentiable hence some care must be taken in deducing (2.2) from the standard (linear) maximum principle. See the Appendix in [AB98] for details. This result has been extended to initial data which are just bounded continuous functions by Gilding,Guedda and Kersner [GGK98]. They make nice use of "Bernstein-type" techniques to show that a unique solution exists, and is classical for positive time.

Naturally, our next goal is to investigate the well-posedness of (1.1) in wider spaces of less regular functions, for instance $L^q(\mathbb{R}^n)$ for suitable exponents q (possibly depending on p). To allow such solutions the equation (1.1) is first cast in the integral form,

$$u(x,t) = \int_{\mathbb{R}^n} G(x-y,t)u_0(y)dy + \mu \int_0^t \int_{\mathbb{R}^n} G(x-y,t-s)|\nabla u(y,s)|^p dy ds, \quad (2.4)$$

where $G(x,t) = (4\pi t)^{-n/2}\exp(-|x|^2/4t)$ is the heat kernel. Taking ∇_x of Eq. (2.4) and using norms of the type $\sup_{t\in(0,T]} t^\alpha \|\nabla u(\cdot,t)\|_{L^r(\mathbb{R}^n)}$ for suitable r,α, as in [We80] one obtains the local (in time) existence of solutions to (2.4) in $L^q(\mathbb{R}^n)$, for certain exponents. Then, by using the regularizing effect of the parabolic equation (2.4) (see also [BGL99] for a direct argument) one shows that the solution $u(\cdot,t) \in C_b^2(\mathbb{R}^n)$ for $t > 0$, hence global existence follows from Theorem 2.1 above. As for uniqueness, we note that the solution was constructed by using "growth norms" of the type $\sup_{t\in(0,T]} t^\alpha \|\nabla u(\cdot,t)\|_{L^r(\mathbb{R}^n)}$. Thus, a contraction argument yields uniqueness using such norms (the "Kato-Fujita condition" [KF62]). However, an alternative approach as in [Br94] gives uniqueness for solutions in classes like $C([0,T]; L^q(\mathbb{R}^n)) \cap C((0,T]; C_b(\mathbb{R}^n))$. The exact exponents are summarized in the following theorem.

Theorem 2.2. *[BSW99] For $1 \leq p < 2$, let $q_c = n\frac{p-1}{2-p}$ and take any $q \geq \max(1, q_c)$, $q < \infty$ (but $q > 1$ if $q_c = 1$). Then, given any $u_0 \in L^q(\mathbb{R}^n)$ (and any $\mu \in \mathbb{R}$), the equation (2.4) has a unique, global (in time) solution $u \in C([0,\infty), L^q(\mathbb{R}^n))$.*

In particular we note that if

$$p > p_c := \frac{n+2}{n+1} \quad (2.5)$$

then $q_c > 1$ and the exponent $q = 1$ is outside the scope of Theorem 2.2. Indeed, as the following claim shows, one cannot expect, for $p > p_c$, to have solutions u of (2.4) for any $u_0 \in L^1(\mathbb{R}^n)$, even under the mildest assumptions on u. In presenting the next claim, there is no attempt at achieving maximal generality.

Claim 2.1. Let $p > p_c = \frac{n+2}{n+1}$ and $\mu = 1$.

Denote, for $0 < \delta < \frac{1}{2}(\frac{n+1}{n+2} - \frac{1}{p})$,

$$v_\delta(x) = \begin{cases} |x|^{-n+\delta}, & |x| < 1, \\ 0 & |x| \geq 1. \end{cases}$$

Then, given any $T > 0$, there is no solution $u(x,t)$ of (2.4) in $(x,t) \in \mathbb{R}^n \times (0,T]$, where $u_0 = v_\delta$ and such that

$$u \in L^p((0,T); W^{1,p}(\mathbb{R}^n)).$$

Proof. Assume the existence of a solution $u(x,t)$ with the above properties. Since $\int_0^T \int_{\mathbb{R}^n} |\nabla u|^p dxdt < \infty$, given $\varepsilon > 0$ there exists a sequence $t_j \to 0$ such that

$$\int_{\mathbb{R}^n} |\nabla u(x,t_j)|^p dx < \varepsilon t_j^{-1}, j = 1, 2, \ldots, \tag{2.6}$$

which implies, by the Sobolev inequality,

$$\int_{\mathbb{R}^n} u(x,t_j)^{p^*} dx \leq C(\varepsilon t_j^{-1})^{\frac{p^*}{p}}, \quad \frac{1}{p^*} = \frac{1}{p} - \frac{1}{n}. \tag{2.7}$$

($u \geq 0$ in view of (2.4)).

Take $0 < \beta < \frac{1}{2}$ (to be specified later) and use Hölder's inequality and (2.7) to get,

$$\int_{|x|<t_j^\beta} u(x,t_j)dx \leq (C\varepsilon t_j^{-1})^{\frac{1}{p}} \cdot (\omega_n t_j^{\beta n})^{1-\frac{1}{p^*}}, \tag{2.8}$$

($\omega_n =$ volume of unit ball).

Now $p > \frac{n+2}{n+1}$ implies $n(1 - \frac{1}{p^*}) > \frac{2(n+1)}{n+2}$, so from (2.8)

$$\int_{|x|<t_j^\beta} u(x,t_j)dx \leq C\varepsilon^{1/p} t_j^{-\frac{1}{p}+\frac{2\beta(n+1)}{n+2}}, \ j = 1, 2, \ldots \tag{2.9}$$

Since $p > \frac{n+2}{n+1}$ we can choose $\beta < \frac{1}{2}$ such that $\eta = -\frac{1}{p} + 2\beta\frac{n+1}{n+2} > 0$, hence

$$\int_{|x|<t_j^\beta} u(x,t_j)dx \leq C\varepsilon^{1/p} t_j^\eta \to 0 \text{ as } j \to \infty. \tag{2.10}$$

Let $\tilde{u}(x,t) = G(x,t) * u_0$ be the solution to the heat equation with the same initial data. Clearly $u(x,t) \geq \tilde{u}(x,t)$. We have, for $t > 0$,

$$\int_{|x|>t^\beta} \tilde{u}(x,t)dx = \int_{|x|>t^\beta} \int_{\mathbb{R}^n} G(x-y,t)u_0(y)dydx$$

$$= \int_{|x|>t^\beta} \int_{|y|<\frac{1}{2}t^\beta} + \int_{|x|>t^\beta} \int_{|y|>\frac{1}{2}t^\beta} G(x-y,t)u_0(y)dydx$$

$$\leq \int_{|\xi|>\frac{1}{2}t^\beta} G(\xi,t)d\xi \cdot \|u_0\|_{L^1(\mathbb{R}^n)} + \int_{|y|>\frac{1}{2}t^\beta} u_0(y)dy \qquad (2.11)$$

Since $\beta < \frac{1}{2}$ we have

$$\int_{|\xi|>\frac{1}{2}t^\beta} G(\xi,t)d\xi = O(t^N) \text{ as } t \to 0, N = 1,2,\dots$$

and, for $u_0 = v_\delta$

$$\int_{|y|>\frac{1}{2}t^\beta} u_0(y)dy = (1 - 2^{-\delta}t^{\beta\delta})\|u_0\|_{L^1(\mathbb{R}^n)}, t < 1, \qquad (2.12)$$

so, since $\|\tilde{u}(\cdot,t)\|_{L^1(\mathbb{R}^n)} = \|u_0\|_{L^1(\mathbb{R}^n)}$, we conclude that

$$\int_{|x|<t^\beta} \tilde{u}(x,t)dx = 2^{-\delta}t^{\beta\delta}\|u_0\|_{L^1(\mathbb{R}^n)} + O(t^N) \text{ as } t \to 0.$$

Setting $t = t_j$ and comparing with (2.10) we get, for $j = 1,2,\dots$,

$$C\varepsilon^{1/p}t_j^\eta \geq 2^{-\delta}t_j^{\beta\delta}\|u_0\|_{L^1(\mathbb{R}^n)} + O(t_j^N) \qquad (2.13)$$

which is a contradiction by the choice of δ, since $\beta < \frac{1}{2}$ can be chosen such that $\beta\delta < \eta$. $\qquad\Box$

Remark 2.1. In view of the last claim, one may ask, in the case $p > p_c, \mu = 1$, what is the set of initial data $u_0(x) \in L^1(\mathbb{R}^n)$ for which a solution to (2.4) does exist. Theorem 2.2 implies that this set contains all $u_0 \in L^1(\mathbb{R}^n) \cap L^q(\mathbb{R}^n)$, $q \geq q_c$, and in particular, all $u_0(x) \in L^1(\mathbb{R}^n) \cap L^\infty(\mathbb{R}^n)$. However, Claim 2.1 says this set is not all of $L^1(\mathbb{R}^n)$. The situation is still not clear for $\mu = -1$. On the other hand, if $\mu = 1$ and $p \geq 2$, Claim 2.1 can be strengthened as follows.

Proposition 2.1. *[BSW99] Let $u(x,t)$ be a classical solution of (1.1), with $\mu = 1, p \geq 2$, in a strip $\mathbb{R}^n \times (0,T)$. Assume that $\lim_{t\to 0} u(\cdot,t) = u_0$ in $L^1_{loc}(\mathbb{R}^n)$. Then $\exp(u_0) \in L^1_{loc}(\mathbb{R}^n)$.*

Finally, while (for $p > p_c, \mu = 1$) existence is not guaranteed for all $u_0 \in L^q(\mathbb{R}^n)$, $1 \leq q < q_c$, uniqueness can also fail, as the following theorem shows.

Theorem 2.3. *[BSW99] Assume $2 > p > p_c$ and let $1 \leq q < q_c$ and $\mu = 1$. Then, for $u_0 = 0$, there exists a positive solution u to (2.4). In fact, u is self-similar,*

$$u(x,t) = t^{-k}U(|x|t^{-\frac{1}{2}}), \quad k = \frac{2-p}{2(p-1)},$$

where $U = U(r) \in C^2([0,\infty))$.

Remark 2.2. The case of a coupled system of equations of the type (1.1) was treated in [AR98].

3 Further Extensions. The Case $\mu = -1$

We consider here some further results for solutions of Eq. (1.1) (or (2.4)) under the assumptions that $\mu = -1$ and $u_0 \geq 0$. The maximum-minimum principle guarantees that the solution u is nonnegative and is majorized by the corresponding solution of the heat equation.

In this case, the subcritical part of Theorem 2.2 has been extended by Benachour and Laurencot [BL99] to include positive bounded measures, as follows.

Theorem 3.1. *[BL99] Let $1 < p < p_c = \frac{n+2}{n+1}$, $\mu = -1$ and $u_0 \in M_b^+(\mathbb{R}^n)$ (= the space of positive bounded Borel measures). Then there exists a unique weak solution (in the sense of (2.4)) u such that, $u \in C((0,\infty); L^1(\mathbb{R}^n)) \cap L^p_{loc}((0,\infty); W^{1,p}(\mathbb{R}^n))$.*

Remark 3.1. (a) We refer to [BL99] for a precise definition of a "weak solution". Also, for the uniqueness a "growth condition" (as $t \to 0$) of the "Kato-Fujita" type is required, as in the discussion preceding Theorem 2.2 above.

(b) The case $p = 1$ (and $\mu = -1$) was treated in [BRV97], by probabilistic methods, producing a spherically symmetric solution for any initial data $u_0(x)$ which is a "profiled" spherically symmetric bounded positive measure.

(c) The more general equation $u_t - \Delta u = -a(x)u^q(\nabla u)^p, u_0 \geq 0$, was treated in [Pi00].

The supercritical case $(p \geq p_c)$ is more difficult. Clearly, the method of proof of Claim 2.1 does not work here and the question whether or not the equation is well-posed in $L^1(\mathbb{R}^n)$ remains an open problem. However, Benachour and Laurencot [BL99] have managed to prove the non-existence of "source-type" solutions, namely, solutions that converge (in the sense of distributions) to a multiple of the Delta-function. The exact formulation of the theorem is as follows.

Theorem 3.2. *[BL99] Let $M, T > 0, p \geq p_c$. There is no $u \in L^\infty((0,T); L^1(\mathbb{R}^n)) \cap L^p((0,T); W^{1,p}(\mathbb{R}^n))$ such that $u_t - \Delta u = -|\nabla u|^p$ in $\mathcal{D}'(\mathbb{R}^n \times (0,T))$ and $\lim\limits_{t \to 0} \int\limits_{\mathbb{R}^n} u(x,t)\Psi dx = M\Psi(0), \forall \Psi \in C_0^\infty(\mathbb{R}^n)$.*

It turns out that this result can be generalized to a class of initial singular measures, which we label as "p-atomic". They include all atomic measures.

Let $n \geq 2$, $p_c = \frac{n+2}{n+1} < p < n$ and $p^* = \frac{n-p}{np}$. Let $\nu \geq 0$ be a Borel measure on $\mathbb{R}^n$.

Definition 3.1. We say that ν is *p-atomic* if there exist constants $C > 0$, $0 < \delta < 1$, such that the following is satisfied: for every $0 < t < 1$ there exist sequences $\{x_k\}_{k=1}^\infty \subset \mathbb{R}^n$, $\{r_k\}_{k=1}^\infty \subset (0,\infty)$, such that

$$(i) \qquad \mathrm{supp}(\nu) \subset \bigcup_{k=1}^\infty B(x_k, r_k) \qquad (B(y,r) = \{x; |x - y| < r\}),$$

$$(ii) \qquad \sum_{k=1}^\infty r_k^{n(1-(1/p^*))} \leq Ct^{1/p},$$

$$(iii) \qquad \sum_{k=1}^\infty \nu(B(x_k, r_k))e^{-(1-\delta)r_k t^{-1/2}} \to 0, \quad \text{as } t \to 0.$$

Note that if, for example, ν is atomic then ν is p-atomic $(p > p_c)$. Indeed one takes $r_k = 2^{-k}t^{1/(n(p-1)+p)}$.

Theorem 3.3. *[BSW00] Let $\mu < 0$, $n \geq 2$, $p_c = \frac{n+2}{n+1} < p < n$ and $\psi \geq 0$, $\psi \not\equiv 0$, be a p-atomic measure. Then there is no local nonnegative solution of (1.1) such that*

$$u(.,t) \to \psi \quad in\ \mathcal{M}\ as\ t \to 0. \tag{3.1}$$

The proof is similar in spirit to that of Claim 2.1, inspecting carefully the heat kernel. We refer to [BSW00] for details.

4   Decay as $t \to +\infty$

Let us go back to classical (say, as in Theorem 2.1) solutions to (1.1), where we assume now that $u_0 \geq 0$ and $\mu = -1$. Then the solution $u(x,t)$ is nonnegative and an integration of (1.1) shows that if, in addition, $u_0 \in L^1(\mathbb{R}^n)$ then $u(\cdot, t) \in L^1$ for all $t \geq 0$ and the nonnegative function $I(t) = \int_{\mathbb{R}^n} u(x,t)dx$ is nonincreasing. Thus, the limit $I_\infty = \lim_{t \to \infty} I(t) \geq 0$ always exists. It is interesting that the question whether or not $I_\infty = 0$ is determined uniquely by $p_c = \frac{n+2}{n+1}$, the same critical value as in the previous sections. We have the following theorem.

Theorem 4.1. *[BK99] Let $0 \leq u_0 \in C_b^2(\mathbb{R}^n) \cap L^1(\mathbb{R}^n), u_0 \neq 0$. Let $u(x,t)$ be the solution to (1.1), with $\mu = -1$. Then*

$$I_\infty > 0 \Leftrightarrow p > p_c = \frac{n+2}{n+1}.$$

Remark 4.1. As was seen in Theorem 2.2, the well-posedness of (1.1) in $L^1(\mathbb{R}^n)$ was also linked to the same critical index p_c. However, there is yet no direct argument connecting this well-posedness (essentially a short-time feature) with the long-time decay as expressed in Theorem 4.1.

Remark 4.2. In the case $p < p_c$ the equation is well-posed in $L^1(\mathbb{R}^n)$. Then, as in the discussion of the preceding Theorem 2.2, if $0 \leq u_0 \in L^1(\mathbb{R}^n)$ (and $\mu = -1$), it follows that $u(\cdot, t) \in C_b^2(\mathbb{R}^n) \cap L^1(\mathbb{R}^n)$ for $t > 0$. Hence, Theorem 4.1 is applicable also, in the subcritical case, to all $0 \leq u_0 \in L^1(\mathbb{R}^n)$.

Remark 4.3. In the case $p \leq p_c$, the rate of decay of $I(t)$ to zero becomes slower as p approaches p_c. More precisely, let $1 < p \leq p_c$ and $\alpha > \frac{2-p}{2(p-1)} - \frac{n}{2}$. Then [BK99] $I(t) \leq Ct^{-\alpha}$ (for all sufficiently large t) implies $u_0 = 0$. In particular, if $p = p_c$ then $I(t)$ cannot decay like $t^{-\alpha}$ for any $\alpha > 0$. On the other hand, if $p = 1$ and u_0 is compactly supported then, for some $A, \theta > 0$ we have

$$\sup_{0 \leq t < \infty} \exp(At^\theta)I(t) < \infty.$$

(see [AB98]).

Acknowledgements

The author is deeply indebted to L. Amour, J. Goodman, H. Koch, A. Levy, Ph. Souplet and F. Weissler for pleasant collaborations and useful discussions. It is also a pleasure to thank S. Benachour for interesting correspondence and exchange of ideas.

References

[Al96]   N. Alaa, Solutions faibles d'équations paraboliques quasilinéaires avec données initiales mesures, Ann. Math. Blaise Pascal **3** (1996), 1–15.

[AW92]   L. Alfonsi and F..B. Weissler, Blow-up in $\mathbb{R}^n$ for a parabolic equation with a damping nonlinear gradient term, in "Progress in Nonlinear Differential Equations", N.G. Lloyd et al (Eds.), Birkhäuser 1992.

[AB98]   L. Amour and M. Ben-Artzi, Global existence and decay for visccus Hamilton-Jacobi equations, Nonlinear Anal. TMA **31** (1998), 621–628.

[AR98]   L. Amour and T. Raoux, L^1 decay properties for a semilinear parabolic system (Preprint 1998).

[BL99]   S. Benachour and Ph. Laurencot, Global solutions to viscous Hamilton-Jacobi equation with irregularinitial data, Comm. PDE **24** (1999), 999-2021.

[BL98]   S. Benachour and Ph. Laurencot, "Solutions très singulières" d'une équation parabolique non linéaire avec absorption, Preprint 1998.

[BRV97]   S. Benachour, B. Roynette and P. Vallois, Asymptotic estimates of solutions of $u_t - \frac{1}{2}\Delta u = -|\nabla u|$ in $\mathbb{R}_+ \times \mathbb{R}^d, d \geq 2$, J. Func. Anal. **144** (1997), 301–324.

[Be92]   M. Ben-Artzi, Global existence and decay for a nonlinear parabolic equation, Nonlinear Anal. TMA **19** (1992), 763–768.

[BGL99]   M. Ben-Artzi, J. Goodman and A. Levy, Remarks on a nonlinear parabolic equation, Trans. AMS **352** (1999),731-751.

[BK99]   M. Ben-Artzi and H. Koch, Decay of mass for a semilinear parabolic equation, Comm. PDE **24** (1999),869-881.

[BSW99]   M. Ben-Artzi, Ph. Souplet and F.B. Weissler, Sur la non-existence et la non-unicité des solutions du problème de Cauchy pour une équation parabolique semi-linéaire, CRAS **329** (1999), 371-376.

[BSW00]   M. Ben-Artzi, Ph. Souplet and F.B. Weissler, The local theory for viscous Hamilton-Jacobi equations in Lebesgue spaces (Preprint).

[Br94]   H. Brezis, Remarks on the preceding paper by M. Ben-Artzi, "Global solutions of two-dimensional Navier-Stokes and Euler Equations", Arch. Rat. Mech. Anal. **128** (1994), 359–360.

[BF83]   H. Brezis and A. Friedman, Nonlinear parabolic equations involving measures as initial conditions, J. Math. Pures Appl. IX, Ser. 62 (1983), 73–97.

[BPT86]   H. Brezis, L. Peletier and D. Terman, A very singular solution of the heat equation with absorption, Arch. Rat. Mech. Anal. **95** (1986), 185–219.

[GGK98]   B. Gilding, M. Guedda and R. Kersner, The Cauchy problem for $u_t = \Delta u + |\nabla|^q$, Preprint 1998.

[GKV98]   M. Guedda, R. Kersner and L. Veron, On self-similar-type solutions to the generalized KPZ equation, Preprint 1998.

[HW82]   A. Haraux and F.B. Weissler, Non-uniqueness for a semilinear initial value problem, Indiana Univ. Math. J. **31** (1982), 167–189.

[KF62]   T. Kato and H. Fujita, On the nonstationary Navier-Stokes system, Rend. Sem. Math. Univ. Padova **32** (1962), 243–260.

[KS88]   J. Krug and H. Spohn, Universality classes for deterministic surface growth, Phys. Rev. A **38** (1988), 4271–4283.

[KS91]   J. Krug and H. Spohn, Kinetic roughening of growing surfaces, in "Solids far from equilibrium", C. Godreche (Ed.), Cambridge Univ. Press, 1991, pp. 479–582.

[Li82]   P.L. Lions, "Generalized solutions of Hamilton-Jacobi Equations", Pitman Research Notes in Mathematics, 69, 1982.

[Pi00]   R.G. Pinsky, Decay of mass for the equation $u_t = \Delta u - a(x)u^p|\nabla u|^q$, J. Diff. Eqs. (to appear).

[STW99]   S. Snoussi, S. Tayachi and F.B. Weissler, Asymptotically self-similar global solutions of semilinear parabolic equations with nonlinear gradient terms, Proc. Royal Soc. Edinburgh **i129A** (1999),1291-1307.

[So95]   Ph. Souplet, Résultats d'explosion en temps fini pour une èquation de la chaleur non linéaire, C.R. Acad. Sci. Paris, Seriè I. **321** (1995), 721–726.

[So99]   Ph. Souplet, Geometry of unbounded domains, Poincaré inequalities and stability in semilinear parabolic equations, Comm. PDE **24** (1999), 951-973.

[SW99]   Ph. Souplet and F.B. Weissler, Poincaré's inequality and global solutions of a nonlinear parabolic equation, Ann. Inst. H. Poincaré, Anal. Nonlin. **16** (1999), 335-372.

[Ta96]   S. Tayachi, Forward self-similar solutions of a semilinear parabolic equation with a nonlinear gradient term, Diff. Integral Eqs. **9** (1996), 1107–1117.

[We80]   F.B. Weissler, Local existence and nonexistence for semilinear parabolic equations in L^p, Indiana Univ. Math. J. **29** (1980), 79–102.

Address

M. BEN-ARTZI, Institute of Mathematics, Hebrew University, Jerusalem 91904, Israel

E-MAIL: mbartzi@math.huji.ac.il

2000 Mathematics Subject Classification. Primary 35K10, 35K55

Operator Theory:
Advances and Applications, Vol. 126
© 2001 Birkhäuser Verlag Basel/Switzerland

Non-Convex Minimization - the Case of Vector Fields

E. Brüning

1   Introduction

Typically minimization problems are solved by applying one or the other version of a so-called "Generalized Weierstraß Theorem": *A weakly sequentially lower semi-continuous function attains its infimum on a bounded and weakly sequentially closed subset of a real reflexive Banach space.* Various other versions are for instance given in Chapter 38 of [Ze85]. In particular these theorems have proven to be extremely powerful for the proof of the existence of solutions of many classes of nonlinear partial differential equations [Br65,Br70,Da89,ET76,Mo66,Ze85]. Nevertheless by the demands of concrete problems, for instance in material sciences, control theory, and optimization, and also as a matter of principle one would like to go beyond these theorems since their basic assumptions are not available in many problems.

Let us recall the main issues of a minimization problem in a real reflexive Banach space E. Let $M \subset E$ be a non-empty subset and $f : M \to \mathbb{R}$ a real-valued function which is bounded from below on M. Then $I = \inf \{f(u) : u \in M\}$ is finite and there are minimizing sequences $(u_n)_n \subset M$, i.e., $I = \lim_{n \to \infty} f(u_n)$. Let us assume (or prove using suitable assumptions on (f, M)) that there are minimizing sequences which are bounded in E. It follows that there are minimizing sequences which converge weakly in E. Furthermore, let us assume (or prove what might be difficult) that there is a minimizing sequence $(u_n)_n$ with weak limit $u \in M$. Then, in order to conclude, the following condition has to be made for this sequence:

$$f(u) \leq \liminf_{n \to \infty} f(u_n). \tag{1.1}$$

Certainly, it is an extremely difficult problem to find **one** minimizing sequence which has all the properties listed above. Therefore, traditionally (see any book on the direct methods in the calculus of variations), minimization problems are solved under the assumption that the function f satisfies condition (1.1) for **all** weakly convergent sequences $(u_n)_n$, i.e., f is sequentially lower semi-continuous with respect to weak convergence in E. However in general there are not too many results available which assure weak lower semi-continuity of a function on a reflexive Banach space. Even for concrete functions such as

$$f(u) = \int_{\Omega} F(x, u(x), Du(x))dx \tag{1.2}$$

on some Sobolev space $E = W^{1,p}(\Omega)$ ($\Omega \subset \mathbb{R}^n$ a domain with Lipschitz boundary) the characterization of weak sequential lower semi-continuity is a very difficult problem which has a long history [Da89,Ze85]. Under certain technical restrictions on the integrand F the functional f is w.s.l.s.c. if and only if the function $y \mapsto F(x, u, y)$ is convex respectively quasi-convex in the case of vector-valued functions $u : \Omega \to \mathbb{R}^m, m > 1$, [Da89,Mo52,Mo66,Ev90] and this then implies that the Euler equation for f, i.e., the equation $f'(u) = 0$, is elliptic.

Certainly, not for all functions of interest (for instance for functions arising in the context of solving nonlinear partial differential equations), this convexity condition is available and one is confronted with the problem of minimization without weak lower semi-continuity. This then amounts to proving condition (1.1) not for all weakly convergent sequences but only for a certain subclass and this is what we intend to do for functions arising in the context of solving systems of nonlinear PDE's for vector fields.

It seems that L. C. Young (1937), guided by an analysis of the Bolza problem, was the first to have addressed the problem of minimization without assuming weak lower semi-continuity of the function in question. He suggested two new strategies which turned out to be rather fruitful later [Yo37,Yo69]: relaxation [Da89,Ek79,Yo69,Ze85] and the use of 'Young' measures [Mü98,Pe99] as solutions. However, often we need to find solutions which have at least locally ' finite energy'. Then Young measures do not solve the problem.

In Section 2 we recall our alternative: A theory "Minimization without weak lower semi-continuity" [Brü94] which has been successfully applied to concrete classes of eigen-/boundary- value problems for (global) quasi-linear partial differential operators (scalar case) [Brü94,Brü97].

2 Minimization Without Weak Lower Semi-Continuity.

Let E be a real Banach space and E' its topological dual with canonical bi-linear form $\langle \cdot, \cdot \rangle$. In the following K will always denote a nonempty subset of E whose weak closure $\overline{K}^\sigma$ contains $0 \in E$. Our basic definitions are as follows (see [Brü94]):

Definition A map $T : E \to E'$ is called

(a) K**-monotone at** $v \in E$ iff for all $a \in K$ $\langle T(v + a) - T(v), a \rangle \geq 0$.

(b) **weakly** K**-monotone at** $v \in E$ iff $\varlimsup_{i \to \infty} \langle T(v + a_i) - T(v), a_i \rangle \geq 0$ for all sequences $a_i \epsilon K$ converging weakly (in E) to zero: $0 = w - \lim_{i \to \infty} a_i$.

(c-d) K**-monotone (weakly** K**-monotone)** iff T is K-monotone (weakly K-monotone) at every $v \in E$.

Clearly whenever a map T is K-monotone (at $v \in E$) then T is also weakly K-monotone (at $v \in E$). We mention some elementary properties of (weakly) K-monotone maps.

Proposition 2.1. *(1) The set of mappings $T : E \to E'$ which are (weakly) K-monotone at $v \in E$ is a convex cone.*

(2) If $T : E \to E'$ is weakly K-monotone at $v \in E$ and if $S : E \to E'$ is completely continuous at v then $T + S$ is weakly K-monotone at $v \in E$.

(3) Let $T : E \to E'$ be weakly K-monotone at $v \in E$ and let $A : E \to E$ be a weakly sequentially continuous linear map with adjoint $A' : E' \to E'$. Then $A' \circ T \circ A$ is weakly $A^{-1}(K)$-monotone at $A^{-1}v$.

(4) If $T : E \to E'$ is weakly K-monotone at $v \in E$ and if K_0 is some subset of K with $0 \in \overline{K_0}^{\sigma}$ then T is weakly K_0-monotone.

The following lemma shows that **a locally bounded map can fail to be weakly K-monotone only in the case of "infinite-dimensional" sets K.**

Lemma 2.1. *Suppose $T : E \to E'$ is locally bounded, i.e. T maps bounded subsets of E onto bounded subsets of E' and suppose that K is a subset of E which is contained in some finite dimensional subspace F of E. Then T is weakly K-monotone.*

Let E be a real reflexive Banach space and $f : E \to \mathbb{R}$ a continuous function on E which has a semi-continuous Gâteaux-derivative $f' : E \to E'$. Then it follows for all $u, v \in E$

$$f(u) - f(v) = \int_0^1 \langle f'(v + t(u - v)), u - v \rangle dt. \tag{2.1}$$

We study the minimization problem for (f, M) for the class of functions f having the following properties:

(a) $f' : E \to E'$ is locally bounded;

(b) f is bounded from below on M;

(c) The pre-image under $f_M = f|M$ of a bounded set in $\mathbb{R}$ is a bounded set in the Banach space E.

(d) Weak limit points of minimizing sequences are contained in M.

(H)

An easy way to ensure condition (d) is to assume that M is weakly closed. In general it can be difficult to prove (d) but successful strategies are known [Brü90].

Theorem 2.1. *Under the assumption (a) - (d) on the pair (f, M) the minimization problem for (f, M) has a solution if, and only if, there is a weakly convergent minimizing sequence $\underline{v} = (v_j)$ such that f' is weakly $K_\varepsilon(\underline{v})$-monotone at $v = w - \lim v_j \in M$ for all $0 < \varepsilon \leq \varepsilon_0$ for some $0 < \varepsilon_0 \leq 1$ where*

$$K_\varepsilon(\underline{v}) = \{u \in E \mid u = t(v_j - v), \varepsilon \leq t \leq 1, \ j = 1, 2, \ldots\} \tag{2.2}$$

Remark: (a) Suppose that the Gâteaux-derivative of f is weakly K-monotone at $v \in M$. Then clearly we can successfully apply the above theorem when we can find a weakly convergent minimizing sequence $\underline{v}$ with $v = w - \lim_{j \to \infty} v_j \in M$ and some $0 < \varepsilon_o < 1$ such that $K_\varepsilon(\underline{v}) \subseteq K$ for all $0 < \varepsilon \le \varepsilon_0$. (b) Suppose f' is weakly E-monotone. Then for any weakly convergent minimizing sequence $\underline{v}$ with weak limit in M Remark (a) applies and thus we conclude that the minimization problem for the pair $(f,\ M)$ has a solution. Since weak E-monotonicity is the same as condition (P) of Browder and Hess which is known to characterize in the context of the above result weak sequential lower semi-continuity of f we have a proof of a slightly stronger version of the generalized Weierstraß Theorem mentioned earlier.

<h3 align="center">Estimates for $K_\varepsilon(\underline{v})$</h3>

For $R > 0$ denote by B_R the open ball in the Banach space E with radius R centered at $u = 0$. For any $0 < \varepsilon \le 1$ and any $v \in M$ introduce the sets

$$D(\varepsilon, R, v) = \{w \in E \mid w = t(u - v),\ \varepsilon \le t \le 1, u \in M,\ u - v \in B_R\} \tag{2.3}$$

that is the set of line segments in E generated by all difference vectors $u - v$ of points u of M in the neighborhood $B_R(v) = v + B_R$ of the point $v \in M$. Now let a weakly convergent minimizing sequence $\underline{v}$ be given with $v = w - \lim_{j \to \infty} v_j \in M$. Then there is some $R > 0$ such that $||v_j - v|| < R$ for all $j \in \mathbb{N}$. Hence it follows $K_\varepsilon(\underline{v}) \subseteq D(\varepsilon, R, v)$. Next we formulate a restriction on the set M.

Definition 2.1. Suppose that $C \subset E$ is some non-empty convex cone in E. We say that M satisfies the *cone condition* for the cone C at a point $v \in M$ if, and only if, for some $R > 0$ one has

$$D(\varepsilon, R, v) \subseteq C \cap B_R. \tag{2.4}$$

If there exists a convex cone $C \subset E$ such that M satisfies the cone condition for this cone at every $v \in M$ we say that M satisfies the cone condition for the cone C.

Remark: a) If $v \in M$ is an interior point M then $D(\varepsilon, R, v)$ contains an open ball in E and thus allows any direction in E. b) If however $v \in M$ is not an interior point, in particular when M has no interior points at all, then $D(\varepsilon, R, v)$ fails to contain an open ball centered at the origin of E and the set of "directions" in $D(\varepsilon, R, v)$ is restricted accordingly. c) Condition (2.4) means that the set of directions in $D(\varepsilon, R, v)$ is contained in some cone $C \subset E$ depending on the point $v \in M$. Intuitively this requires, in the case of a level surface for instance, that the "curvature" of M does not vary too much.

Theorem 2.2. *Suppose that the pair $(f,\ M)$ satisfies conditions (a) - (d) and in addition assume:*

(1) M satisfies the cone condition for a convex cone $C \subset E$.
(2) $f' : E \to E'$ is weakly $C \cap B_R$-monotone for some $R > 0$.

Then the minimization problem for the pair (f, M) has a solution.

3 Non-Quasiconvex Minimization

Now we indicate briefly a concrete situation where a functional of the form (1.2) can be minimized in circumstances where no quasi-convexity is available. When a functional of the form (1.2) is studied for proving the existence of (weak) solution of a system of nonlinear PDE's it is natural to assume that the integrand $F(x, u, P)$ is not only a Caratheodory function but has derivatives with respect to u ($F_{,u}$) and P ($F_{,P}$) which are Caratheodory functions too. Here $P \in M^{m \times n}$, the space of all real $m \times n$ matrices. Then, under growth restrictions as in [Brü94], one obtains for the Gâteaux derivative of f at $u \in W^{1,p}(\Omega; \mathbb{R}^m)$

$$j'(u)(v) = \langle F_{,u}(\cdot, u, Du) \cdot v \rangle + \langle F_{,P}(\cdot, u, Du) \cdot Du \rangle \tag{3.1}$$

for all $v \in E = W^{1,p}(\Omega, \mathbb{R}^m)$. Here we use $\langle \cdot \rangle = \int_\Omega \cdot dx$. Now suppose that $C \subset M^{m \times n}$ is a closed convex cone and define

$$K = \{u \in E \mid Du(x) \in C \text{ for almost all } x \in \Omega\}. \tag{3.2}$$

One verifies that the zero element of E is in the weak closure of K. Certainly, for $C \neq M^{m \times n}$ we have $K \neq E$ and thus weak K-monotonicity does not imply quasi-convexity and therefore not weak lower semi-continuity.

Proposition 3.1. *Suppose that for all $u \in \mathbb{R}^m$, all $P \in M^{m \times n}$, almost all $x \in \Omega$, and all $Q \in C$ we have*

$$[F_{,P}(x, u, P + Q) - F_{,P}(x, u, P)] \cdot Q \geq 0. \tag{3.3}$$

Then, under growth and smoothness restrictions as in [Brü94], the Gâteaux derivative f' is weakly K-monotone at any $v \in M$, for any non-empty subset $M \subset E$ which satisfies the cone condition for the cone K.

The proofs of this proposition and the following theorem are given in [Brü00].

Theorem 3.1. *Consider the functional f in (1.2) on $E = W^{1,p}(\Omega; \mathbb{R}^m)$ and assume the condition (3.3) of restricted monotonicity in addition to our standard assumptions (a) - (d). Then f has a minimum on any set $M \subset E$ which satisfies the cone condition for the cone K, Equation 3.2.*

4 Conclusions

As our discussion shows one can minimize functionals f of vector fields u, for instance of the form (1.2), over certain sets $M \subset E$, even if the integrand F is not quasi-convex in the variable Du whenever the Gâteaux derivative is weakly K-monotone for some subset $K \subset E$. In some detail we have presented the case where K is some cone in the space $E = W^{1,p}(\Omega; \mathbb{R}^m)$ and the set M satisfies the cone condition for this cone. This indicates that several important classes of functionals from material science (nonlinear elasticity, thermo-elasticity, visco-elasticity, plasticity etc.) could be minimized by this approach based on weak K-monotonicity (work in progress).

References

[Br65]   F. E. Browder, *Variational methods for non-linear elliptic eigen-value problems*, Bull. Amer. Math. Soc. **71** (1965), 176-183.

[Br70]   F. E. Browder, *Existence theorems for nonlinear partial differential equations*, pp. 1-60 in: Global Analysis, ed.: S.-S. Chern and S. Smale, Proceedings of Symposia in Pure Mathematics vol. **16**, American Math. Soc., Providence, RI, 1970.

[Br76]   F. E. Browder, *Nonlinear operators and nonlinear equations of evolution in Banach spaces*, Proceedings of Symposia in Pure Mathematics vol. **18.2**, American Math. Soc., Providence, RI, 1976.

[Br70]   F. E. Browder, *Pseudo-monotone operators and the direct method of the calculus of variations*, Arch. Rational Mech. Anal. **38** no. 4 (1970), 268-277.

[Br77]   F. E. Browder, *Nonlinear functional analysis*, in *Mathematical Developments arising from Hilbert Problems*, ed.: F. E. Browder; Proceedings of Symposia in Pure Mathematics Vol. **28** (1977), 68-73; AMS, Providence, RI.

[Brü90] E. Brüning, *On the variational approach to semi-linear elliptic equations with scale covariance*, Journ. Differential Equations **83** (1990), 109-144.

[Brü94] E. Brüning, *Minimization without weak lower semi-continuity*, Applicable Analysis **54** (1994), 91-111.

[Brü97] E. Brüning, *Eigenvalue problems for global quasi-linear partial differential operators*, Commun. Appl. Nonlinear Analysis, vol.4 no. 3 (1997), 55-67.

[Brü00] E. Brüning, Non-convex minimization for vector field problems; preprint, University of Durban-Westville, Durban 2000.

[Da89]   B. Dacorogna, *Direct methods in the calculus of variations*, Applied Mathematical Sciences vol. **78**, Springer-Verlag, Berlin Heidelberg New York, 1989.

[Ek79]   I. Ekeland, *Non-convex minimization problems*, Bull. A. M. S., new series vol. **1** (1979), 443-475.

[ET76]   I. Ekeland and R. Temam, *Convex analysis and variational problems*, Monographs, North Holland, Amsterdam Oxford, 1976.

[Ev90]   L. C. Evans, *Weak convergence metheods for nonlinear partial differential equations*, CBMS - Regional Conference Series in Mathematics vol. **74**, American Mathematical Society, Providence, Rhode Island 1990.

[Mo52] C. B. Morrey, *Quasi-convexity and semi-continuity of multiple integrals*, Pacific J. Math. **2** (1952), 25-53.

[Mo66] C. B. Morrey, *Multiple integrals in the calculus of variations.*, Monographs, Springer, Berlin, 1966.

[Mü98] S. Müller, *Variational models for microstructure and phase transition*, Lecture Notes, Max-Planck-Institut für Mathematik in den Naturwissenschaften Leipzig, vol. **2**, Leipzig 1998.

[OR76] J. T. Oden and J. N. Reddy, *Variational Methods in Theoretical Mechanics*, Springer-Verlag Berlin Heidelberg New York 1976.

[Pe99] P. Pedregal, *Optimization, relaxation and Young measures*, Bulletin of the American Mathematical Society vol. **36** no. 1 (1999), 27-58.

[Yo37] L. C. Young, *Generalized curves and the existence of an attained absolute minimum in the calculus of variations*, Comptes Rendues de la Société des Sciences et des Lettres de Varsovie, class III, vol. **30** (1937), 212-234.

[Yo69] L. C. Young, *Lectures on the calculus of variations and optimal control theory*, W. B. Saunders, Philadelphia, 1969.

[Ze85] E. Zeidler, *Variational methods and optimization, Nonlinear functional analysis and its applications*, vol. **III**, Springer-Verlag, New York Berlin Heidelberg London Paris Tokyo, 1985.

Address

ERWIN BRÜNING, Department of Mathematics & Applied Mathematics, University of Durban-Westville

E-MAIL: ebruning@pixie.udw.ac.za

2000 Mathematics Subject Classification. Primary 35A15, 35G20; Secondary 49K20

Operator Theory:
Advances and Applications, Vol. 126
© 2001 Birkhäuser Verlag Basel/Switzerland

On the One Dimensional Behaviour of Atoms in Intense Homogeneous Magnetic Fields

R. BRUMMELHUIS AND P. DUCLOS

Abstract. We show that the Hamiltonian of an atomic ion with N bosonic electrons which is submitted to an homogeneous magnetic field and when appropriately scaled and zoomed, converges, as the strength of the field tends to infinity and in the norm resolvent sense, to the Hamiltonian of a one dimensional atomic ion where the coulombic interactions have been replaced by "delta potentials".

1 Introduction

By intense magnetic fields we mean magnetic fields which are so strong that the energy to jump between two Landau levels is much larger than the one to ionise a hydrogen atom. The electrons being confined in the lowest Landau level have no freedom in the directions orthogonal to the field and therefore will behave as one dimensional particles. Moreover these particles are expected to interact through delta potentials. This note is a contribution to a rigorous analysis of this phenomena in the particular case of bosonic electrons, see the theorem below.

The strength B of the field, which is necessary for such a situation to occur, must be larger that 10^5 tesla a value which is today beyond the possibilities of the laboratories on Earth but seems to be present on the surface of neutron stars. This is why one can find in the astrophysical literature many discussions about possible formations of new atoms and molecules in this context. Among these new species stable atoms with 2Z electrons are conjectured to exist for $B >> Z^3$ where Z is the nuclear charge, see [LSoY94, p524].

With the help of the theorem below it is sufficient to study this stability question for a one dimensional atom whose particles interact through delta potentials and whose Hamiltonian reads:

$$h = \sum_{j=1}^{N} -\partial_j^2 - Z\delta(x_j) + \sum_{j<k} \delta(x_j - x_k).$$

Unfortunately these Hamiltonians do not seem to have been much studied in the literature except of course for the trivial case $N = 1$ and the case $N = 2$, see [Ro71]. In the latter case it is proven that an L^2-ground state exists as long as $Z > 0.374903$, which shows that the conjecture is true at least up to $N = 2$. This communication is also a call to experts in zero-range interactions to devote efforts to study the above Hamiltonian h.

We are far from being the first to tackle this problem. We recommend the two papers [LSoY94], [BaSoY00], which deal with fermionic atoms. However these authors use a variational approach, which does not directly guarantee that the ground state exists and which does not seem to be able to give information on the structure of this ground state. Two things we can handle with our theorem below (but for bosonic atoms, though).

The Hamiltonian of an N-electron atom in an homogeneous magnetic field with a fixed centre of mass and without Fermi-statistics acts in $L^2(\mathbb{R}^{3N})$ and reads as

$$\mathbb{H} := \sum_{j=1}^{N} H_j + \sum_{j<k} \frac{1}{|r_j - r_k|} \quad \text{where} \quad H_j := (-i\nabla_{r_j} - B\hat{x} \wedge r_j)^2 - \frac{Z}{|r_j|}.$$

Here $r_j = (x_j, y_j, z_j)$ denotes the position of the jth electron with respect to the nucleus, Z is the charge of the nucleus and B is proportional to the strength of the magnetic field which points towards the x-axis; $\hat{x}$ denotes the unit vector on the x-axis of the ambient space $\mathbb{R}^3$. We have set the Planck constant $\hbar = 1$ and the mass of the electron $m_e = 1/2$ since their values are irrelevant for our discussion. If we denote by $L_{x_j} := -i\partial_{z_j} y_j + i\partial_{y_j} z_j$ the angular momentum operator of the jth electron with respect to the x-axis and $\mathbb{L} := \sum_{j=1}^{N} L_{x_j}$, it is straightforward to check that $\mathbb{H}$ commutes with $\mathbb{L}$. Therefore $\mathbb{H}$ is reduced by the spectral projections of $\mathbb{L}$. We shall be concerned here by the part $\mathbb{H}^{(0)}$ of $\mathbb{H}$ which corresponds to a zero total angular momentum with respect to the x-axis. This choice is motivated partly to simplify the analysis and partly because it is believed that the lowest energy is reached in this sector, see [BaSe00] and [AHS81] for the special case $N = 1$. To state the main theorem of this note we introduce the unitary transformation U associated to the following scaling

$$\forall j = 1, \ldots N, \quad r_j \to \left(\frac{x_j}{\log B}, \frac{y_j}{\sqrt{B}}, \frac{z_j}{\sqrt{B}} \right)$$

and the projector $\mathit{\Pi}_0^{(0)}$ on states where all the electrons are in the lowest Landau level and have zero angular momentum with respect to the x-axis. Although these objects are standard we give a precise definition of them in the next section. Then one has the

Theorem. *As B tends to infinity one has, uniformly with respect to ξ in every compact of the resolvent set of h, that:*

$$\left(\frac{1}{(\log B)^2} U \left(\mathbb{H}^{(0)} - 2NB \right) U^{-1} - \xi \right)^{-1} - (h - \xi)^{-1} \otimes \mathit{\Pi}_0^{(0)} = \mathcal{O}((\log B)^{-\frac{1}{3}}).$$

From this theorem we can draw two conclusions. Let Z and N be such that h possesses a true L^2-ground state φ with energy e. Then for B large enough $\mathbb{H}^{(0)}$ has also an L^2-ground state $\Phi(B)$ with energy $E(B)$ such that

$$\lim_{B \to \infty} \frac{E(B) - 2NB}{(\log B)^2} = e,$$

and

$$\lim_{B \to \infty} \| \Phi(B) - \sqrt{\log B}\, \varphi(\log B \,\cdot) \otimes \sqrt{B} \chi_0^{(0)}(\sqrt{B}\,\cdot) \| = 0$$

where $\chi_0^{(0)}$ denotes the product of N lowest Landau level functions with zero angular momentum with respect to the x-axis.

2   Proof of the Theorem

First we give the definition of h. Let $D : \mathbb{R}^{N-1} \to \mathbb{R}^N$ be a regular parametrisation of a plane in $\mathbb{R}^N$ of codimension 1 and $t_D : \mathcal{H}^1(\mathbb{R}^N) \to \mathbb{C}$ be the quadratic form defined by ($\mathcal{H}^1(\mathbb{R}^N)$ stands for the first L^2 Sobolev space)

$$t_D[\psi] := \int_{\mathbb{R}^{N-1}} |\psi \circ D(u)|^2 |D'(u)| du.$$

If D_j and $D_{j,k}$ denote respectively a parametrisation of $x_j = 0$ and $x_j = x_k$ we let $t := t_0 + t_1$ with $t_0, t_1 : \mathcal{H}^1(\mathbb{R}^N) \to \mathbb{C}$ being the quadratic forms defined by $t_0[\psi] := \sum_{j=1}^{N} \|\partial_j \psi\|^2$ and

$$t_1 = -Z \sum_{j=1}^{N} t_{D_j} + \sum_{j<k}^{N} t_{D_{j,k}}.$$

One can check easily that t is closed symmetric, densely defined and bounded below and therefore t defines a selfadjoint operator that we denote by h.

Next we explain what $\mathbb{H}_0^{(0)}$ is. Each H_j is unitarily equivalent to

$$H_j \sim B \bigoplus_{m_j \in \mathbb{Z}} \left(-\partial_{x_j}^2 + H_{\mathrm{osc},j}^{(m_j)} - 2m_j - \lambda Z \frac{1}{|r_j|} \right), \quad \lambda := \frac{1}{\sqrt{B}}$$

where $H_{\mathrm{osc},j} := -\partial_{y_j}^2 - \partial_{z_j}^2 + y_j^2 + z_j^2$ is the harmonic oscillator which describes the Landau levels of the jth electron and $H_{\mathrm{osc},j}^{(m)}$ its restriction to angular momenta in the x-direction of value m. In other words

$$H_{\mathrm{osc},j}^{(m_j)} - 2m_j = \bigoplus_{n_j=0}^{\infty} (2(|m_j| - m_j) + 4n_j + 2)\, \Pi_{j,n_j}^{(m_j)}$$

where $\varPi_{j,n_j}^{(m_j)}$ is the eigenprojector on the n_j^{th} level of this harmonic oscillator. Then

$$\varPi_0^{(0)} := \varPi_{1,0}^{(0)} \otimes \ldots \otimes \varPi_{N,0}^{(0)} \quad \text{and} \quad \hat{\varPi}_0^{(0)} := \varPi^{(0)} - \varPi_0^{(0)}$$

where

$$\varPi^{(0)} := \sum_{m_1+\ldots+m_N=0} \sum_{n\in\mathbb{N}^N} \varPi_{1,n_1}^{(m_1)} \otimes \ldots \otimes \varPi_{N,n_N}^{(m_N)}.$$

With these notations one has: $\mathbb{H}^{(0)} := \varPi^{(0)} \mathbb{H} \varPi^{(0)}$.

Then we introduce new notations. Let

$$a := 2\lambda \log \lambda^{-1} = \frac{\log B}{\sqrt{B}}$$

and

$$(\log B)^{-2} U \left(\mathbb{H}^{(0)} - 2NB \right) U^{-1} =: T^a - \frac{\lambda}{a^2}\mathcal{V}^a, \quad \mathcal{V} := \sum_{j=1}^{N} \frac{1}{|r_j|} - \sum_{j<k} \frac{1}{|r_j - r_k|}$$

where T^a is the kinetic part i.e. $T^a = \sum_{j=1}^{N} -\partial_{x_j}^2 + a^{-2}(H_{\text{oscc},j} - 2 - 2L_{x_j})$ and $\mathcal{V}^a(r) := \mathcal{V}(x/a, y, z)$, both sandwiched between $\varPi^{(0)}$.

We compute the resolvent of this operator at a point $\xi < -NZ^2/4$ since such a point surely belongs to the resolvent set of h. We do this with a Feshbach-type formula using the decomposition of Ran $\varPi^{(0)}$ as Ran $\varPi_s \oplus$ Ran $\varPi_r$ where we have set $\varPi_s := \varPi_0^{(0)}$ and $\varPi_r := \hat{\varPi}_0^{(0)}$, to ease the notations. In the sequel a subscript s or r on an operator indicate that the projection $\varPi_s$ or $\varPi_r$ has been applied on the left or on the right according to the position of the subscript: for example $\mathcal{V}_{s,r}^a := \varPi_s \mathcal{V}^a \varPi_r$, or simply $T_r^a := \varPi_r T^a \varPi_r$ when this projection commutes with this operator. So

$$(T^a - \frac{\lambda}{a^2}\mathcal{V}^a - \xi)^{-1} = \begin{pmatrix} S^a & \frac{\lambda}{a^2}S^a\mathcal{V}_{s,r}^a R^a \\ \frac{\lambda}{a^2}R^a\mathcal{V}_{r,s}^a S^a & R^a + \frac{\lambda^2}{a^4}R^a\mathcal{V}_{r,s}^a S^a\mathcal{V}_{s,r}^a R^a \end{pmatrix}$$

where

$$R^a := (T_r^a - \frac{\lambda}{a^2}\mathcal{V}_{r,r}^a - \xi)^{-1}, \quad S^a := (T_s^a - \frac{\lambda}{a^2}\mathcal{V}_{s,s}^a - \frac{\lambda^2}{a^4}\mathcal{V}_{s,r}^a R^a\mathcal{V}_{r,s}^a - \xi)^{-1}.$$

We inspect all terms step by step.

Bound on $R_0^a := (T_r^a - \xi)^{-1}$. Since $T_r^a \geq 4a^{-2}\varPi_r$ one has $\|R_0^a\| \leq a^2/4$. We denote by $R_0 := (T_r - a^2\xi)^{-1}$ and therefore $\|R_0\| \leq 1/4$. Here T means T^1.

Bound on $|r_j|^{-1}R_0|r_j|^{-1}$. Using the Hardy's inequality, the spectral decomposition of $T := T^1$ and the facts $\sum_j m_j = 0$, $\xi \leq 0$ we get:

$$\begin{aligned}
(|r_j|^{-1}R_0|r_j|^{-1})(|r_j|^{-1}R_0|r_j|^{-1}) &= |r_j|^{-1}R_0|r_j|^{-2}R_0|r_j|^{-1} \\
&\leq |r_j|^{-1}R_0(-4\Delta_{r_j})R_0|r_j|^{-1} \\
&\leq 4|r_j|^{-1}R_0(-\Delta_{r_j} + y_j^2 + z_j^2)R_0|r_j|^{-1} \\
&\leq 6|r_j|^{-1}R_0|r_j|^{-1}
\end{aligned}$$

and solving this quadratic inequality gives

$$\forall j = 1, \ldots, N, \quad \||r_j|^{-1}R_0|r_j|^{-1}\|^{-1} \leq 6.$$

Bound on $\lambda a^{-2}\sqrt{R_0^a}Z|r_j^a|^{-1}\sqrt{R_0^a}$. Here r_j^a denotes $(x_j/a, y_j, z_j)$. Doing the reverse dilation $x_j \to ax_j$, $j = 1, \ldots, N$ we get

$$\begin{aligned}
\|\lambda a^{-2}\sqrt{R_0^a}\,Z|r_j^a|^{-1}\sqrt{R_0^a}\| &= \|\lambda Z\sqrt{R_0}\,|r_j|^{-1}\sqrt{R_0}\| \\
&\leq |\lambda Z|\|\sqrt{R_0}\|\||r_j|^{-1}R_0|r_j|^{-1}\|^{\frac{1}{2}} \\
&\leq |\lambda Z|\frac{1}{2}\sqrt{6} = \mathcal{O}(\lambda).
\end{aligned}$$

Bound on R^a. Let R_{NI}^a be defined as R^a with the interactions between the electrons removed, then $R^a \leq R_{\mathrm{NI}}^a$ and

$$R_{\mathrm{NI}}^a = \sqrt{R_0^a}\left(1 - \frac{\lambda Z}{a^2}\sum_{j=1}^N \sqrt{R_0^a}|r_j^a|^{-1}\sqrt{R_0^a}\right)^{-1}\sqrt{R_0^a}$$

which implies $\|R^a\| = \mathcal{O}(a^2)$.

Bound on $|r_j - r_k|^{-1}R_0|r_j - r_k|^{-1}$. Using the unitary transform associated to the change of variables

$$(r_j, r_k) \to \left(t := \frac{r_j - r_k}{\sqrt{2}}, s := \frac{r_j + r_k}{\sqrt{2}}\right) \tag{2.1}$$

$-\Delta_{r_j} - \Delta_{r_k} + y_j^2 + z_j^2 + y_k^2 + z_k^2$ is changed in $-\Delta_s - \Delta_t + s_2^2 + s_3^2 + t_2^2 + t_3^2$, and $L_{x_j} + L_{x_k}$ in $L_{s_1} + L_{t_1}$. Thus we may use the bound on $|r_j|^{-1}R_0|r_j|^{-1}$ to deduce that

$$\forall j \neq k, \quad \||r_j - r_k|^{-1}R_0|r_j - r_k|^{-1}\| \leq 6$$

Bound on $\lambda^2 a^{-4}\mathcal{V}_{s,r}^a R_0^a \mathcal{V}_{r,s}^a$. Doing the reverse scaling $x_j \to ax_j$, $j = 1, \ldots, N$ and expanding gives :

$$\|\lambda^2 a^{-4}\mathcal{V}_{s,r}^a R_0^a \mathcal{V}_{r,s}^a\| = \|\lambda^2 a^{-2}\mathcal{V}_{s,r}R_0\mathcal{V}_{r,s}\|$$

$$\leq 2N \sum_j \|\lambda^2 a^{-2} |r_j|^{-1} R_0 |r_j|^{-1}\|$$

$$+ N(N-1) \sum_{j \neq k} \|\lambda^2 a^{-2} |r_j - r_k|^{-1} R_0 |r_j - r_k|^{-1}\|$$

$$= \mathcal{O}\left(\frac{\lambda^2}{a^2}\right).$$

Bound on $\lambda^2 a^{-4} \mathcal{V}^a_{s,r} R^a \mathcal{V}^a_{r,s}$. Using $R^a \leq R^a_{\mathrm{NI}}$, the formula for R^a_{NI} in term of R^a_0 and the previous bound gives:

$$\lambda^2 a^{-4} \mathcal{V}^a_{s,r} R^a \mathcal{V}^a_{r,s} = \mathcal{O}\left(\frac{\lambda^2}{a^2}\right).$$

Now we need the following

Lemma. *As B tends to infinity one has*

$$\left((\log B)^{-2} I\!I_s U (I\!H^{(0)} - 2NB) U^{-1} I\!I_s - \xi \right)^{-1} - (h - \xi)^{-1} = \mathcal{O}((\log B)^{-\frac{1}{3}}).$$

Proof. Let W be the following one dimensional potential

$$W(x) := \int_0^\infty \frac{e^{-u}}{\sqrt{x^2 + u}} du.$$

It is proven in [BRu99] that: $B^{-1} I\!I_s U_{B^{-\frac{1}{2}}} \left(I\!H^{(0)} - 2NB \right) U^{-1}_{B^{-\frac{1}{2}}} I\!I_s$ is equal to

$$\sum_{j=1}^N -\partial^2_{x_j} - \lambda Z W(x_j) + \frac{\lambda}{\sqrt{2}} \sum_{j<k} W(\frac{x_j - x_k}{\sqrt{2}});$$

where $U_{B^{-\frac{1}{2}}}$ denotes the unitary implementation of the scaling $r_j \to B^{-\frac{1}{2}} r_j$, $j = 1, \ldots, N$. This shows that we have to face a weak coupling limit problem since we are interested in the limit $\lambda \to 0$. We use the approach described in the next section. We perform an extra scaling $x_j \to x_j/a$, $j = 1, \ldots, N$ which finally gives:

$$h(B) := (\log B)^{-2} I\!I_s U_{B^{-\frac{1}{2}}} (I\!H^{(0)} - 2NB) U^{-1}_{B^{-\frac{1}{2}}} I\!I_s =$$

$$\sum_{j=1}^N -\partial^2_{x_j} - \frac{\lambda}{a^2} Z W(\frac{x_j}{a}) + \frac{\lambda}{a^2 \sqrt{2}} \sum_{j<k} W(\frac{x_j - x_k}{a\sqrt{2}}).$$

One can check easily that

$$\int_{|x|<R} W(x) dx = 2\log(2R) + \gamma + \mathcal{O}(R^{-2}), \quad \text{as} \quad R \to \infty$$

where γ is the Euler constant. It is shown in the next section that under the above asymptotic behaviour one has[1]:

$$\|\frac{\lambda}{a^2}W(\frac{\cdot}{a}) - \delta\|_{-1,1} = \mathcal{O}((\log B)^{-\frac{1}{3}}) \quad \text{as} \quad B \to \infty$$

where $\|\cdot\|_{-1,1}$ means the operator norm from $\mathcal{H}^1(\mathbb{R})$ to $\mathcal{H}^{-1}(\mathbb{R})$; here δ must be understood as the operator $\psi \to \psi(0)\delta$. Obviously this implies

$$\|1\ldots\otimes 1\left(\frac{\lambda}{a^2}W(x_j/a) - \delta(x_j)\right)1\otimes\ldots 1\|_{-1,1} = \mathcal{O}((\log B)^{-\frac{1}{3}})$$

as $B \to \infty$ and as an operator from $\mathcal{H}^1(\mathbb{R}^n)$ to $\mathcal{H}^{-1}(\mathbb{R}^n)$, for all $j = 1,\ldots N$.

The analogous statement, for all $j \neq k$,

$$\|1\ldots 1\otimes \left(\frac{\lambda}{a^2\sqrt{2}}W_{0,0}(\frac{x_j - x_k}{a\sqrt{2}}) - \delta(x_j - x_k)\right)\otimes 1\ldots 1\|_{-1,1} = \mathcal{O}((\log B)^{-\frac{1}{3}})$$

as $B \to \infty$ follows by using a change of variables of the type (2.1). These two properties are sufficient to imply the statement of the lemma since

$$(h(B) - \xi)^{-1} = \sqrt{(h - \xi)^{-1}}(1 + K)^{-1}\sqrt{(h - \xi)^{-1}}$$

with

$$K := \sqrt{(h - \xi)^{-1}}\left(-Z\sum_j \left(\frac{\lambda}{a^2}W(\frac{x_j}{a}) - \delta(x_j)\right) + \right.$$

$$\left.\sum_{j<k}\left(\frac{\lambda}{a^2\sqrt{2}}W(\frac{x_j - x_k}{\sqrt{2}a}) - \delta(x_j - x_k)\right)\right)\sqrt{(h - \xi)^{-1}}$$

and $K \to 0$ as $B \to \infty$. ∎

Convergence of S^a to $(h - \xi)^{-1}$. From the above lemma one has: $(T_s^c - \frac{\lambda}{a^2}V_{s,s}^a - \xi)^{-1} - (h - \xi)^{-1} = \mathcal{O}\left((\log B)^{-\frac{1}{3}}\right)$ whereas using the second resolvent equation one gets $S^a - (T_s^a - \frac{\lambda}{a^2}V_{s,s}^a - \xi)^{-1} = \mathcal{O}\left(\lambda^2 a^{-2}\right) = \mathcal{O}((\log B)^2)$.

The last terms. It is now easy to verify that $V_{s,r}^a R^a = \mathcal{O}(a^2)$ and thus

$$\frac{\lambda}{a^2}S^a V_{s,r}^a R^a = \mathcal{O}(\lambda), \quad \frac{\lambda^2}{a^4}R^a V_{r,s}^a S^a V_{s,r}^a R^a = \mathcal{O}(\lambda^2) = \mathcal{O}(B(\log B)^{-1}).$$

The theorem is proven.

---

[1] we warn the reader that the parameters a and v of the next section are presently changed as $a \to 2a$ and $v \to v/2$; a change which does not affect the following limit as one can see with a simple scaling argument.

3   The Weak Coupling Limit Revisited

In this section we address the following question. Let $V : \mathbb{R} \to \mathbb{R}_-$ be a measurable function such that (with $\langle x \rangle := \sqrt{1 + x^2}$)

$$\langle x \rangle V \in L^\infty(\mathbb{R}) \tag{H1}.$$

How much V is attractive is made more precise by

$$\exists v > 0, \quad \mathcal{V}(R) := - \int_{|x|<R} V(x)dx = v \log(R) + \mathcal{O}(1) \quad \text{as} \quad R \to \infty \tag{H2}.$$

Let $H_\lambda := -\frac{d^2}{dx^2} + \lambda V$. *Does H_λ possess negative eigenvalues as λ tends to 0?* The answer is yes as one can see with an easy variational argument. If V obeys slightly stronger assumptions, a perturbation expansion for the lowest eigenvalue in terms of λ is derived in [AHS81] which uses a modified Birman-Schwinger technic. We shall rederive part of this result with a different method better suited to this note. This method is based on the following property: H_λ, when appropriately scaled and zoomed, converges in the norm resolvent sense to $h_0 := -\frac{d^2}{dx^2} - v\delta$. Let

$$h(a, \lambda) := -\frac{d^2}{dx^2} + \frac{\lambda}{a^2} V(\frac{\cdot}{a}) = a^{-2} U_{\frac{1}{a}} H_\lambda U_{\frac{1}{a}}^{-1}$$

where $U_{\frac{1}{a}}$ denotes the unitary scaling associated to the change of variable $x \to x/a$, $a > 0$.

Lemma. *Let $V : \mathbb{R} \to \mathbb{R}_-$ satisfy (H1,2) and $a(\lambda) := \lambda \log \lambda^{-1}$. Then $h(a, \lambda)$ converges in the norm resolvent sense to h_0 as $\lambda \downarrow 0$. More precisely for every ξ in the resolvent set of h_0*

$$(h(a, \lambda) - \xi)^{-1} - (h_0 - \xi)^{-1} = \mathcal{O}\left(\left(\log \lambda^{-1} \right)^{\frac{1}{3}} \right) \quad \text{as} \quad \lambda \to 0.$$

Proof. We assume without loss of generality that $v = 1$. We split V in two parts:

$$\forall R > 0, \quad V_R(x) := \begin{cases} V(x) & \text{if } |x| < R \\ 0 & \text{otherwise} \end{cases}, \quad \text{and} \quad \tilde{V}_R := V - V_R.$$

and we impose the following relation between the three parameters a, R and λ:

$$a = -\lambda \int_{|x|<R} V(x)dx \iff a = \lambda \mathcal{V}(R) \tag{3.1}$$

so that

$$\frac{\lambda}{a^2} \int_{\mathbb{R}} V_R(\frac{x}{a})dx = \frac{\lambda}{a} \int_{\mathbb{R}} V_R(x)dx = -1.$$

Let $r_0 := (h_0 + 5/4)^{-1}$ thus $\|r_0\| \le 1$; clearly the norm resolvent convergence of $h(a,\lambda)$ to h_0 follows from the convergence to 0 of

$$\sqrt{r_0}\left(\frac{\lambda}{a^2}V(\tfrac{\cdot}{a}) + \delta\right)\sqrt{r_0} = \sqrt{r_0}\left(\frac{\lambda}{a^2}V_R(\tfrac{\cdot}{a}) + \delta\right)\sqrt{r_0} + \sqrt{r_0}\frac{\lambda}{a^2}\tilde{V}_R(\tfrac{\cdot}{a})\sqrt{r_0}.$$

The second term is estimated as follows (we use (3.1) for the last inequality):

$$\|\sqrt{r_0}\frac{\lambda}{a^2}\tilde{V}_R(\tfrac{\cdot}{a})\sqrt{r_0}\| \le \frac{\lambda}{a^2}\|\tilde{V}_R(\tfrac{\cdot}{a})\|_\infty \le \frac{\lambda}{a^2}\sup\{|V(x)|, |x \ge R\}$$

$$\le \frac{\lambda}{a^2}\frac{1}{\sqrt{1+R^2}}\|\langle x\rangle V\|_\infty \le \frac{\|\langle x\rangle V\|_\infty}{\lambda R\mathcal{V}(R)^2}.$$

For the first term we have, using also (3.1) and with u, v in $\mathcal{H}^1(\mathbb{R})$ and $\varphi := u\bar{v}$:

$$\left|\left\langle\left(\frac{\lambda}{a^2}V_R(\tfrac{\cdot}{a}) + \delta\right)u, v\right\rangle_{-1,1}\right| = \left|\int_\mathbb{R}dx\frac{\lambda}{a^2}V_R(\tfrac{x}{a})(\varphi(x) - \varphi(0))\right|$$

$$= \left|\int_\mathbb{R}dx\frac{\lambda}{a}V_R(x)(\varphi(ax) - \varphi(0))\right|$$

$$\le \frac{\lambda}{a}\int_{|x|<R}dx|V(x)|\int_0^{ax}dy|\varphi'(y)|$$

$$\le \frac{\lambda}{a}\sqrt{aR}\int_{|x|<R}dx|V(x)|\|\varphi'\|$$

$$\le 2\lambda\sqrt{\frac{R}{a}}\int_{|x|<R}dx|V(x)|\|u\|_1\|v\|_1$$

and consequently

$$\|\sqrt{r_0}\left(\frac{\lambda}{a^2}V_R(\tfrac{\cdot}{a}) + \delta\right)\sqrt{r_0}\| \le 2\lambda\sqrt{\frac{R}{a}}\mathcal{V}(R) = 2\sqrt{\lambda R\mathcal{V}(R)};$$

$\langle\cdot,\cdot\rangle_{-1,1}$ denotes the duality brackets between $\mathcal{H}^1(\mathbb{R})$ and $\mathcal{H}^{-1}(\mathbb{R})$ and $\|\cdot\|_1$ the norm of $\mathcal{H}^1(\mathbb{R})$. Thus with $C_V := \|\langle x\rangle V\|_\infty$

$$\|\sqrt{r_0}\left(\frac{\lambda}{a^2}V_R(\tfrac{\cdot}{a}) + \delta\right)\sqrt{r_0}\| \le 2\sqrt{\lambda R\mathcal{V}(R)} + \frac{C_V}{\lambda R\mathcal{V}(R)^2}.$$

We choose R as a function of λ so that

$$2\sqrt{\lambda R\mathcal{V}(R)} = \frac{C_V}{\lambda R\mathcal{V}(R)^2} \iff \lambda = \left(\frac{C_V}{2}\right)^{\frac{2}{3}}\frac{1}{R\mathcal{V}(R)^{\frac{5}{3}}} =: \frac{1}{G(R)}.$$

Let $R_0 \ge 0$ be the infimum of the essential support of V in $\mathbb{R}_+$, then the map $G : [R_0, \infty[\to \mathbb{R}$ is strictly increasing since

$$(C_V/2)^{\frac{2}{3}}G'(R) = \mathcal{V}(R)^{\frac{2}{3}}(5/3(-V(R) - V(-R))R + \mathcal{V}(R)).$$

Therefore for every λ in $]0, 1/G(R_0)]$ there exists a unique R solving $\lambda = 1/G(R)$. With such a choice of R we get

$$\left\| \sqrt{r_0} \left(\frac{\lambda}{a^2} V_R(\frac{\cdot}{a}) + \delta \right) \sqrt{r_0} \right\| \leq \left(\frac{4C_V}{\mathcal{V}(R)} \right)^{\frac{1}{3}}.$$

Using (H3) it is straightforward though tedious to find that

$$R_\lambda = \frac{C}{\lambda(\log \lambda^{-1})^{\frac{5}{3}}} \left(1 + \mathcal{O}\left(\frac{\log_2 \lambda^{-1}}{\log \lambda^{-1}} \right) \right), \quad a = \lambda \log \lambda^{-1} \left(1 + \mathcal{O}\left(\frac{\log_2 \lambda^{-1}}{\log \lambda^{-1}} \right) \right).$$

and consequently $\mathcal{V}(R) = \mathcal{O}((\log \lambda)^{-1})$ as $\lambda \to 0$. Finally an easy argument based on the scaling properties of δ shows that one can replace a by any other function b such that $a/b \to 1$ as $\lambda \to 0$. Thus we are allowed to take $a = \lambda \log \lambda^{-1}$. $\blacksquare$

Once the norm resolvent convergence is proven it is standard to show that the infimum of the spectrum is a simple isolated eigenvalue for λ small enough and to derive the perturbation expansion of this eigenvalue in terms of λ.

Acknowledgements

This problem was suggested to both of us by Mary Beth Ruskai with whom we have had many fruitfull discussions. We are very greatfull to the referee whose very carefull reading has helped to improve the quality of the manuscript.

References

[AHS81]   J.E. Avron, I. Herbst, B. Simon: *Schrödinger Operators with Magnetic Fields III. Atoms in Homogeneous Magnetic Field* Commun. Math. Phys. **79** 529-572 (1981)

[BaSe00]   B. Baumgartner, R. Seiringer: *Atoms with bosonic "electrons" in strong magnetic fields* Preprint mp_arc n°00-283

[BaSoY00]   B. Baumgartner, J.-Ph. Solovej, J. Yngvason: *Atoms in strong magnetic fields: The high field limit at fixed nuclear charge*, Commun. Math. Physics 212 (3), 703 - 724 (2000)

[BRu99]   R. Brummelhuis, B. Ruskai: *A one-dimensional Model for many-electrons atoms in extremely strong magnetic fields: maximum negative ionisation* J. Phys. A, **32**, 2567 - 2582 (1999)

[LSoY94]   Lieb H.L., Solovej J. Ph., Yngvason J.: *Asymptotics of Heavy Atoms in High Magnetic Fields: I. Lowest landau band Regions.* Commun. Math. Phys. **47**, 513-591 (1994)

[Ro71]   C.M. Rosenthal: *Solution of Delta Function Model for Heliumlike Ions* Journ. Chem. Phys. **35**(5) 2474-2483 (1971)

Addresses

R. BRUMMELHUIS, Université de Reims Champagne Ardennes, Reims, France

E-MAIL: raymond.brummelhuis@univ-reims.fr

P. DUCLOS, Université de Toulon et du Var et Centre de Physique Théorique de Marseille, France

E-MAIL: duclos@univ-tln.fr

2000 Mathematics Subject Classification. Primary 81V55 ;Secondary 81Q15

Operator Theory:
Advances and Applications, Vol. 126
© 2001 Birkhäuser Verlag Basel/Switzerland

Semi-Classical Resolvent Estimates and Spectral Asymptotics for Trapping Perturbations

VINCENT BRUNEAU AND VESSELIN PETKOV

Abstract. We study semi-classical asymptotics of the Spectral Shift Function (SSF) for long-range trapping perturbations of $-h^2\Delta$. Without any assumption on the behaviour of the classical trajectories we prove the estimates $\|R(\lambda+i\tau)\|_{s,-s} \leq C \exp(Ch^{-p})$ for the resolvent $R(\lambda+i\tau)$. We apply this estimates to obtain a new semi-classical representation formula of the SSF.

1 Introduction

We consider a self-adjoint operator $L = L(h)$ depending on $h \in]0, h_0]$ and satisfying the long-range "black box" assumptions (introduced in [SZ91], [S97] and [S98]). This framework allows a treatment of the scattering phenomena without having a precise information of the perturbation in a compact set. The operator L is defined in a domain $\mathcal{D} \subset \mathcal{H}$ of a complex Hilbert space $\mathcal{H}$ with an orthogonal decomposition

$$\mathcal{H} = \mathcal{H}_{R_0} \oplus L^2(\mathbb{R}^n \setminus B(0, R_0)), \ B(0, R_0) = \{x \in \mathbb{R}^n : |x| \leq R_0\}, \ n \geq 2\,.$$

Concerning the "black box" we assume that $\mathbf{1}_{B(0,R_0)}(L + i)^{-1}$ is compact and on $\mathbb{R}^n \setminus B(0, R_0)$, L becomes a second order elliptic differential operator which is a long-range perturbation of $-h^2\Delta$ (see [S97], [S98], [BP00], [BP01] for more details).

An example of an operator satisfying our assumptions is the Schrödinger operator $L(h) = -h^2\Delta + V(x)$, with $V \in C^\infty(\mathbb{R}^n, \mathbb{R})$ satisfying for any $\alpha \in \mathbb{N}^n$:

$$|\partial_x^\alpha V(x)| \leq C_\alpha \langle x \rangle^{-\delta-|\alpha|}, \ \delta > 0, \ \langle x \rangle := (1 + |x|^2)^{\frac{1}{2}} \tag{1.1}$$

The first purpose of this note is to obtain semi-classical estimates for the resolvent $R(z) = (L(h) - z)^{-1}$, $z \in \mathbb{C} \setminus \mathbb{R}$, depending on $h \in]0, h_0]$, for $z \in \mathcal{B}_\pm = \{z = \lambda + i\tau \in \mathbb{C} : \lambda \in J, \pm\tau \in]0, 1]\}$, $J =]\mu_0, \mu_1[\subset\subset \mathbb{R}^+$ containing no eigenvalues. We introduce the spaces

$$\mathcal{H}^{0,s} = \mathcal{H}_{R_0} \oplus L^2(\mathbb{R}^n \setminus B(0, R_0), \langle x \rangle^s dx).$$

First we obtain semi-classical resolvent estimates

$$\|R(\lambda \pm i\tau)\|_{s,-s} \leq Cr(h), \ \lambda \in J, \ \tau \in]0, 1], \ h \in]0, h_0], \tag{1.2}$$

where $\|.\|_{s,-s}$ denotes the norm in $\mathcal{L}(\mathcal{H}^{0,s}, \mathcal{H}^{0,-s})$. Such estimates have been obtained by many authors in great generality in the case when every $\lambda \in J$ is a *non-trapping energy level* for the principal symbol $l_0(x, \xi)$ of a differential operator L (see [RT87] and the survey article [R98] for other references). Roughly speaking, we call these operators *non-trapping* perturbations and in this case we have $r(h) = h^{-1}$ in (1.2).

The case of *trapping energy levels* is more complicated. In the special case of a Schrödinger operator $-h^2\Delta + V(x)$, with potential $V(x)$ having the form of a "well in an island", the results in [GMR89] imply (1.2) with $r(h) = e^{Ch^{-1}}$. In the general case some results are known concerning the cut-off resolvents which depend on the distribution of resonances ([TZ98],[B98]). In particular for long-range trapping perturbations in the exterior of a connected obstacle Burq [B98] established that

$$\|\chi R(\lambda \pm i0)\chi\| \leq Ce^{ch^{-p}}, \ \lambda \in J. \tag{1.3}$$

with $p = 1$. Here we assume the estimate (1.3) with $p \geq 1$, provided $\chi \in C_0^\infty(\mathbb{R}^n)$ is equal to 1 on a sufficiently large region. Then we prove the estimate (1.2) with $r(h) = \exp(Ch^{-p})$ (see Theorem 2.1).

The second purpose of this note is to obtain a semi-classical representation for the derivative of the *Spectral Shift Function* (SSF) related to two self-adjoint operators $L_1 = L_1(h)$, $L_2 = L_2(h)$ on $\mathbb{R}^n$, $n \geq 2$, satisfying long-range "black box" assumptions and such that L_2 is a short-range perturbation of L_1. As an example we can consider the Schrödinger operators on $L^2(\mathbb{R}^n)$, $L_j(h) = -h^2\Delta + V_j(x)$, $j = 1, 2$, with $V_1, V_2 \in C^\infty(\mathbb{R}^n, \mathbb{R})$ satisfying (1.1) and for any $\alpha \in \mathbb{N}^n$:

$$|\partial_x^\alpha (V_1 - V_2)(x)| \leq C_\alpha \langle x \rangle^{-\bar{n}-|\alpha|}, \ \bar{n} > n. \tag{1.4}$$

For such operators the SSF $\xi(\lambda) \in S'(\mathbb{R})$ is defined as the temperate distribution

$$\langle \xi, f' \rangle_{S',S} = -\mathrm{tr}(f(L_2) - f(L_1)).$$

For long-range perturbations $P_j = P_j(h)$ acting in a dense set of $L^2(\mathbb{R}^n)$, Robert [R94] established a representation formula for the derivative of the SSF. The representation in [R94] involves some remainder terms related to the technical constructions of long times parametrices for the propagators $U_j(t) = e^{ith^{-1}P_j}$. In this paper we treat the problem of the representation of the derivative $\xi'(\lambda)$ of SSF in great generality covering the case of "black box" semi-classical scattering ([S97], [S98], [TZ98]) without the constructions of parametrices. Moreover, we can obtain (see [BP01]) a Weyl type asymptotic of $\xi(\lambda)$ generalizing to "black box" semi-classical setup, the results of [C98] and [R94].

2 Resolvent Estimates

Let us denote by $l_0(x, \xi)$ the principal symbol of the operator L, for the Schrödinger operator $l_0(x, \xi) = |\xi|^2 + V(x)$.

Theorem 2.1. *Let $\rho_0 > R_0$ be a constant, depending on J, such that all bounded trajectories of the Hamilton field $H_{l_0} := (\partial_\xi l_0, -\partial_x l_0)$ are included in $\{x : |x| \leq \rho_0/2\}$. Assume the estimate (1.3) is fulfilled for $\lambda \in J_1 \supset\supset J$ with $\chi \in C_0^\infty(\{x : |x| \leq \rho_1\})$, $\chi = 1$ for $|x| \leq \rho_0$. Then for each $s > \frac{1}{2}$, we have the inequality (1.2) with $r(h) = \exp(Ch^{-p})$.*

The idea of our proof is to construct a self-adjoint operator $\tilde{L} = \tilde{L}(h)$ on $L^2(\mathbb{R}^n)$ such that each $\lambda \in J$ is a non-trapping energy level for $\tilde{L}$ and $L(h)\psi = \tilde{L}(h)\psi$ for any $\psi \in C_0^\infty(\mathbb{R}^n)$ supported away from a ball $B(0, \rho_1)$. For this purpose we show that the bounded trajectories related to the symbol of $L(h)$ and to the energy levels $\lambda \in J$ are included in a compact set $K(J) \subset \mathbb{R}^{2n}$ depending on J. Next we represent $R(z)$ by a sum of terms involving the resolvent $\tilde{R}(z)$ of $\tilde{L}$ and an application of the semi-classical estimates for non-trapping energy levels reduces the problem to the estimation of $\|\chi(x)R(z)\chi(x)\|_{\mathcal{H} \mapsto \mathcal{H}}$.

3 Representation of the SSF

In the following $E_j'(\lambda)$ denote the spectral projectors for L_j given by the Stone formula $E_j'(\lambda) = \frac{1}{2\pi i}(R_j(\lambda+i0) - R_j(\lambda-i0))$, $\lambda \in J$, and we write $[a_j]_{j=1}^2 = a_2 - a_1$.

Theorem 3.1. *Let $\rho_0 > R_0$ be as in Theorem 2.1 and let $\chi \in C_0^\infty(\{x : |x| \leq \rho_1\})$ be equal to 1 for $|x| \leq 2\rho_0$, $\rho_1 > 2\rho_0$. Then for any $\lambda \in J$, the operator $\chi E'(\lambda)\chi$ is trace class one and we have*

$$\xi'(\lambda) = \left[\mathrm{tr}\left(\chi E_j'(\lambda)\chi\right)\right]_{j=1}^2 + h^{-n}\sum_{k \geq 0} d_k(\lambda)h^k \\ + \mathrm{tr}\,T^+(\lambda + i0) - \mathrm{tr}\,T^-(\lambda - i0),$$ (3.1)

where $d_k(\lambda)$ are C^∞ functions on J, while $T^\pm(z)$ are trace class operators for $\pm\Im z > 0$ such that for any $\tilde{\chi} \in C_0^\infty(\mathbb{R}^n)$, $\tilde{\chi} = 1$ on $\mathrm{supp}\chi$ we have

$$\|T^\pm(z)\|_{\mathrm{tr}} \leq \mathcal{O}(h^\infty)\left(1 + \|\tilde{\chi}R_1(z)\tilde{\chi}\|_{\mathcal{H} \to \mathcal{H}} + \|\tilde{\chi}R_2(z)\tilde{\chi}\|_{\mathcal{H} \to \mathcal{H}}\right), \quad \pm\Im z > 0,$$

uniformly with respect to $z \in \mathcal{B}_\pm = \{z \in \mathbb{C} : (\Re z, \pm\Im z) \in J \times]0,1]\}$ and $h \in]0, h_0]$. Moreover, the application $z \mapsto \mathrm{tr}(T^\pm(z))$ is holomorphic on $\mathcal{B}_\pm$.

The idea of the proof of Theorem 3.1 is to use an approximation of $\xi(\lambda)$ by a sequence of locally integrable functions $\xi_\rho'(\lambda) = \left[\mathrm{tr}\left(\chi_\rho E_j'(\lambda)\chi_\rho\right)\right]_{j=1}^2$, where $\chi_\rho \in C_0^\infty(B(0, 2\rho))$, $\chi_\rho = 1$ on $B(0, \rho)$. The operators $E_j'(\lambda)$ are not trace class so we decompose $\xi_\rho'(\lambda)$ in a sum of three terms and we examine the corresponding limits separately. The analysis of the term $\mathrm{tr}(T^\pm(z))$ is based on the localization in the incoming and outgoing regions and the ideas of Robert [R94] to combine the semi-classical estimates in these regions with the cyclicity of the trace.

References

[BP00]   V. Bruneau, V. Petkov. Semiclassical resolvent estimates for trapping perturbations, *Commun. Math. Phys.* **213**, 413-432, 2000.

[BP01]   V. Bruneau, V. Petkov. Representation of the spectral shift function and spectral asymptotics for trapping perturbations, Preprint, 2000.

[B98]   N. Burq. Absence de résonances près du réel pour l'opérateur de Schrödinger, Exposé XVII, *Séminaire EDP, Ecole Polytechnique*, 1997/1998.

[C98]   T. Christiansen. Spectral asymptotics for general compactly supported perturbations of the Laplacian on $\mathbb{R}^n$, *Comm. P.D.E.* **23**, 933-947, 1998.

[GMR89]   C. Gérard, A. Martinez and D. Robert. Breit-Wigner formulas for the scattering poles and total scattering cross-section in the semi-classical limit, *Commun. Math. Phys.* **121**, 323-336, 1989.

[R94]   D. Robert. Relative time-delay for perturbations of elliptic operators and semiclassical asymptotics, *J. Funct. Anal.* **126**, 36-82, 1994.

[R98]   D. Robert. Semiclassical approximation in quantum mechanics. A survey of old and recent mathematical results, *Helv. Phys. Acta*, **71**, 44-116, 1998.

[RT87]   D. Robert and H. Tamura. Semiclassical estimates for resolvents and asymptotics for total scattering cross-sections, *Ann. Inst. H. Poincaré (Physique théorique)*, **46**, 415-442, 1987.

[S97]   J. Sjöstrand. A trace formula and review of some estimates for resonances, in *Microlocal analysis and spectral theory (Lucca, 1996)*, 377–437, NATO Adv. Sci. Inst. Ser. C Math. Phys. Sci., 490, Dordrecht, Kluwer Acad. Publ. 1997.

[S98]   J. Sjöstrand. Resonances for bottles and trace formulae, *Math. Nachrichten*, (to appear). Preprint Ecole Polytechnique 98.

[SZ91]   J. Sjöstrand and M. Zworski. Complex scaling and the distribution of scattering poles, *J. Amer. Math. Soc.*, **4**, 729-769, 1991.

[TZ98]   S. Tang and M. Zworski. From quasimodes to resonances, *Math. Res. Lett.*, **5**, 261-272, 1998.

Address

VINCENT BRUNEAU AND VESSELIN PETKOV, Université Bordeaux I, 351 cours de la Libération, F-33405 Talence

E-MAIL: vbruneau@math.u-bordeaux.fr, petkov@math.u-bordeaux.fr

2000 Mathematics Subject Classification. Primary 35J10; Secondary 35P25

Operator Theory:
Advances and Applications, Vol. 126
© 2001 Birkhäuser Verlag Basel/Switzerland

Semiclassical Pseudodifferential Operators with Double Discontinuous Symbols and their Application to Problems of Quantum Statistical Physics

A.M. Budylin* and V.S. Buslaev

1 Integral Operator and its Resolvent

1.1 Integral Operator

Two-particle correlation function $g(t, s, \beta)$ for Bose gas of one-dimensional impenetrable particles is represented by the determinant $g = \det(I - V)$ where the integral operator V on $\mathbb{R}$ has the kernel (see, for example, [IIKV92],[KBI93])

$$V(x, y) = \theta(x) \frac{i}{\pi} \frac{e^{it(x^2 - y^2)} v_t(y) - v_t(x) e^{-it(x^2 - y^2)}}{x - y} \theta(y) .$$

Here

$$\theta(x) = \theta_1(x + s/2t), \quad \theta_1 = (1 + e^{\lambda^2 - \beta})^{-1/2} , \tag{1.1}$$

and $v_t(x)$ is equal to

$$v_t(x) = v(\sqrt{t}x), \quad v(x) = c \int_0^x e^{2iy^2} \, dy .$$

In latter definition c is the constant:

$$c = -2e^{-i\frac{\pi}{4}} \sqrt{\frac{2}{\pi}} . \tag{1.2}$$

Notice that

$$v(x) \sim -\mathrm{sgn}(x) - c \frac{e^{2ix^2}}{4ix} , \qquad x \to \infty . \tag{1.3}$$

In the above formulas t has the sense of the time coordinate, s has the sense of the space coordinate and β is the so-called normalized chemical potential. We are interested in the asymptotic behavior of $g(t, s, \beta)$ when $t \to +\infty$ so that

$$0 < C_1 < \frac{s}{t} < C_2 < +\infty . \tag{1.4}$$

* Partially supported by Russian Fund of Fundamental Reseach 98-01-01091

The asymptotic behavior of g is in fact known, see [IIKV92],[KBI93]. Our goal here is to describe an independent approach based on a direct asymptotic inversion of the so-called semiclassical pseudodifferential operators with double discontinuous symbols. Earlier we have applied analogous approaches to some other problems, in particular, to study the large time asymptotic behavior of solutions of completely integrable wave equations, see [BB91],[BB92],[BB94],[BB96a],[BB96b]. Our approach does not depend on the explicit form of θ. We suppose only that θ is smooth, vanishes quickly at infinity and satisfies the estimate

$$|\theta(x)| < 1/\sqrt{2}\,. \tag{1.5}$$

1.2 Notations

Let
P be the operator of multiplication by the characteristic function $\mathrm{P}(x)$ of the semiaxis $(0, \infty)$,
$\mathrm{Q}(x) = 1 - \mathrm{P}(x)\,, \quad Q = I - P\,, \quad S = P - Q\,,$
$\mathcal{D}$ be the operator of differentiation $\mathcal{D} = \frac{d}{dx}\,,$
$\mathbf{x}$ be the operator of multiplication by x,
θ be the operator of multiplication by the function $\theta(x)$,
$\mathbf{v}$ be the operator of multiplication by the function $v_t(x)$,
T_1 be the operator of multiplication by the function $\mathrm{T}_1(x) = e^{itx^2}\,,$
T_2 be the operator of multiplication by the function $\mathrm{T}_2(x) = v_t(x)\mathrm{T}_1^*(x)\,.$
Introduce also the commutator of operators A and B: $[A, B] = AB - BA$.

1.3 Pseudodifferential Operators

The integral operator V will be treated as a pseudodifferential operator (PDO), i.e. we shall deal sooner with its symbol than with its kernel. In terms of the symbol $\sigma(x, y, \xi)$ (more precisely, amplitude) the kernel $A(x, y)$ of the corresponding "$1/t$"-PDO A is defined by the formula

$$A(x, y) = \left(\frac{t}{2\pi}\right)^{1/2} \int_{\mathbb{R}} e^{it(x-y)\xi}\sigma(x, y, \xi)d\xi\,.$$

If $\sigma = \mathrm{P}(x)\,, \mathrm{Q}(x)$ then $A = P\,, Q$, respectively. If $\sigma = \mathrm{P}(\xi)\,, \mathrm{Q}(\xi)$ then $A = p\,, q$, $p + q = I$, where p, q are the mutually orthogonal projection operators in $L_2(\mathbb{R})$ on the subspaces of functions analytical in upper and lower complex half-planes, respectively:

$$p = \frac{-1}{2\pi i} \int_{\mathbb{R}} \frac{f(y)}{x - y + i0}\, dy\,,$$

$$q = \frac{1}{2\pi i} \int_{\mathbb{R}} \frac{f(y)}{x - y - i0}\, dy\,.$$

The operator $s = p - q$ is the Hilbert transformi, given by:

$$(sf)(x) = \frac{i}{\pi}\, \text{v.p.}\!\int_{\mathbb{R}} \frac{f(y)}{x - y}\, dy\,.$$

Suppose that one considers the operator A with the symbol σ. Consider the operator $T_1 A T_1^*$. Its symbol σ_1 is connected with σ by the formula:

$$\sigma_1(x, \xi) = \sigma(x, \xi - 2x)\,.$$

In other words, the transformation $A \mapsto T_1 A T_1^*$ is generated by the classical canonical transformation $(x, \xi) \mapsto (x, \xi - 2x)$ of the phase plane (x, ξ). The transformation $A \mapsto T_1^* A T_1$ allows a natural analogous interpretation: it is generated by the canonical transformation $(x, \xi) \mapsto (x, \xi + 2x)$.

The operator V can be represented by the formula

$$V = \theta(T_1 s T_2 - T_2 s T_1)\theta = \theta(T_1 s \mathbf{v} T_1^* - T_1^* \mathbf{v} s T_1)\theta\,.$$

1.4   Resolvent

Further we shall seek the asymptotic behavior of the resolvent $R_z = A^{-1}$, $A = I - zV$, $z \in \mathbb{C}$, of the operator V, as $t \to \infty$. The asymptotics of the determinant $\det(I - V)$ can be computed by means of the well known formula

$$\ln \det(I - V) = -\int_0^1 \mathrm{tr}\big[R_z - I\big]\frac{dz}{z}\,. \tag{1.6}$$

Let us notice that if one replaces the function v_t by its leading asymptotic term, $v_t(x) \mapsto -\mathrm{sgn}(x)$ then the symbol of V becomes discontinuous with respect to both phase variables x and ξ.

1.5   Decomposition

The properties of the symbol of the operator A are typical for semiclassical pseudodifferential equations with double discontinuous symbols, see [BB91], [BB92], [BB94], [BB96a], [BB96b]. The following decomposition of the operator A is a key to the computation of the asymptotics of the resolvent R_z:

$$A = A_0 - V_1 - V_2\,,$$

where

$$A_0 = I + 2z\theta^2 S\mathbf{v},$$
$$V_1 = z\theta T_1(S + s)\mathbf{v}T_1^*\theta,$$
$$V_2 = z\theta T_1^*\mathbf{v}(S - s)T_1\theta.$$

Note that the operator A_0 is an operator of multiplication by a function.

1.6 Inversion of the Components

There are two reasons for introducing the decomposition. The first is the fact that two operators $A_1 = A_0 - V_1$ and $A_2 = A_0 - V_2$ are explicitly invertible. Indeed,

$$A_1 = T_1 \left[A_0 - z\theta(S + s)\mathbf{v}\theta \right] T_1^*,$$

$$A_2 = T_1^* \left[A_0 - z\theta\mathbf{v}(S - s)\theta \right] T_1.$$

Therefore, both the operators A_1 and A_2 are elementary similar to classical singular integral operators of the form

$$C = I - \mathbf{a}p\mathbf{b}.$$

We have to suppose that a and b are smooth, fast vanishing at infinity functions. If the index of the function $1 - ab$ on the extended axis $\overline{\mathbb{R}}$ is equal to zero, the operator $I - \mathbf{a}p\mathbf{b}$ is a bounded bijection of $L_2(\mathbb{R})$. In addition, the inverse has the form

$$(I - \mathbf{a}p\mathbf{b})^{-1} = I + \mathbf{a}c_+ p c_- \mathbf{b}.$$

Here $\mathbf{c}_\pm$ are operators of multiplication by two smooth functions $c_\pm(x)$ that allow analytical continuation to the upper and the lower complex half-planes, respectively, and that satisfy the factorization relation

$$c_+(x)c_-(x) = c(x), \qquad c(x) = \frac{1}{1 - a(x)b(x)}. \tag{1.7}$$

Notice that

$$A_0 - z\theta(S + s)\mathbf{v}\theta = \left[I - 2z\theta p\mathbf{v}(1 + 2z\theta^2 P\mathbf{v})^{-1}\theta \right] (I + 2z\theta^2 P\mathbf{v}).$$

It means that in our case the function c in (1.7) is determined by the equality:

$$c(x) = \frac{1 + 2z\theta^2(x)\mathrm{P}(x)v_t(x)}{1 - 2z\theta^2(x)\mathrm{Q}(x)v_t(x)}$$
$$= [1 + 2z\theta^2(x)v_t(x)] \cdot \mathrm{P}(x) + \frac{1}{1 - 2z\theta^2(x)v_t(x)} \cdot \mathrm{Q}(x).$$

The function c is smooth since v is smooth and $v(0) = 0$.

Applying the general formula for the inversion of a singular operator one obtains:

$$A_1^{-1} = (1 + 2z\theta^2 S\mathbf{v})^{-1}$$
$$+ z\theta T_1 (1 + 2z\theta^2 P\mathbf{v})^{-1}\mathbf{c}_+(S + s)\mathbf{c}_+^{-1}(1 - 2z\theta^2 Q\mathbf{v})^{-1}T_2\theta\,.$$

Analogously,

$$A_0 - z\theta\mathbf{v}s\theta = (I - 2z\theta^2 Q\mathbf{v})\cdot[I + 2z\theta(1 - 2z\theta^2 Q\mathbf{v})^{-1}\mathbf{v}p\theta]\,.$$

and

$$A_2^{-1} = (1 + 2z\theta^2 S\mathbf{v})^{-1}$$
$$+ z\theta T_2 (1 + 2z\theta^2 P\mathbf{v})^{-1}\mathbf{c}_-(S - s)\mathbf{c}_-^{-1}(1 - 2z\theta^2 Q\mathbf{v})^{-1}T_1\theta\,.$$

1.7 Supports of the Components

The second reason of the effectiveness of the decomposition is the fact that the supports of the symbols of V_1 and V_2 do have an intersection only on a relatively poor, i.e. one-dimensional, set, namely on the axis $x = 0$. Indeed, it is obvious that the support of the symbol of the operator $s + S$ is the unification of two quadrants $x \geqslant 0, \xi \geqslant 0$ and $x \leqslant 0, \xi \leqslant 0$, therefore, the support of the symbol is the vertical angle between the straight-lines $x = 0$ and $\xi = 2x$. Analogously, the support of the symbol of the operator $s - S$ is the unification of two quadrants $x \leqslant 0, \xi \geqslant 0$ and $x \geqslant 0, \xi \leqslant 0$, and, therefore, the support of the symbol of V_2 is a vertical angle between the straight-lines $x = 0$ and $\xi = -2x$. This means that the supports of the symbols of V_1 and V_2 intersect only along the straight-line $x = 0$. As a result, asymptotically, when $t \to \infty$, the slow variable x of integration in products $V_1 V_2$ and $V_2 V_1$, i.e. the variable x, which is not accompanied by the factor $\sqrt{t}$, can be put equal to zero.

1.8 Alternating Method

To justify this idea rigorously we have to apply an appropriately modified alternating Schwarz method, see [BB92],[BB94],[BB96a].

Let us introduce the "reflection" operators Γ_1 and Γ_2 by the formulas

$$A_1^{-1} = A_0^{-1}(I - \Gamma_1)\,, \quad A_2^{-1} = A_0^{-1}(I - \Gamma_2)\,,$$

that is

$$\Gamma_1 = -z\theta T_1(1 - 2z\theta^2 Q\mathbf{v})\mathbf{c}_+(S+s)\mathbf{c}_+^{-1}(1 - 2z\theta^2 Q\mathbf{v})^{-1}T_2\theta\,,$$
$$\Gamma_2 = -z\theta T_2(1 - 2z\theta^2 Q\mathbf{v})\mathbf{c}_-(S-s)\mathbf{c}_-^{-1}(1 - 2z\theta^2 Q\mathbf{v})^{-1}T_1\theta\,. \tag{1.8}$$

Then

$$R_z = A^{-1} = (A_0 - A_1 - A_2)^{-1} = A_0^{-1}(I - \Gamma)\,,$$

where

$$I - \Gamma = (I - \Gamma_1)(I - \Gamma_1\Gamma_2)^{-1}(I - \Gamma_2)\,.$$

1.9 Model Operator

Simultaneously with A we shall consider the operator B,

$$B = I - z\theta_0^2(T_1 s T_2 - T_2 s T_1)$$

where $\theta_0 = \theta(0)$. We can say that B is obtained from A by replacing the slow variables x and y in the kernel of A by 0. The operator B depends on t, $B = B(t)$, but it is unitary similar to $B(1)$:

$$B(t) = WB(1)W^*, \qquad Wf(x) = t^{1/4}f(\sqrt{t}x)\,.$$

This implies that the asymptotic inversion of $B(t)$ must be, in fact, exact.

Side by side with the above computations we obtain

$$B = B_0 - U_1 - U_2\,,$$

where

$$B_0 = I + 2z\theta_0^2 S\mathbf{v}\,,$$
$$U_1 = z\theta_0^2 T_1(S+s)T_2\,,$$
$$U_2 = z\theta_0^2 T_2(S-s)T_1\,.$$

Then

$$B^{-1} = (B_0 - U_1 - U_2)^{-1} = B_0^{-1}(I - \Delta)\,,$$

where

$$I - \Delta = (I - \Delta_1)(I - \Delta_1\Delta_2)^{-1}(I - \Delta_2)$$

and

$$\Delta_1 = z\theta_0^2 T_1(1 - 2z\theta_0^2 Q\mathbf{v})\mathbf{d}_+(S+s)\mathbf{d}_+^{-1}(1 - 2z\theta_0^2 Q\mathbf{v})^{-1}T_2\,,$$
$$\Delta_2 = z\theta_0^2 T_2(1 - 2z\theta_0^2 Q\mathbf{v})\mathbf{d}_-(S-s)\mathbf{d}_-^{-1}(1 - 2z\theta_0^2 Q\mathbf{v})^{-1}T_1\,.$$

Here $\mathbf{d}_\pm$ are the operators of multiplication by two functions $d_\pm$ allowing analytic continuation to the upper and the lower complex half-planes of the variable x, respectively, and satisfy the factorization relation

$$d_+(x)d_-(x) = d(x),$$

where

$$d(x) = \frac{1 + 2z\theta_0^2 \mathrm{P}(x)v_t(x)}{1 - 2z\theta_0^2 \mathrm{Q}(x)v_t(x)}$$

$$= [1 + 2z\theta_0^2 v_t(x)] \cdot \mathrm{P}(x) + \frac{1}{1 - 2z\theta_0^2 v_t(x)} \cdot \mathrm{Q}(x).$$

Of course, the functions $d, d_\pm$ depend only on the "fast" variable.

To some extent the operator B is a model for the operator A.

1.10 Asymptotic Behavior of the Resolvent

Follow the idea suggested in the subsection 1.7 we can put $x = 0$ in all the internal integrations in the formula for $I - \Gamma$. After simple combinatorial reconstructions we obtain:

$$R_z = A_0^{-1}(I - \Gamma_1 - \Gamma_2) - B_0^{-1}(\Delta - \Delta_1 - \Delta_2) + O(t^{-\frac{1}{2}} \ln t). \tag{1.9}$$

More precisely, we can prove the following theorem.

Theorem 1.1. *If $|z| \leqslant 1$ and $t \to \infty$, the resolvent R_z allows the asymptotic representation (1.9) in the Hilbert-Schmidt norm with respect to $L_2(\mathbb{R})$.*

Let us notice that we have constructed explicitly all the components of the asymptotic representation except Δ. To find Δ and to find B^{-1} are equivalent problems.

To compute the asymptotic behavior of the correlation function we have to compute the asymptotic behavior of the trace $\mathrm{tr}(R_z - I)$. Improving the claim of the previous theorem, we can prove the second theorem.

Theorem 1.2. *If $|z| \leqslant 1$ and $t \to \infty$, the following asymptotic expansion*

$$\mathrm{tr}(R_z - I) = R_1 - R_2 + O(t^{-\frac{1}{2}} \ln t), \tag{1.10}$$

$$R_1 = \mathrm{tr}[A_0^{-1}(I - \Gamma_1 - \Gamma_2) - I], \quad R_2 = \mathrm{tr}[B_0^{-1}(\Delta - \Delta_1 - \Delta_2)]$$

holds.

2 Asymptotic Behavior of the Correlation Function

2.1 Model Operator

We define

$$W = T_1 s T_2 - T_2 s T_1 \,.$$

One can easily verify that

$$[\mathcal{D}^2 + 4\mathbf{x}^2, W] = 0 \,.$$

Since the eigenvalues and eigenfunctions of $\mathcal{D}^2 + 4\mathbf{x}^2$ are well known, due to the last formula we can explicitly find the spectrum and the eigenvectors of W.

The operator W preserves the parities of the function, therefore we can try to find its even or odd eigenfunctions:

$$
\begin{aligned}
u_e(x,\nu) &= D_\nu(2e^{i\frac{\pi}{4}}x) + D_\nu(-2e^{i\frac{\pi}{4}}x)\,, \\
u_o(x,\nu) &= D_\nu(2e^{i\frac{\pi}{4}}x) - D_\nu(-2e^{i\frac{\pi}{4}}x)
\end{aligned}
\tag{2.1}
$$

where $D_\nu(z)$ is the Weber parabolic cylinder function and

$$\nu = -\frac{1}{2} + i\sigma\,, \qquad \sigma \in \mathbb{R}\,.$$

The corresponding eigenvalues are respectively:

$$\xi_\nu = 1 - i\tan\frac{\pi\nu}{2}$$

for u_e and

$$\eta_\nu = 1 + i\cot\frac{\pi\nu}{2}$$

for u_o.

Now it is easy to see that the operator $W - I$ is a unitary operator. Its spectrum is simple, continuous and covers the whole circle. The functions $u_e(x,\nu)$ and $u_o(x,\nu)$ constitute an orthogonal and complete system of eigenfunctions of B. The upper spectral semicircle of the operator $W - I$ corresponds to the even eigenfunctions, and the lower semicircle corresponds to the odd eigenfunctions.

To complete the asymptotic decomposition of B we have to find the spectral function. In fact, it was computed by Cherry, see [Ch49], [BE53]. From Cherry's result it follows that

$$\delta(x - y) = \frac{i}{4\pi} \int_{-\frac{1}{2}-i\infty}^{-\frac{1}{2}+i\infty} d\nu \, \frac{e^{i\frac{\pi}{2}(\nu+\frac{1}{2})}}{\sin \pi\nu} \{u_e(x,\nu)u_e(y,\nu) + u_o(x,\nu)u_o(y,\nu)\}\,. \tag{2.2}$$

Knowing the spectral resolution of W, we can explicitly compute the kernel of the operator $B^{-1} = (I - z\theta_0^2 W)^{-1}$.

2.2 Asymptotic Behavior of the Function g

Come back to the formula of Theorem 1.2.

The explicit formulas for $A_0^{-1}, \Gamma_1, \Gamma_2$ directly lead to the following asymptotic formula for R_1:

$$R_1 = k_1 t + k_2 \ln t + b + (t^{-\frac{1}{2}} \ln t), \tag{2.3}$$

where

$$k_1 = 2 \int_{\mathbb{R}} |x| \nu(x)\, dx, \tag{2.4}$$

$$k_2 = \frac{\nu_0^2}{2}, \tag{2.5}$$

$$b = 2 \int_{\mathbb{R}} |x|(\eta(x) - \nu_0)\, dx \tag{2.6}$$

$$-\frac{1}{4} \int_{\mathbb{R}^2} \operatorname{sgn}(x)\operatorname{sgn}(y) \ln|x - y| \cdot \left[\frac{d\nu(x)}{dx}\frac{d\nu(y)}{dy} + \frac{d\eta(x)}{dx}\frac{d\eta(y)}{dy} \right] dx dy$$

$$-\nu_0 \int_{\mathbb{R}} \operatorname{sgn}(x) \ln|x| \cdot \frac{d\nu(x)}{dx}\, dx - \frac{1}{2i} \int_{\mathbb{R}} \operatorname{sgn}(x)\eta(x)\frac{d\ln v(x)}{dx}\, dx.$$

Here we used the following notations:

$$\nu(x) = \frac{1}{\pi} \ln[1 - 2\theta^2(x)],$$

$$\nu_0 = \frac{1}{\pi} \ln(1 - 2\theta_0^2), \tag{2.7}$$

$$\eta(x) = \frac{1}{\pi} \ln[1 + 2\theta_0^2 \operatorname{sgn}(x)v(x)].$$

A little bit more difficult, but still direct computations give R_2:

$$R_2 = \frac{i\nu_0}{2} + \nu_0^2 \ln 2 + \frac{\nu_0^2}{2} + \int_0^{i\nu_0} \left[\psi(\lambda) + \frac{\pi}{2} \cot\frac{\pi\lambda}{2} \right] \lambda\, d\lambda - 2 \int_{\mathbb{R}} |x|(\eta(x) - \nu_0)\, dx$$

$$+\frac{1}{4} \int_{\mathbb{R}^2} \operatorname{sgn}(x)\operatorname{sgn}(y) \ln|x - y| \cdot \frac{d\eta(x)}{dx}\frac{d\eta(y)}{dy}\, dx dy$$

$$+\frac{1}{2i} \int_{\mathbb{R}} \operatorname{sgn}(x)\eta(x)\frac{d\ln v(x)}{dx}\, dx.$$

In deriving the formula we have to use the following identity for Euler's ψ–function:

$$\frac{1}{4i} \int\limits_{\frac{1}{2}-i\infty}^{\frac{1}{2}+i\infty} \left\{ \frac{1}{\cot\frac{\pi\lambda}{2} - \cot\frac{\pi\nu}{2}} \cdot \frac{1}{\sin^2\frac{\pi\nu}{2}} \right.$$

$$\left. + \frac{1}{\cot\frac{\pi\lambda}{2} + \tan\frac{\pi\nu}{2}} \cdot \frac{1}{\cos^2\frac{\pi\nu}{2}} \right\} \psi(\nu)\,d\nu = \lambda\psi(\lambda) - \lambda + 1 \,.$$

2.3 Final Result

Summarizing R_1 and R_2 we arrive to the following final result:

$$g(t,s,\beta) = \frac{2t}{\pi} \int\limits_{\mathbb{R}} |x|\ln[1 - 2\theta^2(x)]\,dx + \frac{\nu_0^2}{2} \cdot \ln t$$

$$- \frac{1}{4} \int\limits_{\mathbb{R}^2}\!\!\int \mathrm{sgn}(x)\mathrm{sgn}(y) \ln|x-y| \cdot \frac{d\nu(x)}{dx}\,\frac{d\nu(y)}{dy}\,dxdy$$

$$- \nu_0 \int\limits_{\mathbb{R}} \mathrm{sgn}(x)\ln|x| \cdot \frac{d\nu(x)}{dx}\,dx$$

$$+ \frac{i\nu_0}{2} + \nu_0^2 \ln 2 + \frac{\nu_0^2}{2} + \int\limits_0^{i\nu_0} \left[\psi(\lambda) + \frac{\pi}{2}\cot\frac{\pi\lambda}{2} \right]\lambda\,d\lambda + O(t^{-\frac{1}{2}}\ln t) \,,$$

that coincides with the result of [IIKV92],[KBI93].

References

[BE53]   Bateman H., Erdélyi A., *Higher transcendental functions*, vol. 2, Mc Graw-Hill Book Company, Inc., NY, Toronto, London, 1953.

[BB91]   Budylin A.M., Buslaev V.S., *Reflection operator and their applications to asymptotic investigations of semiclassical integral equations*, Adv. Sov. Math., Amer. Math. Soc. **7** (1991), 107–157.

[BB92]   Budylin A.M., Buslaev V.S., *Quasiclassical integral equations*, Soviet Math. Dokl. **44** (1992), no. 1, 127–131.

[BB94]   Budylin A.M., Buslaev V.S., *Quasiclassical integral equations with slowly decreasing kernel on bounded domains*, St.Petersburg Math.J. **5** (1994), no. 1, 141–158.

[BB96a]   Budylin A.M., Buslaev V.S., *Semiclassical asymptotics of the resolvent of the integral convolution operator with the sine-kernel on a finite interval.*, St.Petersburg Math.J. **7** (1996), no. 6, 925–942.

[BB96b]   Budylin A.M., Buslaev V.S., *Semiclassical integral equations and asymptotic behavior of the Korteweg–de Vries equation on t → ∞*, Dokl.Rus.Akad.Nauk **348** (1996), no. 4, 455–458.

[Ch49]   Cherry T.M., *Expansion in terms of Parabolic Cylinder Functions*, Proc. Edinburgh Math. Soc. **2** (1949), no. 8, 50–65.

[IIKV92]   Its A.R., Izergin A.G., Korepin V.E., VarzuginG.G., *Large time and distance asymptotics of field correlation function of impenetrable bosons at finite temperature*, **54** (1992), 351–395.

[KBI93]   Korepin V.E., Bogolyubov N.M., Izergin A.G., *Quantum inverse scattering method and correlation functions*, Cambridge Monagraphs on Math. Phys., Cambridge Univ. Press, Cambridge, 1993.

Address

A.M. BUDYLIN AND V.S. BUSLAEV, Dept. of Mathematical Physics, Inst. for Physics, St. Petersburg State University, 2, Ul'yanovskaya str, St. Petersburg-Petrodvorets, 198904, Russia

E-MAIL: budylin@mph.phys.spbu.ru, buslaev@mph.phys.spbu.ru

2000 Mathematics Subject Classification. Primary 46N55; Secondary 47G30, 45E10

Operator Theory:
Advances and Applications, Vol. 126
© 2001 Birkhäuser Verlag Basel/Switzerland

A Free Boundary Value Problem Arisen in Unsteady Compressible Flow

SHUXING CHEN

Abstract. We study a free boundary value problem of multidimensional unsteady potential flow equation. The problem is defined in a domain bounded by two conical surfaces, one is given, and other is to be determined. In self-similar coordinates the problem can be reduced to a free boundary value problem of an elliptic equation. Then the existence of the solution is proved by using partial hodograph transformation and nonlinear alternating iteration. This result indicates that the structure of the weak solution with discontinuity is stable under small perturbation of data.

1   Introduction

The paper concerns a free boundary value problem arisen in multidimensional unsteady compressible flow. Consider a given static compressible gas filling the whole space except a body located at the origin with negligible size. Assume that at initial time the body at the origin expands like explosion, then there will appear a shock front around the body moving into the static gas. Ahead of the shock front the state of the gas is unchanged, but the location of the moving shock and the flow field in between the body and the shock front is to be determined. The problem is a multidimensional version of the well known piston problem in one space dimensional case, it is a fundamental prototype in the theory of hyperbolic conservation laws (see [CF48,Li94,Wh74]).

We will use the unsteady potential flow equation to describe the motion of the gas. This equation is a good description of the flow, if the possible shock is weak, or the speed of the expansion of the body is small comparing to the sonic speed. Assume that the problem is not axi-symmetric, namely the velocity of the expanded body, which will also be called as piston in this paper, varies in different direction, and is independent of time, then we can consider the problem in the self-similar coordinates. In such a coordinate system the equation is elliptic due to the fact that the normal component of relative velocity of the flow behind the shock front is subsonic. Therefore, the problem is reduced to a free boundary value problem of an elliptic equation. To deal with the free boundary problem we use partial hodograph transformation to fix the free boundary. Moreover, we also introduce the method of nonlinear alternative iteration, which will prevent the appearance of any new free boundary in the process of coordinate transformation, and establish a convergent sequence of approximate solutions as well. The limit of the sequence gives the existence of the solution to the original nonlinear problem, meanwhile, the result also indicates the solution is stable under small perturbation of the data – the velocity of the expansion of the body.

2   Main Result

We use the unsteady potential flow equation to describe the motion of the compressible flow (see [MT87,Mo94]). In two space dimension the equasion is

$$\frac{\partial}{\partial t}H + \frac{\partial}{\partial x}(\Phi_x H) + \frac{\partial}{\partial y}(\Phi_y H) = 0 \tag{2.1}$$

where Φ is the velocty potential satisfying $\nabla\Phi = (u,v)$, H stands for the density. For polytropic gas it can be expressed as a function of derivatives of Φ:

$$H = H(-\Phi_t - \frac{1}{2}|\nabla\Phi|^2) = (\frac{\gamma - 1}{\gamma}(-\Phi_t - \frac{1}{2}|\nabla\Phi|^2))^{\frac{1}{\gamma-1}}.$$

On any shock front, the R-H condition and the entropy condition should be satisfied. If Σ is a shock front, on which the parameters of the flow has jump, then the R-H conditin on Σ is

$$\begin{aligned}&\Phi \text{ is continuous,}\\&n_t[H] + n_x[\Phi_x H] + n_y[\Phi_y H] = 0\end{aligned} \tag{2.2}$$

where (n_t, n_x, n_y) is the vector normal to Σ, $[\cdot]$ stands for the jump of the function in the bracket. Besides, the entropy condition means that the relative normal velocity is supersonic ahead of shock, and is subsonic behind shock.

Assume that the state of the gas at the initial time is characterized by $\rho = \rho_0, (u,v) = (0,0)$. If the piston is located at the origin, and starting from $t = 0$ it expands with velocity depending on $\theta = \arctan y/x$, and being independent of time t, then the path of the piston can be described by

$$B: \quad x = h_1(\theta)t, \quad y = h_2(\theta)t, \tag{2.3}$$

which is a conical surface in (t, x, y) space. Then the boundary value condition on the path of the piston is

$$\Phi_x = h_1(\theta), \quad \Phi_y = h_2(\theta) \tag{2.4}$$

Since the whole problem is invariant under the dilation $t \to \alpha t, x \to \alpha x, y \to \alpha y$, we can only consider the self-similar solution of (2.1). Take $\xi = x/t, \eta = y/t$,

$\Phi(t,x,y) = t\psi(x/t, y/t)$, and using $r = (\xi^2 + \eta^2)^{\frac{1}{2}}, \theta = \arctan \eta/\xi$, the problem can be reduced to

$$(a^2 - (\psi_r - r)^2)\psi_{rr} - 2(\psi_r - r)\frac{\psi_\theta}{r^2}\psi_{r\theta} + (a^2 - \frac{\psi_\theta^2}{r^2})\frac{1}{r^2}\psi_{\theta\theta} + \frac{\psi_r}{r}(a^2 + \frac{\psi_\theta^2}{r^2}) - \frac{2}{r^2}\psi_\theta^2 = 0. \tag{2.5}$$

Correspondingly, if Σ and B are denoted by $r = s(\theta)$ and $r = b(\theta)$, the boundary condition are

$$\psi_r = b(\theta) \qquad \text{on} \quad r = b(\theta), \tag{2.6}$$

$$\psi = \psi_0, \quad [(\psi_r - r)H] = 0 \qquad \text{on} \quad r = s(\theta), \tag{2.7}$$

The main result in this paper is the existence of solution with a shock front away from the path of the piston, provided $b(\theta)$ is a small perturbation of a constant b_0. More precisely, we have

Theorem 2.1. *Assume that the path $b(\theta)$ of the piston satisfies*

$$\|b(\theta) - b_0\|_{C^{2+\alpha}} \leq \epsilon_0. \tag{2.8}$$

where $b_0 = \min b(\theta)$, and ϵ_0 is sufficiently small, then we can find a function $s(\theta)$ defined in $0 \leq \theta \leq 2\pi$, a function $\psi(r, \theta)$ defined in $b(\theta) \leq r \leq s(\theta), 0 \leq \theta \leq 2\pi$, such that (2.5)-(2.7) is satisfied. Moreover, if we denote by $\psi_B(r)$ the background solution of the problem with $b(\theta) = b_0$, and by s_0 the corresponding right end of the interval where $\psi_B(r)$ is defined, then

$$\|s(\theta) - s_0\|_{C^{2+\alpha}} \leq C\epsilon_0 \tag{2.9}$$

$$\|\psi(r, \theta) - \psi_B(r)\|_{C^{2+\alpha}} \leq C\epsilon_0 \tag{2.10}$$

3   Symmetric Case

In symmetric case, $b(\theta) = b_0$, then $\psi(r, \theta)$ and $s(\theta)$ will also be independent of θ. Therefore, (2.9) becomes an ordinary differential equation

$$(a^2 - (\psi_r - r)^2)\psi_{rr} + \frac{a^2}{r}\psi_r = 0 \quad r \in (b_0, s_0), \tag{3.1}$$

while the boundary conditions are

$$\psi_r = b_0 \qquad \text{on} \quad r = b_0, \tag{3.2}$$

$$\psi = \psi_0, \quad [(\psi_r - r)H] = 0 \qquad \text{on} \quad r = s_0. \tag{3.3}$$

where s_0 is unknown, and will be determined together with ψ. Besides, according to the entropy condition we have

$$\psi_r > 0, a > r - \psi_r > 0 \tag{3.4}$$

on $r = s_0$. The Rankine-Hugoniot condition gives a restriction to the state behind the shock front. To describe it we have following assertions:

Lemma 3.1. *For all possible state ψ, which can be connected with the state $\psi = \psi_0$ by a shock front moving outward, the corresponding (r, ψ_r) forms a curve Γ called shock polar. The shock polar locates below the diagonal $r = \psi_r$. It is increasing, and takes diagonal as its asymptote.*

For any point P_s on Γ, we denote $s_0 = r(P_s), \chi_0 = \psi_r(P_s) > 0$. Taking initial data as

$$\psi|_{r=s_0} = \psi_0, \quad \psi_r|_{r=s_0} = \chi_0$$

and integrating (3.1) in $r < s_0$, we obtain a solution of (3.1).

Lemma 3.2. *There exists an integral curve of (3.1) with initial data $\psi = \psi_0$ and ψ_r determined by the location of the starting point on the shock polar Γ. The integral curve intersect with the diagonal $r = \psi_r$.*

Based on these two lemmas, we establish the existence of the problem (3.1)-(3.3).

Theorem 3.1. *For any point $b_0 > 0$ there is a unique solution $\psi(r)$ of (3.1)-(3.3). Corresponding to each solution the curve $(r, \psi_r(r))$ is decreasing in (b_0, s_0), it intersects with the diagonal $r = \psi_r$ at $r = b_0$ and with Γ at $r = s_0$.*

Proof. The left end point of Γ is $(a_0, 0)$. Through the end point the solution of (3.1) satisfying $\psi(a_0) = \psi_0$ is $\psi \equiv \psi_0$, which is the interval $(0, a_0)$ on the r-axis. Starting from any point P_s on Γ, we have a solution of (3.1) satisfying $\psi(r_{P_s}) = \psi_0$. The integral curve ℓ intersects with the diagonal $r = \psi_r$ at P_b. By the property of ordinary differential equation the coordinates of P_b is a continuous function of P_s. When P_s runs to the point $(a_0, 0)$, the corresponding integral curve sweep the domain bounded by Γ, r-axis, the diagonal and the integral curve ℓ. Besides, we confirm that for each point in between O and P_b on the diagonal there is one and only one integral curve passing through the point. In fact, if two integrals ℓ_1 and ℓ_2 of (3.1) intersect at P_b. Since on the diagonal the equation (3.1) becomes

$$a^2 \psi_{rr} + \frac{1}{r} a^2 \psi_r = 0, \tag{3.5}$$

then ψ_{rr} on ℓ_1 and ℓ_2 takes same value at P_b.

On the other hand, by differentiating (3.1) we can obtain a second order differential equation for $\psi_r(r)$ (or denoted by $\chi(r)$)

$$(a^2 - (\chi - r)^2)\chi_{rr} - 2(\chi - r)\chi_r^2 + \frac{a^2}{r}\chi_r + (\chi_r + \frac{\chi}{r})2a(a_\psi\chi + a_\chi\chi_r) = 0 \quad (3.6)$$

By uniqueness of solution to (3.6) with initial data

$$\chi(r_{P_b}) = \chi_{P_b}, \quad \chi_r(r_{P_b}) = -\frac{1}{r_{P_b}}\chi_{P_b}$$

ℓ_1 and ℓ_2 must coincide. This also means the one to one correspondance of P_b and P_s. Since P_s can be any point on Γ and the slope of all integrals of (3.1) on (r, ψ_r) plane is bounded , then for any positive number b_0 we can find s_C and a solution of (3.1), satisfying the boundary conditions (3.2) and (3.3). Hence the theorem is proved.

4 Partial Hodograph Transformation and Domain Composition

Consider the problem (2.5)-(2.7) in non-symmetric case. Assume that $b_0 = \min b(\theta)$, and $\|b(\theta) - b_0\|_{C^{2+\alpha}} < \epsilon_0$ with ϵ_0 being sufficiently small, we expect the solution of (2.5)-(2.7) is also a perturbation of the problem (3.1)-(3.3). In the sequel we call the solution of (3.1)-(3.3) as background solution, and denote it by ψ_B.

The problem (2.5)-(2.7) is a free boundary value problem. Since (2.5) is elliptic for ψ_B, then it is also elliptic for the small perturbation of ψ_B. To avoid the difficulty caused by the moving boundary $r = s(\theta)$, we introduce a partial hodograph transformation to fix it (see [MT87,Mn97]). The transformation is

$$T : \begin{cases} \sigma = \theta, \\ p = -\psi(r, \theta). \end{cases} \quad (4.1)$$

which changes the position of the unknown function ψ and the variable r. Since ψ equals a constant ψ_0 on the shock front, then T transforms the shock front to a fixed boundary $p = -\psi_0$. The inverse of T is

$$T^{-1} : \begin{cases} \theta = \sigma, \\ r = u(p, \sigma). \end{cases} \quad (4.2)$$

The transform T changes the equation (2.5) to the new form

$$E_{11}u_{pp} + 2E_{12}u_{p\sigma} + E_{22}u_{\sigma\sigma} + Q(u, u_p, u_\sigma) = 0 \quad (4.3)$$

where

$$E_{11} = (a^2 - (\frac{1}{u_p} + u)^2) - 2(\frac{1}{u_p} + u)(\frac{u_\sigma^2}{u^2 u_p}) + (a^2 - \frac{1}{u^2}\frac{u_\sigma^2}{u_p^2})\frac{u_\sigma^2}{u^2},$$

$$E_{12} = -(\frac{1}{u_p} + u)\frac{u_\sigma}{u^2} - (a^2 - \frac{1}{u^2}\frac{u_\sigma^2}{u_p^2})\frac{u_\sigma u_p}{u^2},$$

$$E_{22} = (a^2 - \frac{1}{u^2}\frac{u_\sigma^2}{u_p^2})\frac{u_p^2}{u^2},$$

$$Q(u, u_p, u_\sigma) = -\frac{1}{u}(a^2 u_p^2 + \frac{u_\sigma^2}{u^2}) - \frac{2}{u^2}u_\sigma^2 u.$$

The boundary conditions will also have new forms in the new coordinates. Denote the path of the expanding piston by $p = g(\sigma)$, the boundary condition on it is

$$u = b(\sigma), \quad u_p = -\frac{1}{b(\sigma)}, \tag{4.4}$$

while on the shock front $p = \psi_0$, the condition is

$$(\frac{1}{u_p} + u)H - u\rho_0 = 0. \tag{4.5}$$

Let us call the problem (2.5)-(2.7) as (NL), and call the problem (4.3)-(4.5) as $(NL)^*$. Obviously, these two problems are equivalent in fact.

Since the boundary $p = g(\sigma)$ is unknown, the problem $(NL)^*$ is also a free boundary value problem. To avoid the appearance of any new free boundary, we try to only consider (2.9) near $r = b(\theta)$, and only consider (4.3) near $p = -\psi_0$. To this end we decompose the annular domain $b(\theta) < r < s(\theta)$ to overlapped annuluses. Then we introduce a set of sub boundary value problems in these overlapped domains and try to construct the solution of the original nonlinear problem through these auxiliary boundary value problems. For instance, we choose constants r_1, r_2 satisfying $b_0 < r_2 < r_1 < s_0$, and introduce

$$(NL)^{(a)} : \begin{cases} \text{equation}(2.5) \quad \text{in} \quad \Omega_a : b(\theta) < r < r_1, \\ \text{boundary condition} \quad (2.6) \quad \text{on} \quad r = b(\theta), \\ \psi = d(\theta) \quad \text{on} \quad r = r_1, \end{cases} \tag{4.6}$$

$$(NL)^{(b)} : \begin{cases} \text{equation}(4.3) \quad \text{in} \quad \Omega_b : -\psi_B(r_1) > p > -\psi_0, \\ u = q(\sigma) \quad \text{on} \quad p = -\psi_B(r_2), \\ \text{boundary condition}(4.5) \quad \text{on} \quad p = -\psi_0. \end{cases} \tag{4.7}$$

Both $(NL)^{(a)}$ and $(NL)^{(b)}$ are defined in a domain with fixed boundary. They will also be denoted by $(NL)^{(a)}\{b(\theta), d(\theta)\}$, $(NL)^{(b)}\{q(\sigma)\}$ respectively. As we will see in the following sections, the solvability and the corresponding estimates will lead us to obtain the solution of the problem (NL) (and $(NL)^*$).

Remark 4.1. Generally, to ensure the existence and uniqueness of all auxiliary problems defined in overlapped annuluses, we have to let each annulus be sufficiently narrow. Correspondingly, the whole domain $b(\theta) < r < s(\theta)$ should be decomposed into many annuluses, overlepped with each other. But to emphasize the main idea and simplify analysis we only deal with the case of decomposition of whole domain into two annuluses in the sequel.

5 The Solution to $(NL)^{(a)}$ and $(NL)^{(b)}$

According to the explination in the remark, we discuss the property of solution to problem $(NL)^{(a)}$ and $(NL)^{(b)}$. By using the Schauder estimates for the linearized problem of $(NL)^{(a)}$ and the implicit function theorem the following propositions on exisence and corresponding estimates of the solution to $(NL)^{(a)}$ can be established.

Lemma 5.1. *Assume that δ, ϵ are sufficiently small, and ϵ is small with respect to δ. $|r_1 - b_0| < \delta, \|b(\theta) - b_0\|_{C^{2+\alpha}(0,2\pi)} < \epsilon, \|d(\theta) - \psi_{10}\|_{C^{2+\alpha}(0,2\pi)} < \epsilon$ with $\psi_{10} = \psi_B(r_1)$, then the problem $(NL)^{(a)}\{b(\theta), d(\theta)\}$ has unique solution $\psi(r,\theta)$. Moreover,*

$$\|\psi(r,\theta) - \psi_B(r)\|_{C^{2+\alpha}(\Omega_a)} \to 0 \quad \text{when} \quad \epsilon \to 0 \tag{5.1}$$

Lemma 5.2. *Assume that δ, ϵ are sufficiently small, and ϵ is small with respect to $\delta, |r_1-b_0| < \delta, \|b(\theta)-b_0\|_{C^{2+\alpha}(0,2\pi)} < \epsilon, \|d_j(\theta)-\psi_{10}\|_{C^{2+\alpha}(0,2\pi)} < \epsilon \ (j=1,2)$, and $\psi_j(r,\theta)$ is the solution of the problem $(NL)^{(a)}\{b(\theta), d_j(\theta)\}$, then the comparison principle is valid, i.e. $d_2 \geq d_1$ implies $\psi_2 \geq \psi_1$.*

Lemma 5.3. *Assume that δ, ϵ are small quantities described in the above lemma. $b_0 = \min b(\theta), \tilde{b}_0 = \max b(\theta), \ |b(\theta) - b_0\|_{C^{2+\alpha}(0,2\pi)} < \epsilon, \ \psi(r,\theta) = (NL)^{(a)}\{b(\theta), d(\theta)\}, \tilde{\psi}_B(r)$ is the solution of (3.1)-(3.3) with b_0 replaced by $\tilde{b}_0, \tilde{\psi}_B(r_1) \leq d(\theta) \leq \psi_B(r_1)$, then*

$$\tilde{\psi}_B(r) \leq \psi(r,\theta) \leq \psi_B(r) \quad \text{in} \quad \Omega_a.$$

Besides

$$\|\psi(r,\theta) - \psi_B(r)\|_{C^{2+\alpha}(\Omega_a^-)} \leq C(\|d(\theta) - \psi_{10}\|_{C(0,2\pi)} + \epsilon) \tag{5.2}$$

where $\Omega_a^- = \{(r,\theta) : b(\theta) \leq r \leq r_1 - \frac{1}{10}\delta, \ 0 \leq \theta \leq 2\pi\}$

Similarly, we can also establish corresponding propositions for problem $(NL)^{(b)}$. That is

Lemma 5.4. *Assume that δ, ϵ are small quantities as above, $\|q(\sigma) - r_2\|_{C^{2+\alpha}(0,2\pi)} < \epsilon$, then $(NL)^{(b)}\{q(\sigma)\}$ has a unique solution. Moreover,*

$$\|u(p,\sigma) - u_B(p)\|_{C^{2+\alpha}(\Omega_b)} \to 0 \quad \text{when} \quad \epsilon \to 0. \tag{5.3}$$

Lemma 5.5. *Assume that δ, ϵ are small quantites as above, and assume $\|q^{(j)}(\sigma) - r_2\|_{C^{2+\alpha}(0,2\pi)} < \epsilon$, $u^{(j)}(p,\sigma) = NL^{(b)}\{q^{(j)}(\sigma)\}$, then*

$$q^{(2)}(\sigma) \geq q^{(1)}(\sigma) \Longrightarrow u^{(2)}(p,\sigma) \geq u^{(1)}(p,\sigma). \tag{5.4}$$

Besides

$$\|u^{(j)}(p,\sigma) - u_B(p)\|_{C^{2+\alpha}(\Omega_b^-)} \leq C(\|q^{(j)}(\sigma) - r_2\|_{C^0(0,2\pi)} + \epsilon). \tag{5.5}$$

6   Solution to (NL) and $(NL)^*$

Based on the lemmas given in the last section, we establish a sequence of approximate solutions to (NL) and $(NL)^*$ by alternatively solving $(NL)^{(a)}$ and $(NL)^{(b)}$. First, we take $\psi^{(0)} = \psi_B$, $u^{(0)} = u_B$, then we define $\psi^{(1)}(r,\theta)$ as the solution of the problem $(NL)^{(a)}\{b(\theta), \psi_B(r_1)\}$. For $n \geq 1$ we define

$$u^{(n)}(p,\sigma) = (NL)^{(b)}\{(\psi^{(n)})^{-1}(\psi_B(r_2),\sigma)$$

$$\psi^{(n+1)}(r,\theta) = NL^{(a)}\{b(\theta), -(u^{(n)})^{-1}(r_1,\theta)\}$$

inductively.

Lemma 6.1. *If $\|b(\theta) - b_0\|_{C^{2+\alpha}} < \epsilon$ with ϵ being sufficiently small, $\tilde{b}_0 = \max b(\theta)$, $\psi_B(r), \tilde{\psi}_B(r), u_B(p), \tilde{u}_B(p)$ are the corresponding background solutions for (NL) and $(NL)^*$ in symmetric case. Then the sequences $\{\psi^{(n)}\}, \{u^{(n)}\}$ are well defined, which satisfy*

$$\begin{aligned}
\tilde{\psi}_B(r) \leq \psi^{(n)}(r,\theta) \leq \psi_B(r) &\quad \text{in} \quad \Omega_a^-, \\
\|\psi^{(n)}(r,\theta) - \psi_B(r)\|_{C^{2+\alpha}(\Omega_a^-)} &\leq C\epsilon,
\end{aligned} \tag{6.1}$$

and

$$\begin{aligned}
\tilde{u}_B(p) \geq u^{(n)}(p,\sigma) \geq u_B(p) &\quad \text{in} \quad \Omega_b^-, \\
\|u^{(n)}(p,\sigma) - u_B(p)\|_{C^{2+\alpha}(\Omega_b^-)} &\leq C\epsilon,
\end{aligned} \tag{6.2}$$

Moreover, $\{\psi^{(n)}\}$ is monotone decreasing with respect to n, $\{u^{(n)}\}$ is monotone increasing with respect to n.

The proof of Theorem 2.1   As we proved in Lemma 6.1, the sequences $\{u^{(n)}(p,\sigma)\}$, and $\{\psi^{(n)}(r,\theta)\}$ are bounded and monotone with respect to n, then these sequences are convergent. Denote their limits by $u(p,\sigma), \psi(r,\theta)$ respectively. Since the $C^{2+\alpha}$ norm of $\psi^{(n)}$ on $r = r_2$ is dominated by its C^0 norm on $r = r_1$, the $C^{2+\alpha}$ norm of $u^{(n)}$ in Ω_b^- is dominated by their C^0 norm on $p = -\psi_B(r_1)$, so the $C^{2+\alpha}$ norm of $\psi^{(n)}, u^{(n)}$ are uniformly bounded in corresponding domains. The fact also implies ψ, u are $C^{2+\alpha}$ function there.

On the other hand, ψ satisfies (2.9),(2.10) and $\psi(r_1,\theta) = -(u)^{-1}(r_1,\theta)$. Notice that (4.3) is an equivalent form of the equation (2.9) in the coordinates (p,σ), then both $\psi(r,\theta)$ and $-(u_1)^{-1}(r,\theta)$ satisfy the equation (2.9) in the overlapped domain $\Omega_a \cap T^{(-1)}(\Omega_b)$. Besides, these two functions coinside on the boundaries $r = r_1$ and $r = r_2$. Since the domain Ω_a is chosen small enough, so that the problem $(NL)^{(a)}$ and its linearization has a unique solution in Ω_a. Therefore, it is easy to see that $\psi(r,\theta)$ and $u^{-1}(r,\theta)$ coincide on the domain $\Omega_a \cap T^{(-1)}(\Omega_b)$. Namely, viewing functions $u^{-1}(r,\theta)$ as extensions of $\psi(r,\theta)$, we obtain the solution of (NL) in the whole domain $b(\theta) < r < s(\theta)$ with $\psi(s(\theta),\theta) = \psi_0$.

As for the stability, (2.10) can be obtained by taking limit in (6.1),(6.2). Besides, in view of $s(\theta) = \psi^{-1}(-v_0,\theta)$, (2.9) can also be obtained.

Acknowledgements

This work is partially supported by NNSF of China and Doctoral Foundation of State Educational Ministry.

References

[Al89]   Alinhac, S., *Existence d'ondes de rarefaction pour des systèmes quasi-linéaires hyperboliques multidimensionnels*, Commu. P.D.E., **14**, 173-230, 1989.

[CK98]   Canic, S. and Kerfitz, B. *Riemann problems for the two-dimensional unsteady transonic small disturbance equation*, SIAM J. Appl. Math. **58** 636-665, 1998.

[CKL99] Canic,S. , Kerfitz, B. and Lieberman, G. *A proof of perturbed steady transonic shocks via a free boundary problem*, Comm. Pure Appl. Math. **52**. 1999.

[CKK00] Canic, S., Kerfitz, B. and Kim, E.H. *Free boundary problems for the unsteady transonic small disturbance equation: transonic regular reflection* (to appear).

[CG97]   Chen, G. Q. *Remarks on spherically symmetric solutions of the compressible Euler equations* Proceedings of the Royal Socialy of Edinburgh, **127A** 243-259, 1997.

[CS99]   Chen, S.X. *Non-symmetric conical supersonic flow*, International Series of Numerical Mathematics, **129** Birkhauser Verlag, Basel/Switzerland, 140-158, 1999.

[CL00]   Chen, S.X. and Li, D.N. *Supersonic flow past a symmetrically curved cone*, Indiana Univ. Math. Jour., 2000.

[CF48]   Courant, R. and Friedrichs, K.O. *Supersonic Flow and Shock Waves*, Interscience Publishers Inc., New York, 1948.

[Da00]   Dafermos, C.M. *Hyperbolic Conservation Laws in Continuum Physics*, Springer-Verlag, Berlin Heiderberg New York, 2000.

[Li94]   Li, T.T. *Global classical solutions for quasilinear hyperbolic systems*, Wiley, Masson, New York, Paris, 1994.

[LL99]   Lien, W.C. and Liu, T.P. *Nonlinear stability of a self-similar 3-d gas flow*, Comm. Math. Phys., **304**,524-549, 1999.

[Ma83a]   Majda, A. *The stability of multi-dimensional shock fronts*, Memoir Amer. Math. Soc. **273**,1983.

[Ma83b]   Majda, A. *The existence of multi-dimensional shock fronts*, Memoir Amer. Math. Soc. **281**, 1983.

[Ma91]   Majda, A. *One perspective on open problems in multi-dimensional conservation laws*, IMA vol. Math. Appl. **29** 217-237, 1991.

[MT87]   Majda, A. and Thomann, E. *Multi-dimensional shock fronts for second order wave equations*, Comm. P.D.E., **12**, 777-828, 1987.

[Me91]   Metivier, G. *Ondes Soniques*, Jour. Math. Pure Appl., **70**, 197-268, 1991

[Mn97]   Mnif, M. *Probleme de Riemann pour une loi de conservation scalaire hyperbolique d'order deux*, Comm. P.D.E., 22, 1589-1627,1997.

[Mo94]   Morawetz, C.S. *Potential theory for regular and Mach reflection of a shock at a wedge*, Comm. Pure Appl. Math. **47**, 593-624, 1994.

[Sc76]   Schaeffer,D.G. *Supersonic flow past a nearly straight wedge* , Duke Math. J. **43**, 637-670, 1976.

[Sm83]   Smoller, J., *Shock waves and reaction-diffusion equations*, Springer-Verlag, New York, 1983.

[Wh74]   Whitham, G.B. *Linear and Nonlinear Waves*, John Wiley and Son, New York, London, Sydney, Toronto, 1974.

Address

SHUXING CHEN, Institut of Mathematics Fudan University, Shanghai, 200433, China

E-MAIL: sxchen@fudan.ac.cn

2000 Mathematics Subject Classification. Primary 35L67, 35L65; Secondary 35L25,35J60

Operator Theory:
Advances and Applications, Vol. 126
© 2001 Birkhäuser Verlag Basel/Switzerland

Some New Results on the Nonlinear Singular Partial Differential Equations

CHEN HUA, LUO ZHUANGCHU, AND TAHARA HIDETOSHI

Abstract. In this paper, we calculate the formal Gevrey index of the formal solution of a class of nonlinear first order totally characteristic type partial differential equations with irregular singular in the space variable.

1 Introduction

Let $(t,x) \in C_t \times C_x$, $N = \{1,2,\ldots\}$, $Z_+ = \{0,1,2,\ldots\}$, and denote by $C[[t,x]]$ (resp. by $C[[x]]$) the ring of formal power series in the variables (t,x) (resp. in the variable x).

Let us consider the following nonlinear singular first order partial differential equation

$$t\frac{\partial u}{\partial t} = F\left(t,x,u,\frac{\partial u}{\partial x}\right), \tag{1.1}$$

where $u = u(t,x)$ is an unknown function, and $F(t,x,u,v)$ is a function defined in an open polydisc Δ centered at the origin of $C_t \times C_x \times C_u \times C_v$. Set $\Delta_0 = \Delta \cap \{t = 0, u = 0 \text{ and } v = 0\}$. We impose the following condition on $F(t,x,u,v)$:

(F1) $F(t,x,u,v)$ is a holomorphic function on Δ;

(F2) $F(0,x,0,0) \equiv 0$ on Δ_0

Then by the Taylor expansion in (t,u,v) we can express $F(t,x,u,v)$ in the form

$$F(t,x,u,v) = a(x)t + b(x)u + \gamma(x)v + \sum_{i+j+\alpha \geq 2} a_{i,j,\alpha}(x)t^i u^j v^\alpha,$$

and $a(x), b(x), \gamma(x), a_{i,j,\alpha}(x)$ are all holomorphic functions on Δ_0.

If for $\gamma(x) \equiv 0$ on Δ_0, the equation (1.1) is called a non-linear Fuchsian type PDE (or is called a "Briot-Bouquet type PDE (cf. [[GT90]-[GT96]])' ; this situation has been discussed. If $\gamma(0) \neq 0$, we can solve $\partial u/\partial x$ from the equation (1.1) and then we can apply the Cauchy-Kowalewski theorem. If $\gamma(x) \not\equiv 0$ and $\gamma(0) = 0$, the indicial operator $C(\lambda,x,\partial/\partial x) = \lambda - b(x) - \gamma(x)\partial/\partial x$ is a singular differential operator; in this situation the equation (1.1) has been called a totally characteristic type PDE(cf.[[CT00]-[CL01]])). Thus, in this paper we assume:

(F3) $\gamma(x) = x^p c(x)$ for $p \in N$ and $c(0) \neq 0$.

In the case $p = 1$ we already have the following result.

Theorem 1.1. (Chen-Tahara). *Assume $p = 1$ and $|i - b(0) - jc(0)| \neq 0$ for any $(i, j) \in \mathbf{N} \times \mathbf{Z}_+$. Then we have:*

(1) The equation (1.1) has a unique formal solution $u(t, x) \in \mathbf{C}[[t, x]]$ with $u(0, x) \equiv 0$.

(2) Moreover, if $c(0) \in \mathbf{C} \setminus [0, \infty)$ holds the unique formal solution in (1) is convergent in a neighborhood of $(0, 0) \in \mathbf{C}_t \times \mathbf{C}_x$.

In this paper we shall discuss the case $p \geq 2$. In this case the indicial operator $C(\lambda, x, \partial/\partial x) = \lambda - b(x) - x^p c(x) \partial/\partial x$ has a irregular singularity at $x = 0 \in \mathbf{C}$ and the formal power series solution of (1.1) is not convergent in general; but still it belongs to a formal Gevrey class.

Definition 1.1. Let $s \geq 1$ and $\sigma \geq 1$. We say that a formal power series $f(t, x) = \sum_{i \geq 0, j \geq 0} f_{i,j} t^i x^j \in \mathbf{C}[[t, x]]$ belongs to the formal Gevrey class $G\{t, x\}_{(s, \sigma)}$ if the power series

$$\sum_{i \geq 0, j \geq 0} \frac{f_{i,j}}{(i!)^{s-1}(j!)^{\sigma-1}} t^i x^j$$

is convergent in a neighborhood of $(0, 0) \in \mathbf{C}_t \times \mathbf{C}_x$.

The following result is a consequence of the main theroem (Theorem 2.1) of this paper.

Theorem 1.2. *Assume $p \geq 2$ and $b(0) \notin \mathbf{N}$. Then:*

(1) The equation (1.1) has a unique formal solution $u(t, x) \in \mathbf{C}[[t, x]]$ with $u(0, x) \equiv 0$.

(2) Moreover, it belongs to the formal Gevrey class $G\{t, x\}_{(s, \sigma)}$ for any $s \geq p/(p-1)$ and $\sigma \geq p/(p-1)$.

The result of this type is often called a Maillet's type theorem.

2   Main Results

We discuss the same equation (1.1) as in §1 under the conditions (F1),(F2),(F3), and $p \geq 2$.

Our equation is written as

$$\left(t \frac{\partial}{\partial t} - b(x) - x^p c(x) \frac{\partial}{\partial x} \right) u = a(x)t + \sum_{i+j+\alpha \geq 2} a_{i,j,\alpha}(x) t^i u^j \left(\frac{\partial u}{\partial x} \right)^\alpha \tag{2.1}$$

where $a(x), b(x), c(x), a_{i,j,\alpha}(x)$ are all holomorphic functions on Δ_c, $c(0) \neq 0$, and the right hand side is a holomorphic function on Δ with $v = \partial u / \partial x$.

Set
$$J = \left\{ (i, j, \alpha); i + j + \alpha \geq 2, \alpha > 0, \text{ and } a_{i,j,\alpha}(0) \neq 0 \right\}.$$

We have:

Theorem 2.1. *Assume (F1),(F2),(F3), $p \geq 2$ and $b(0) \notin \mathbf{N}$. Then, the equation (2.1) has a unique formal solution $u(t, x) \in \mathbf{C}[[t, x]]$ with $u(0, x) \equiv 0$ and it belongs to the formal Gevrey class $G\{t, x\}_{(s,\sigma)}$ for any (s, σ) satisfying*

$$s \geq 1 + \max \left[0, \sup_{(i,j,\alpha) \in J} \left(\frac{1}{(p-1)(i+j+\alpha-1)} \right) \right] \tag{2.2}$$

and $\sigma \geq p/(p-1)$.

Note that

$$1 + \frac{1}{(p-1)(i+j+\alpha-1)} \leq 1 + \frac{1}{(p-1)(2-1)} = \frac{p}{p-1}$$

and therefore $s \geq p/(p-1)$ implies the condition (2.2). Thus, Theorem 1.2 follows from Theorem 2.1.

As a particular case, we have:

Corollary 2.1. *If $J = \emptyset$, the unique formal solution $u(t, x)$ belongs to the class $G\{t, x\}_{(1, p/(p-1))}$.*

This implies that the formal solution is holomorphic in the variable t.

For $f(x) = \sum_{j \geq 0} f_j x^j \in \mathbf{C}[[x]]$ we write $f(x) \gg 0$ if $f_j \geq 0$ holds for all $j \geq 0$. The following proposition asserts that our condition (2.2) is the best possible result in a generic case.

Proposition 2.1. *Assume (F1),(F2),(F3), $p \geq 2$ and $b(0) \notin \mathbf{N}$. Moreover, assume the following additional conditions:*

c1) $a(0) > 0$, $(\partial a/\partial x)(0) > 0$ and $a(x) \gg 0$; c2) $b(0) < 1$ and $(b(x) - b(0)) \gg 0$;

c3) $c(0) > 0$ and $c(x) \gg 0$; c4) $a_{i,j,\alpha}(x) \gg 0$ (for $i + j + \alpha \geq 2$).

Then, the unique formal solution $u(t, x)$ in Theorem 2.1 belongs to the class $G\{t, x\}_{(s,\sigma)}$ if and only if (s, σ) satisfies (2.2) and $\sigma \geq p/(p-1)$.

Thus, we may say that the index (s_0, σ_0) defined by

$$s_0 = 1 + \max\left[0, \ \sup_{(i,j,\alpha)\in J}\left(\frac{1}{(p-1)(i+j+\alpha-1)}\right)\right], \qquad \sigma_0 = \frac{p}{p-1} \qquad (2.3)$$

is *the formal Gevrey index of the equation (2.1)*.

Example 2.1. Let $p, q, l, m, n \in \boldsymbol{Z}_+$ satisfying $p \geq 2$, $n \geq 1$ and $l + m + n \geq 2$. Let us consider

$$t\frac{\partial u}{\partial t} = (1+x)t + x^p\frac{\partial u}{\partial x} + x^q t^l u^m\left(\frac{\partial u}{\partial x}\right)^n. \qquad (2.4)$$

We have

1) (2.4) has a unique formal solution $u(t,x) \in \boldsymbol{C}[[t,x]]$ with $u(0,x) \equiv 0$.

2) When $q \geq 1$, $u(t,x)$ belongs to the class $G\{t,x\}_{(s,\sigma)}$ if and only if

$$s \geq 1 \quad \text{and} \quad \sigma \geq \frac{p}{p-1}.$$

3) When $q = 0$, $u(t,x)$ belongs to the class $G\{t,x\}_{(s,\sigma)}$ if and only if

$$s \geq 1 + \frac{1}{(p-1)(l+m+n-1)} \quad \text{and} \quad \sigma \geq \frac{p}{p-1}.$$

3   Proof of Theorem 2.1

We set $\sigma = p/(p-1)$; then the condition (2.2) is written as

$$s \geq 1 + \max\left[0, \ \sup_{(i,j,\alpha)\in J}\left(\frac{\sigma-1}{i+j+\alpha-1}\right)\right]. \qquad (3.1)$$

Since $b(0) \notin \boldsymbol{N}$ is assumed, we can find a $\rho > 0$ such that $|k - b(0)| \geq \rho k$ holds for all $k \in \boldsymbol{N}$.

For $f(x) = \sum_{j\geq 0} f_j x^j \in \boldsymbol{C}[[x]]$, we write

$$|f|(x) = \sum_{j\geq 0}|f_j|\, x^j, \qquad S(f)(x) = \sum_{j\geq 0} f_{j+1}x^j = \frac{f(x)-f(0)}{x},$$

$$B_\sigma(f)(x) = \sum_{j\geq 0}\frac{f_j}{(j!)^{\sigma-1}}x^j, \quad \sigma > 1.$$

$B_\sigma(f)(x)$ is a variation of the Borel transform of $f(x)$. For $f(x) = \sum_{j\geq 0} f_j x^j$, $g(x) = \sum_{j\geq 0} g_j x^j$ we write $f(x) \ll g(x)$ if $|f_j| \leq g_j$ holds for all $j \geq 0$.

Next, we write $u(t, x)$ as

$$u(t, x) = \sum_{k \geq 1} u_k(x) t^k, \tag{3.2}$$

Under (3.2) the equation (2.1) is decomposed into the following recurrent family:

$$\left(1 - b(x) - x^p c(x) \frac{\partial}{\partial x}\right) u_1 = a(x), \tag{3.3}$$

and for $k \geq 2$

$$\left(k - b(x) - x^p c(x) \frac{\partial}{\partial x}\right) u_k$$
$$= \sum_{2 \leq i+j+\alpha \leq k} a_{i,j,\alpha}(x) \left[\sum_{i+|m|+|n|=k} u_{m_1} \cdots u_{m_j} \times \frac{\partial u_{n_1}}{\partial x} \cdots \frac{\partial u_{n_\alpha}}{\partial x}\right], \tag{3.4}$$

where $|m| = m_1 + \cdots + m_j$ and $|n| = n_1 + \cdots + n_\alpha$. Therefore, if $b(0) \notin \mathbf{N}$, we can determine uniquely $u_k(x) \in G\{x\}_\sigma$ $(k = 1, 2, \ldots)$ inductively on k from (3.3) and (3.4). Thus, we have obtained a unique formal solution $u(t, x)$ in (3.2).

Next, let us prove that this formal solution $u(t, x)$ belongs to the formal Gevrey class $G\{t, x\}_{(s,\sigma)}$ if s satisfies the condition (3.1). To do so, we set

$$w_k(x) = S(u_k)(x) \in G\{x\}_\sigma, \quad k = 1, 2, \ldots.$$

Then we have $u_k(x) = u_k(0) + x w_k(x)$ and by (3.3),(3.4) we have

$$(1 - b(0)) u_1(0) = a(0), \tag{3.5}$$

$$\left(1 - b(x) - x^{p-1} c(x) - x^p c(x) \frac{\partial}{\partial x}\right) w_1 = S(b)(x) u_1(0) + S(a)(x), \tag{3.6}$$

and for $k \geq 2$

$$(k - b(0)) u_k(0)$$
$$= \sum_{2 \leq i+j+\alpha \leq k} a_{i,j,\alpha}(0) \left[\sum_{i+|m|+|n|=k} u_{m_1}(0) \cdots u_{m_j}(0) w_{n_1}(0) \cdots w_{n_\alpha}(0)\right], \tag{3.7}$$

$$\left(k - b(x) - x^{p-1}c(x) - x^p c(x)\frac{\partial}{\partial x}\right) w_k$$
$$= S(b)(x)u_k(0)$$

$$+ \sum_{2 \le i+j+\alpha \le k} S(a_{i,j,\alpha})(x) \left[\sum_{i+|m|+|n|=k} \left(u_{m_1}(0) + xw_{m_1}\right) \times \cdots \times \right.$$
$$\times \left(u_{m_j}(0) + xw_{m_j}\right) \times \left(w_{n_1} + x\frac{\partial w_{n_1}}{\partial x}\right) \cdots \left(w_{n_\alpha} + x\frac{\partial w_{n_\alpha}}{\partial x}\right) \Bigg]$$
$$+ \sum_{2 \le i+j+\alpha \le k} a_{i,j,\alpha}(0) \left[\sum_{i+|m|+|n|=k} \left(\frac{1}{x}\left\{\left(u_{m_1}(0) + xw_{m_1}\right) \times \cdots \times \right.\right.\right.$$
$$\times \left(u_{m_j}(0) + xw_{m_j}\right) \times \left(w_{n_1} + x\frac{\partial w_{n_1}}{\partial x}\right) \cdots \left(w_{n_\alpha} + x\frac{\partial w_{n_\alpha}}{\partial x}\right)$$
$$\left. - u_{m_1}(0) \cdots u_{m_j}(0)\, w_{n_1} \cdots w_{n_\alpha}\right\}$$
$$\left.\left.+ u_{m_1}(0) \cdots u_{m_j}(0)\, S\left(w_{n_1} \cdots w_{n_\alpha}\right) \right) \right]. \tag{3.8}$$

Choose $0 < R < 1$ and $A > 0$ so that $|u_1(0)| \le A$ and $B_\sigma\left(w_1\right)(x) \ll \frac{A}{(R-x)^\sigma}$. Put $\Phi(x) = xB_\sigma(|S(b)|)(x) + 2x^{p-1}B_\sigma(|c|)(x)$ and take $B > 0$ such that

$$\frac{B_\sigma(|S(b)|)(x)}{\rho - \Phi(x)} \ll \frac{B}{(R-x)^\sigma}.$$

Similarly, choose $A_{i,j,\alpha}^{(0)} \ge 0$ and $A_{i,j,\alpha} \ge 0$ so that $|a_{i,j,\alpha}(0)| \le A_{i,j,\alpha}^{(0)}$,

$$\frac{|a_{i,j,\alpha}(0)|}{\rho - \Phi(x)} \ll \frac{A_{i,j,\alpha}^{(0)}}{(R-x)^\sigma}, \qquad \frac{B_\sigma(|S(a_{i,j,\alpha})|)(x)}{\rho - \Phi(x)} \ll \frac{A_{i,j,\alpha}}{(R-x)^\sigma}$$

and that

$$\sum_{i+j+\alpha \ge 2} A_{i,j,\alpha}^{(0)} t^i u^j v^\alpha \quad \text{and} \quad \sum_{i+j+\alpha \ge 2} A_{i,j,\alpha} t^i u^j v^\alpha$$

are convergent in a neighborhood of the origin of $\boldsymbol{C}_t \times \boldsymbol{C}_u \times \boldsymbol{C}_v$. We may assume that $A_{i,j,\alpha}^{(0)} = 0$ if $a_{i,j,\alpha}(0) = 0$.

Using these constants, let us consider the following functional equation with respect to Y:

$$Y = \frac{A}{(R-x)^{2\sigma}} t + \frac{1}{(R-x)^\sigma} \sum_{i+j+\alpha \ge 2} \frac{C_{i,j,\alpha}}{(R-x)^{\sigma(4i+2j+2\alpha-3)}} t^i (2Y)^j (2\beta Y)^\alpha \tag{3.9}$$

where $\beta = (4e\sigma)^\sigma$ and

$$C_{i,j,\alpha} = \left((1 + B/\rho)A^{(0)}_{i,j,\alpha} + A_{i,j,\alpha}\right)(i+j+\alpha)^{\sigma-1} + A^{(0)}_{i,j,\alpha}\left(e^{i+j+\alpha}\right)^{s-1}.$$

Note that by $i + j + \alpha \geq 2$ we have $4i + 2j + 2\alpha - 3 \geq 1$.

Since (3.9) is an analytic functional equation with respect to Y, by the implicit function theorem we see that (3.9) has a unique holomorphic solution $Y = Y(t, x)$ in a neighborhood of the origin of $\boldsymbol{C}_t \times \boldsymbol{C}_x$ with $Y(0, x) \equiv 0$. If we expand Y into the form

$$Y(t, x) = \sum_{k \geq 1} Y_k(x) t^k$$

we see that the coefficients $Y_k(x)$ $(k \geq 1)$ are determined by the following recurrent formula:

$$Y_1 = \frac{A}{(R - x)^{2\sigma}},$$

and for $k \geq 2$

$$Y_k = \frac{1}{(R - x)^\sigma} \sum_{2 \leq i+j+\alpha \leq k} \frac{C_{i,j,\alpha}}{(R - x)^{\sigma(4i+2j+2\alpha-3)}} \left[\sum_{i+|m|+|n|=z} \left(2Y_m\right) \right.$$
$$\left. \times \cdots \times \left(2Y_{m_j}\right)\left(2\beta Y_{n_1}\right) \cdots \left(2\beta Y_{n_\alpha}\right) \right].$$

Moreover we can prove by induction on k that $Y_k(x)$ has the form

$$Y_k(x) = \frac{M_k}{(R - x)^{\sigma(4k-2)}} \qquad \text{(for } k \geq 1\text{)}$$

with constants $M_1 = A$ and $M_k \geq 0$ (for $k \geq 2$).

In addition, we have the following lemma.

Lemma 3.1. *Let* $\beta = (4e\sigma)^\sigma$, *and let* $u(t, x)$ *be the unique formal solution in (3.2). If* s *satisfies the condition (3.1) we have the following estimates for all* $k \in \boldsymbol{N}$:

$$|u_k(0)| \ll \frac{(k - 1)!^{s-1}}{k^\sigma} Y_k(x), \qquad B_\sigma(|w_k|)(x) \ll \frac{(k - 1)!^{s-1}}{k^\sigma} Y_k(x),$$

$$B_\sigma\left(x\frac{\partial}{\partial x}|w_k|\right)(x) \ll \frac{(k - 1)!^{s-1}}{k^{\sigma-1}} \beta Y_k(x),$$

$$B_\sigma\left(\frac{\partial}{\partial x}|w_k|\right)(x) \ll \frac{(k - 1)!^{s-1}}{1} \beta Y_k(x).$$

Then, by Lemma 3.1 we have

$$\sum_{k \geq 1} \frac{B_\sigma(|u_k|)(x)}{(k-1)!^{s-1}}\, t^k \ll \sum_{k \geq 1} \frac{|u_k(0)|}{(k-1)!^{s-1}}\, t^k + x \sum_{k \geq 1} \frac{B_\sigma(|w_k|)(x)}{(k-1)!^{s-1}}\, t^k$$

$$\ll (1+x) \sum_{k \geq 1} Y_k(x) t^k = (1+x) Y(t,x).$$

This implies that our formal solution $u(t,x)$ in (3.2) belongs to the class $G\{t,x\}_{(s,\sigma)}$.

Theorem 2.1 is proved.

Acknowledgements

The work was supported by the National Natural Science Foundation of China.

References

[CT00]   H. Chen and H. Tahara, On the holomorphic solution of non-linear totally characteristic equations, to appear in *Mathematische Nachrichten*, Germany.

[CT99]   H. Chen and H. Tahara, On totally characteristic type non-linear partial differential equations in the complex domain, *Publ. RIMS, Kyoto Univ.*, 26:621-636, 1999.

[CL01]   H. Chen and Z. Luo, On the holomorphic solution of nonlinear totally characteristic equations with several space variables, *Preprint*.

[GT90]   R. Gérard and H. Tahara, Nonlinear singular first order partial differential equations of Briot-Bouquet type, *Proc. Japan Acad.*, 66:72-74, 1990.

[GT290]   R. Gérard and H. Tahara, Holomorphic and singular solution of nonlinear singular first order partial differential equations, *Publ. RIMS, Kyoto Univ.*, 26:979-1000, 1990.

[GT96]   R. Gérard and H. Tahara, Singular nonlinear partial differential equations, *Aspects of Mathematics*, E 28, Vieweg, 1996.

Addresses

CHEN HUA AND LUO ZHUANGCHU, Institute of Mathematics, Wuhan University, Wuhan, 430072, P.R.China

E-MAIL: chenhua@whu.edu.cn

TAHARA HIDETOSHI, Department of Mathematics, Sophia University, Japan

E-MAIL: h-tahara@hoffman.cc.sophia.ac.jp

2000 Mathematics Subject Classification. Primary 35A07; Secondary 35A10, 35A20

Operator Theory:
Advances and Applications, Vol. 126
© 2001 Birkhäuser Verlag Basel/Switzerland

A Unified Approach to the Theory of Fundamental Solutions, Non-degenerate Case

M. Y. Chi and Qi Minyou

In PDE theory, there are two different approaches of construction of fundamental solutions. The earlier one was due to J. Hadamard ([Ha32]) which is applicable to second order linear equations with real analytic coefficients

$$L(u) \equiv \sum_{i,j=1}^{n} a_{ij}(x)\frac{\partial^2 u}{\partial x_i \partial x_j} + \sum_{i=1}^{n} b_i(x)\frac{\partial u}{\partial x_i} + c(x)u = 0, \qquad (1.1)$$

and its principal symbol

$$L_2(x,\xi) \equiv \sum_{i,j=1}^{n} a_{ij}(x)\xi_i\xi_j$$

is assumed to be non-degenerate, i.e.,

$$\det(a_{ij}(x)) = \mathrm{Hess}_\xi L_2(x,\xi) \neq 0.$$

Hadamard defined the fundamental solutions to be those solutions with specific singularity. For this end he constructed a class of characteristic surfaces, namely, the characteristic conoid which has a conic vertex at say, $x = 0$. It can be written as

$$k(x) = 0$$

where $k(x)$ satisfies the following equation

$$L_2(x,k_x) \equiv \sum_{i,j=1}^{n} a_{ij}(x)k_{x_i}k_{x_j} = 4k. \qquad (1.2)$$

Then Hadamard tried to find a solution of (1.1) in the form

$$u(x) = \sum_{h=0}^{\infty} U_h(x)k^{p+h}/\Gamma(p+h+1), \qquad (1.3)$$

where p is to be determined later. He obtained a system of ODE's of the Fuchsian type for $U_h(x)$

$$t\frac{dU_h}{dt} + (p+h+n/2+t\Phi)\,U_h = L(U_{h-1},U_{h-2})$$

In order that bounded and non-vanishing at $x = 0$ solutions exist, p should take specific values.

Finally, he proved the convergence of (1.3) by majoration method, and this is the main reason for him to stay in the real analytic frame.

Another approach of constructing fundamental solutions was initiated by I. G. Petrovsky ([Pe38]). He used systematically Fourier-Laplace transfoms which is one of the main sources of the distribution theory. But Fourier-Laplace transforms are applicable mainly in case of PDE's with constant coefficients

$$P(D)u = 0.$$

Now, fundamental solutions are defined to be solutions for

$$P(D)u = \delta. \tag{1.4}$$

Using Fourier transform, the fundamental solution is formally

$$u(x) = (2\pi)^{-n} \int \frac{e^{<\xi \cdot x>} d\xi}{P(\xi)}.$$

Thus studying the algebraic variety

$$P_m(\xi) = 0$$

is vital in this approach.

In this paper, we want to generalize Hadamard's approach to cover linear C^∞ or real analytic partial differential equations of arbitrary order m which is non-degenerate, i. e.,

$$P(x, \partial_x)u = 0, \tag{1.5}$$

such that the Hessian determinant

$$\mathrm{Hess}_\xi H(x, \xi) \neq 0 \qquad x \in \Omega, \quad \xi \neq 0, \tag{1.6}$$

where Ω is a neighborhood of $x = 0$ small enough, and

$$H(x, \xi) = P_m(x, \xi) = \sum_{|\alpha|=m} a_\alpha(x)\xi^\alpha$$

is the principal symbol of $P(x, \partial_x)$. Here and after, we use ξ to denote ∂_x instead of $-i\partial_x$, because we do not use Fourier transforms.

We first generalize the eikonal equation (1.2). As we mentioned above, Hadamard started from the construction of the characteristic conoid composed of all bicharacteristic curves through the origin. For this end, he introduced a pseudo-Riemannian metric,

$$ds^2 = \sum_{i,j=1}^{n} A_{ij}(x)dx_i dx_j$$

where

$$(A_{ij}) = (a_{ij})^{-1},$$

a_{ij} are coefficients in (1.1), and considered the minimization of the integral

$$j(x) = \int_0^x ds = \int [\sum_{i,j=1}^n A_{ij}\dot{x}_i\dot{x}_j]^{\frac{1}{2}} dt.$$

Since the integrand is ds, we see that Hadamard was using Fermat's principle, i.e., he was working in the Lagrangian frame. The merit of this approach lies in that the integrand is homogeneous in $\dot{x}_i$ of first order, the general form of the variation is particularly simple. But the relation between the extremals of this integral (the eikonal integral) and the bicharacteristics is not very clear. So he also introduced the energy integral

$$J(x) = \int_0^x \sum_{i,j=1}^n a_{ij}\xi_i\xi_j dt$$

and pointed out that they possess the same extremals but parametrized differently.

For general PDO , how can we find the analog of the Fermat's principle? We simply switch to the Hamiltonian frame, where the energy integral is well-defined:

$$J(x) = \int_o^x H(x,\xi)dt.$$

Since the non-degeneracy condition is satisfied, we can apply Legendre's transform $\dot{x} = H_\xi$ and obtain the Lagrangian

$$L(x,\dot{x}) = \dot{x}\xi - H(x,\xi) = (m-1)H(x,\xi),$$

(Euler's formula). It is easy to see that $L(x,\dot{x})$ is homogeneous in $\dot{x}$ of order $\frac{m}{m-1}$. Hence we use another integral as the eikonal integral in our case:

$$j(x) = \int_o^x [L(x,\dot{x})]^{\frac{m-1}{m}} dt = (m-1)^{\frac{m-1}{m}} \int_o^x [H(x,\xi)]^{\frac{m-1}{n}} dt$$

which has the dimension of a length. So the fact that the $x-$components of the Hamiltonian orbits for $H(x,\xi)$ are the extremals of $j(x)$ is just the generalization of Fermat's principle. Since the integrand of $j(x)$ is homogeneous in $\dot{x}$ of first order, to compute its variation is particularly simple. Just like Hadamard, we introduce

$$k(x) = [j(x)]^{\frac{m}{m-1}}$$

which has the dimension of $[\tau \cdot \text{action}]$, we obtain immediately the following result of Leray([Le62]).

Theorem 1.1 (Leray). *Let $P(x, \partial_x)$ of (1.5) be a linear PDO with C^∞ or real analytic coefficients of order $m \geq 2$ which satisfies the non-degeneracy condition (1.6), then the characteristic conoid with vertex at $x = 0$ can be written as*

$$k(x) = 0$$

where $k(x)$ is C^∞ or real analytic near $x = 0$ but $x \neq 0$, and

$$H(x, k_x) = \frac{k}{m-1}, \qquad\qquad (1.7)$$
$$k_x = \tau\xi, \qquad \tau = t^{1/(m-1)}.$$

(1.7) is just the eikonal equation we need.

Now we should find a solution for the equation

$$P(x, \partial_x)u = c(x)k_+^\lambda / \Gamma(\lambda + 1), \qquad\qquad (1.8)$$

where $c(x)$ is C^∞ or real analytic near $x = 0$, λ is a complex parameter. That we can use (1.8) to replace (1.4) is because that $\delta(x)$ can either be expressed as

$$\delta(x) = (2\pi)^{-n} \int e^{i<\xi \cdot x>} d\xi,$$

which is the basis of Fourier transform and physically corresponds to the uncertainty principle, or can be thought of as the residue at $\lambda = -1$ of the distribution-valued meromorphic function of λ

$$\delta(x) = \operatorname{Res}\Big|_{\lambda = -1} [x_+]^\lambda.$$

This point of view can be traced back to M. Riesz (his Riemann-Liouville integrals) and even to Hadamard himself (partie fini of divergent integrals) and is well exposed in Gelfand and Shilov's work ([GS58]).

The next step is to find a solution for the equation (1.8) in the form

$$u(x) = \sum_{h=0}^{\infty} \left[\tau^h U_h(\tau, \eta)\right] k_+^\lambda / \Gamma(\lambda + 1). \qquad\qquad (1.9)$$

We should determine p and find the transport equations for U_h.

Actually Hadamard tried to look for a solution of the form (1.3). Difference in form between Hadamard's and ours means that we switch to another coodinates (τ, η) than x. This is because, the Hamiltonian system

$$\dot{x} = H_\xi(x, \xi), \qquad x\big|_{t=0} = 0;$$
$$\dot{\xi} = -H_x(x, \xi), \qquad \xi\big|_{t=0} = \eta, \quad \|\eta\| = 1,$$

defines a mapping

$$x = x(\tau, \eta), \qquad \tau = t^{\frac{1}{m-1}}, \tag{1.10}$$

which is a diffeomorphism near $x = 0$ when $m = 2$, but is singular at $x = 0$ when $m > 2$. $k(x)$ in Hadamard's case of $m = 2$ is smooth and can be reduced to a quadratic by Morse's lemma, while in general case of $m > 2$ $k(x) = O(1)\tau^m = O(1)\|x\|^{\frac{m}{m-1}}$. Mapping (1.10) is very close to Leray's projéction caractéristique ([Le62]). It is in essence a blow-up of the singularity at $x = 0$ to a domain σ on the hypersurface $\tau = 0$ in (τ, η) space.

In (τ, η) space the transport equations are particularly simple.

Theorem 1.2. *In (τ, η) space the transport equation for U_0 is*

$$\tau \frac{d}{d\tau} U_0 + \left[\lambda + \frac{n}{2}(m - 1)\right] U_0 = c(\zeta), \tag{1.11}$$

$(p = \lambda + m - 1)$ and those for U_h are

$$\tau \frac{d}{d\tau} U_h + \left[\lambda + h + \frac{n}{2}(m - 1)\right] U_h = L_h(U_{h-1}, \cdots, U_{h-n}), \tag{1.12}$$

where L_h is a linear expression of U_{h-j} and their derivatives of order up to degree j with C^∞ or real analytic coefficients, and $U_{h-j} \equiv 0$ when $h < j$.

Theorem 1.3. *Equation (1.8) has the formal solution (1.9), where $p = \lambda + m - 1$ and the coefficeints U_h are determined by the transport equations (1 11) and (1.12).*

Up to now, all the conclusions are valid both for C^∞ and real analytic cases. But as to convergence of the formal solution (1.9), we can prove it by majoration method in real analytic case, but it is very unlikely that one can obtain general results in C^∞ case as the existence would imply the local solvability for C^∞ linear PDE. We have only asymptotic results by the Borel technique.

Theorem 1.4. *In real analytic case, the series (1.9) is convergent near $\tau = 0$; while in C^∞ case, it gives only an asymptotic solution for the equation (1.8).*

Finally, we will show, how (1.9) can give fundamental solutions. The crucial point is that although t_+^λ is a distribution, the structure of k_+^λ as a distribution is tied closely to the singularity of $k(x)$. In (τ, η) space,

$$k(x) = \tau^m H(0, \eta).$$

It consists of two strata, one is a $0-$dimensional stratum, namely the point $x = 0$, the other is an $(n - 1)-$dimensional stratum $H(0, \eta) = 0$, which is a manifold in non-degenerate case. In this case distribution H_+^λ and $\delta(H)$ as its residue are well studied. (See [GS58]). Such structure of the algebraic variety $P_m(0, \tau_i) = 0$ leads to an enlargement of the notion of fundamental solutions, and we give

Definition 1.1. The fundamental solutions for the non-degenerate PDO $P(x, \partial_x)$ are the solutions for

$$P(x, \partial_x)u = \text{Dirac-type distributions.}$$

If the right hand side is a linear expresssion of $\delta(\tau)$ and its derivatives, it will be called a Hadamard-Dirac (H-D) fundamental solution; if it is a linear expresssion of $\delta(H)$ and its derivatives, it will be called a Hadamard-Huygens (H-H) fundamental solution; if it is a sum of both, it will be called a Hadamard-mixed (H-M) fundamental solution.

Since the right hand side of (1.8) is meromorphic in λ, so are its solutions (denoted hence as $u(x, \lambda)$). If $\lambda = \lambda_0$ is a pole of the right hand side of (1.8) with a Dirac-type distribution as residue, then

$$P(x, \partial_x)\text{Res}\,|_{\lambda=\lambda_0}[u(x, \lambda_0)] = \text{Dirac-type distribution}$$

is the corresponding fundamental solution. But the right hand side of (1.8) as a distribution in (τ, η) space is

$$A(\tau, \eta)\tau^{m(\lambda+n)-(n+1)} \otimes H_+^\lambda(0, \eta)/\Gamma(\lambda + 1),$$

and the denominator $\Gamma(\lambda + 1)$ is also meromorphic in λ, so sometimes λ_0 as a whole may be a regular point of the right hand side and the first Taylor coefficient may contain Dirac-type distributions. Thus we should distinguish three cases as follows:

Theorem 1.5. *The right hand side of (1.8) is a distribution-valued meromorphic function of λ divided by another meromorphic function $\Gamma(\lambda + 1)$, and*

1. $\lambda \in P_1 = \{\lambda : -m(\lambda + n) + (n + 1) \in I\!N, but\ \lambda \notin -I\!N\}$ are simple poles with residue
$$B(\tau, \eta)\delta^{(l)}(\tau) \otimes H^{n(1-m)/m-(l-1)/m}.$$

2. $\lambda \in P_2 = \{-l,\ \ l \in I\!N, m(\lambda + n) - (n + 1) \notin I\!N\}$ are regular values with first Taylor coefficient
$$B(\tau, \eta)\tau^{m(n-l)-(n+1)} \otimes \delta^{(l-1)}(H).$$

3. $\lambda \in P_3 = \{\lambda = -l_1 = n(1 - m)/m - (l_2 - 1)/m,\ \ l_1, l_2 \in I\!N\}$ are regular values with first Taylor coefficient
$$B(\tau, \eta)\delta^{(l_1)}(\tau) \otimes \delta^{(l_2-1)}(H).$$

$B(\tau, \eta)$ are suitable functions.

Now we give our final result.

Theorem 1.6. *The procedure above gives distributional fundamental solutions. More precisely,*

1. $\lambda \in P_1$ give H-D solutions;

2. $\lambda \in P_2$ give H-H solutions;

3. $\lambda \in P_3$ give H-M solutions.

References

[GS58]   I. M. Gelfand and Shilov, Generalized Functions, Vol. I, 1958, Moscow.

[Ha32]   J. Hadamard, Le Problème de Cauchy, 1932, Hermann, Paris.

[Le62]   J. Leray, Problème de Cauchy, IV, Bull. Soc. Math. France, t. 90, (1962), 39-156.

[Pe38]   I. G. Petrovsky, On the Cauchy problems for systems of partial differential equations in the domain of non-analytic functions, Bull. Mosc. Univ., Math. Mech. I, (1938), 1-72.

Address

M. Y. CHI AND QI MINYOU, Institute of Mathematics, Wuhan University, Wuhan, 430072, P.R. of China

Operator Theory:
Advances and Applications, Vol. 126
© 2001 Birkhäuser Verlag Basel/Switzerland

Fourier Integral Operators in SG Classes: Classical Operators

SANDRO CORIASCO * AND PAOLO PANARESE

Abstract. We continue the investigation of the calculus of Fourier Integral Operators (FIOs) in the class of symbols with exit behaviour (**SG** symbols). Here we analyse what happens when one restricts the choice of amplitude and phase functions to the subclass of the classical **SG** symbols. It turns out that the main composition theorem, obtained in the environment of general **SG** classes, has a "classical" counterpart. We also analyse the Cauchy problem for classical hyperbolic operators of order $(1,1)$; for such operators we refine the known results about the analogous problem for general **SG** hyperbolic operators. The material contained here will be used in a forthcoming paper to obtain a Weyl formula for a class of operators defined on manifolds with cylindrical ends, improving the results obtained in [MP99].

1   Introduction

The calculus of FIOs developed in [Co98a] is based on the class of (general) **SG** symbols and amplitudes, i.e., the classes of all $a \in C^\infty(\mathbb{R}^n \times \mathbb{R}^n \times \mathbb{R}^n; \mathbb{C})$ satisfying $\forall \alpha, \beta, \gamma \in \mathbb{N}^n \; \exists C_{\alpha\beta\gamma} \geq 0$ such that

$$|\partial_\xi^\alpha \partial_\beta^x \partial_\gamma^y a(x,y,\xi)| \leq C_{\alpha\beta\gamma} \, \langle\xi\rangle^{m_1-|\alpha|} \, \langle x\rangle^{m_2-|\beta|} \, \langle y\rangle^{m_3-|\gamma|} \tag{1.1}$$

where, as usual, $|\alpha| = \alpha_1 + \alpha_2 + \ldots + \alpha_n$ for all $\alpha \in \mathbb{N}^n$, $\langle x\rangle = \sqrt{1-|x|^2}$ for all $x \in \mathbb{R}^n$ and (x,y,ξ) runs through all $\mathbb{R}^n \times \mathbb{R}^n \times \mathbb{R}^n$. When (1.1) is fulfilled, we say that a belongs to the class of **SG** amplitudes of order $m = (m_1, m_2, m_3)$. A **SG** (left) symbol is a **SG** amplitude which does not depend on y. In such a case, the vanishing third component of the order is dropped, and we will use the double order $m = (m_1, m_2)$, denoting the set of **SG** symbols by $\mathbf{SG}_l^m$. With some requirement on the real-valued phase function $\varphi \in \mathbf{SG}_l^{(1,1)}$ (see below) it turns out that FIOs of the type

$$Au(x) = A_{\varphi,a} u(x) = \frac{1}{(2\pi)^n} \int e^{i\varphi(x,\xi)} \, a(x,\xi) \, \hat{u}(\xi) \, d\xi \tag{1.2}$$

make sense for $u \in \mathcal{S}(\mathbb{R}^n)$, $a \in \mathbf{SG}_l^m$, $m \in \mathbb{R}^2$. More precisely, the operator in (1.2) is linear and continuos from $\mathcal{S}(\mathbb{R}^n)$ in itself and extendable to a linear continous operator from $\mathcal{S}'(\mathbb{R}^n)$ in itself. These and other results (in particular, about the

* Supported by the EU Research and Training Network "Geometric Analysis'

compositions of these FIOs with the corresponding **SG** pseudo-differential operators (ψdos)) were obtained in [Co98a], and subsequently applied to the study of **SG**-hyperbolic Cauchy problems in [Co98b] and [CR99]. In section 2 we fix the notations and recall basic facts about the **SG** ψdos calculus[1], together with a short resume of the calculus of the **SG** FIOs, developed in [Co98a]. The classical **SG** calculus is illustrated, e.g., in Egorov and Schulze [ES97] (see also Maniccia and Panarese [MP99] and Witt [Wi98]). A very brief account of it is given in section 3. We will denote the subclass of **SG** classical (left) symbols of order $m \in \mathbb{R}^2$ by $\mathbf{SG}^m_{l,\mathrm{cl}}$. Here we wish to illustrate what happens when amplitude and phase functions of the **SG** FIOs defined by (1.2) are classical **SG** symbols[2]. In such a case, we will speak of classical **SG** FIOs. We prove the following theorem.

Theorem 1.1. *Let $P = \mathrm{Op}\,(p)$ be a classical* **SG** *ψdo and let $A = A_{\varphi,a}$ be a classical* **SG** *FIO with $p \in \mathbf{SG}^r_{l,\mathrm{cl}}$, $a \in \mathbf{SG}^s_{l,\mathrm{cl}}$. Then, the composed operators PA and AP are classical* **SG** *FIOs with the same phase function and amplitude h such that $h \in \mathbf{SG}^{r+s}_{l,\mathrm{cl}}$.*

Section 4 is devoted to the precise definition of classical FIOs. We will give there a sketch of the proof of the theorem above, while in section 5 we will discuss the hyperbolic first order Cauchy problems associated to classical **SG** operators, along the same lines of [Co98b]. Using Theorem 1.1, it is possible to prove the following result.

Theorem 1.2. *Assume that λ is a classical* **SG** *symbol of order $(1,1)$, depending smoothly on a parameter $t \in J = [-T,T]$, $T > 0$. Assume also that the corresponding operator is hyperbolic[3]. Then, the operator $A_{\varphi(t),a(t)}$, approximating the solution operator of*

$$\begin{cases} (\partial_t - i\Lambda(t))u(t) = 0, \, t \in J \\ u(0) = u_0 \end{cases} \tag{1.3}$$

is, modulo smoothing operators, a classical **SG** *FIO.*

For space reasons, the detailed proofs of Lemma 4.2 and of Propositions 5.1 and 5.2, intermediate steps of the proofs of the above theorems, will appear elsewhere.

---

[1] Refer to Cordes [C95], Parenti [Pa72], Schrohe [Sc86] and Egorov and Schulze [ES97], for details and further development of the theory.

[2] The additional requirements for the phase function will be stated in section 4.

[3] See below for the precise definitions.

2 SG Classes of Symbols and Operators. SG Sobolev Spaces

We set, from now on, $d_x\varphi = \left(\dfrac{\partial\varphi}{\partial x^1}, \ldots, \dfrac{\partial\varphi}{\partial x^n}\right) = (\partial_1\varphi, \ldots, \partial_n\varphi) = (\partial_1^x\varphi, \ldots, \partial_n^x\varphi)$
and $\nabla_\xi\varphi = \left(\dfrac{\partial\varphi}{\partial\xi_1}, \ldots, \dfrac{\partial\varphi}{\partial\xi_n}\right) = (\partial^1\varphi, \ldots, \partial^n\varphi) = \left(\partial_\xi^1\varphi, \ldots, \partial_\xi^n\varphi\right)$. For convenience, when dealing with orders of symbols, we will often use the obvious notations $e = (1,1)$, $e_1 = (1,0)$ and $e_2 = (0,1)$. In general, ψdos will be denoted by capital letters and their symbols or amplitudes by the corresponding small letter (i.e., $P = \mathrm{Op}\,(p)$, $q = \mathrm{Sym}\,(Q)$, etc.). $A = A_{\varphi,a}$ will be, unless otherwise stated, a **SG** FIO with phase function φ and amplitude a.

Definition 2.1. For $m = (m_1, m_2, m_3) \in \mathbb{R}^3$ we denote by $\mathbf{SG}^n = \mathbf{SG}^m(\mathbb{R}^n)$ the space of all amplitudes functions $a \in C^\infty(\mathbb{R}^n \times \mathbb{R}^n \times \mathbb{R}^n)$ which satisfy (1.1). $\mathbf{SG}^m(\mathbb{R}^n)$ is given the usual Fréchet topology based upon the seminorms implicit in (1.1). When a is vector or matrix valued, $a \in \mathbf{SG}^m$ means that the estimates (1.1) are fulfilled component by component. For $m = (m_1, m_2) \in \mathbb{R}^2$ denote by $\mathbf{SG}_l^m = \mathbf{SG}_l^m(\mathbb{R}^n)$ the double-order symbol space of functions $a \in \mathbf{SG}^{(m_1, m_2, 0)}$ which are independent of y.

The following lemma is a result about compositions of elements of the **SG** classes of symbols and amplitudes. It is a basic tool in the proof of the theorems of composition among pseudodifferential and Fourier Integral operators.

Lemma 2.1. *Let* $f \in \mathbf{SG}^m$ *and* g *vector valued in* $\mathbb{R}^n$ *such that* $g \in \mathbf{SG}^{e_1}$ *and* $\exists C > 0 \mid C^{-1}\langle\xi\rangle \leq \langle g(x,y,\xi)\rangle \leq C\langle\xi\rangle$. *Then* $f(x,y,g(x,y,\xi)) \in \mathbf{SG}^m$.

Definition 2.2. With each amplitude $p \in \mathbf{SG}^m$ associate a linear operator $P = \mathrm{Op}\,(p) : \mathcal{S}(\mathbb{R}^n) \to \mathcal{S}(\mathbb{R}^n)$ defined as

$$Pu(x) = \mathrm{Op}\,(p)\,u(x) = \int\int e^{i\langle x-y|\xi\rangle}\, p(x,y,\xi)\, u(y)\,dy\,d\!\!\!\!\!/\,\xi. \tag{2.1}$$

Here, as usual, $d\!\!\!\!\!/\,\xi = (2\pi)^{-n}d\xi$. If $r \in \mathbb{R}^2$ and $q \in \mathbf{SG}_l^r$ holds[4], (2.1) reduces to

$$Qu(x) = Op(q)u(x) = \int e^{i\langle x|\xi\rangle}\, q(x,\xi)\, \hat{u}(\xi)\,d\!\!\!\!\!/\,\xi, \tag{2.2}$$

where $\hat{u}(\xi) = \mathcal{F}_{x\to\xi}(u)(\xi)$ is the Fourier transform of u. We denote by $\mathbf{LG}^r$ the space of all the operators defined as in (2.2). An element $P \in \mathbf{LG}^r$ is called a **SG** ψdo , of order less or equal to r.

---

[4] That is, the amplitude is, indeed, a symbol.

The ψdos in $\mathbf{LG} = \bigcup_{r \in \mathbb{R}^2} \mathbf{LG}^r$ form an algebra of linear continuous operators from $\mathcal{S}(\mathbb{R}^n)$ to $\mathcal{S}(\mathbb{R}^n)$, extendable as linear continuous operators from $\mathcal{S}'(\mathbb{R}^n)$ to $\mathcal{S}'(\mathbb{R}^n)$. The usual property about the order of the composed operators holds normally, simply understanding the sum of orders as a sum of vectors in $\mathbb{R}^2$. The residual elements of this ψdos algebra (smoothing operators) are the integral operators with kernel in $\mathcal{S}(\mathbb{R}^n \times \mathbb{R}^n)$, whose set we denote by $\mathcal{K}$. It is possible to prove that the $\mathbf{SG}$ elliptic operators[5] admit, as expected, a parametrix modulo $\mathcal{K}$. Moreover, it is also possible to prove that every operator defined by (2.1) with $p \in \mathbf{SG}^m$, $m = (m_1, m_2, m_3) \in \mathbb{R}^3$, can be represented, modulo $\mathcal{K}$, as an operator of the form (2.2) with $q \in \mathbf{SG}_l^r$, $r = (m_1, m_2 + m_3)$. For $P \in \mathbf{LG}^m$ we denote by $p = \mathrm{Sym}\,(P) \in \mathbf{SG}_l^m$ the symbol of P, that is $P = \mathrm{Op}\,(p)$. Moreover, we denote by $\mathrm{Sym}_{\mathrm{p}}\,(P)$ a principal symbol of P, that is a $p' \in \mathbf{SG}_l^m$ such that $p - p' \in \mathbf{SG}_l^{m-e}$.

In the present situation, the notion of asymptotic expansion is the following: for

$$a \in \mathbf{SG}_l^m,\ a_j \in \mathbf{SG}_l^{m-je},\ a \sim \sum_{j \in \mathbb{N}} a_j \Leftrightarrow \forall N \in \mathbb{N}\ a - \sum_{j=0}^{N} a_j \in \mathbf{SG}_l^{m-(N+1)e}. \text{ It is a}$$

fact that with any sequence of symbols with orders diverging to $(-\infty, -\infty)$ it is possible to associate an asymptotic sum, which is unique modulo $\mathcal{S}(\mathbb{R}^n \times \mathbb{R}^n)$.

In the following definition, we describe the notion of weighted Sobolev spaces "adapted" to the $\mathbf{SG}$ calculus.

Definition 2.3. For $s = (s_1, s_2) \in \mathbb{R}^2$, let π_s denote the product $\pi_s(x, \xi) = \langle \xi \rangle^{s_1} \langle x \rangle^{s_2}$ and $\Pi_s = \mathrm{Op}\,(\pi_s)$ the corresponding operator. The weighted Sobolev spaces $H^s(\mathbb{R}^n) = H^s$, $s = (s_1, s_2) \in \mathbb{R}^2$, are defined by:

$$H^s = \left\{ u \in \mathcal{S}'(\mathbb{R}^n) \mid \Pi_s u \in L^2(\mathbb{R}^n) \right\}, \tag{2.3}$$

with the natural Hilbert norm $\|u\|_s = \|\Pi_s u\|_{L^2} = \|\Pi_s u\|_0$.

The $\mathbf{SG}$ ψdos act continously on the spaces H^s, that is, $P \in \mathbf{LG}^r$ implies that P is linear and continuous from H^s to H^{s-r} for all $r, s \in \mathbb{R}^2$.

We recall now the definition of $\mathbf{SG}$ FIO, and we state the main composition result among general $\mathbf{SG}$ ψdos and FIOs, which corresponds to the first statement of Theorem 1.1 in the non-classical situation. The proof, together with a complete analysis of the properties of $\mathbf{SG}$ FIOs, can be found in [Co98a] or [Co98c].

Definition 2.4. We will call phase function (or simply phase) any real valued $\varphi \in \mathbf{SG}_l^e$ satisfying $C^{-1} \langle x \rangle \le \langle \nabla_\xi \varphi(x, \xi) \rangle \le C \langle x \rangle$ and $C^{-1} \langle \xi \rangle \le \langle d_x \varphi(x, \xi) \rangle \le C \langle \xi \rangle$ for a suitable constant $C > 0$. We will denote by $\mathcal{P}$ the set of all such phases. Moreover, we define the set $\mathcal{P}^\epsilon$, $\epsilon > 0$ of all regular phases as follows

$$\mathcal{P}^\epsilon = \left\{ \varphi \in \mathcal{P} \mid \forall x, \xi\ :\ \left| \det \left(\partial_i^x \partial_\xi^j \varphi \right) \right| \ge \epsilon \right\}.$$

---

[5] See again [C95], [Pa72], [Sc86], [ES97] for the precise notion of ellipticity and hypoellipticity in this context.

Definition 2.5. For any choice of $\varphi \in \mathcal{P}$, $a \in \mathbf{SG}_l^m$ and $u \in \mathcal{S}(\mathbb{R}^n)$, we define the corresponding **SG** FIO as in (1.2).

Remark 2.1. The **SG** FIOs defined in Definition 2.5 above are linear and continuous from $\mathcal{S}(\mathbb{R}^n)$ in itself and extendable as linear continuous operators from $\mathcal{S}'(\mathbb{R}^n)$ in itself. Moreover, if the phase is regular and the amplitude a is in $\mathbf{SG}_l^m$, they are also linear and continuous from H^s to H^{s-m} for all $s, m \in \mathbb{R}^2$. This remains true even if the estimates of Definition 2.4 hold only outside a set of the type $|x| + |\xi| \geq R > 0$ (see [Co98a] and [Co98c]).

Theorem 2.1. *Let be given a FIO $A = A_{\varphi,a}$ such that $\varphi \in \mathcal{P}$ and $a \in \mathbf{SG}_l^m$ and a ψdo $P = \mathrm{Op}\,(p)$ with $p \in \mathbf{SG}_l^t$. Then, the composed operator $H = PA$ is, modulo smoothing operators, a FIO. In fact, $H = H_{\varphi,h}$ where φ is the same phase function and the amplitude $h \in \mathbf{SG}_l^{m+t}$ admits the following asymptotic expansion:*

$$h(x,\xi) \sim \sum_{\alpha \in \mathbb{N}^n} \frac{1}{\alpha!}(\partial_\xi^\alpha p)(x, d_x\varphi(x,\xi)) D_\alpha^y \left[e^{i\psi(x,y,\xi)} a(y,\xi) \right]^{-}_{y=x}. \qquad (2.4)$$

Here

$$\psi(x,y,\xi) = \varphi(y,\xi) - \varphi(x,\xi) - \langle y - x | d_x\varphi(x,\xi)\rangle, \qquad (2.5)$$

and, as usual, $D_\alpha^y = (-i)^{|\alpha|}\partial_\alpha^y$. We will write $h = p \circ_\varphi a$ as a short form of (2.4).

The similar result for the composition AP can be easily deduced from Theorem 2.1 (see [Co98a] and [Co98c]).

Remark 2.2. General FIO calculi, including the **SG** FIOs calculus, already existed (see, e.g., [Mo83] and the references quoted therein). We preferred, in view of our applications, a more transparent approach, completely in the **SG** environment. At our knowledge, however, the present classical **SG** FIOs have not yet been studied.

3   Classical SG Symbols and Pseudo-Differential Operators

A definition of classical **SG** symbol can be given as follows[6].

Definition 3.1. Let S^m, $m \in \mathbb{R}$, denote the space of global classical symbols in one variable. This means that $a \in S^m$ if $a = a(\xi)$ is smooth on $\mathbb{R}^n$ satisfies estimates like (1.1) in the only variable ξ and there exist functions $a_{(m-j)} \in$

---

[6] See [Wi98].

$C^\infty(\mathbb{R}^n \setminus \{0\})$, $j \in \mathbb{N}$, homogeneous of degree $m - j$, such that, for some excision function[7] ω, we have

$$a(\xi) \sim \sum_{j=0}^{\infty} \omega(\xi) a_j(\xi).$$

Then, for $m = (m_1, m_2) \in \mathbb{R}^2$, $\mathbf{SG}_{l,\mathrm{cl}}^m = S_\xi^{m_1} \hat{\otimes}_\pi S_x^{m_2}$.

An equivalent definition by means of asymptotic expansions in terms of sub-classes of homogeneous $\mathbf{SG}$ symbols can be found in [ES97], which we refer to for most of the notations concerning the classical $\mathbf{SG}$ symbols[8]. It easily turns out that the classical symbols are closed under sums and products. Moreover, to any $a \in \mathbf{SG}_{l,\mathrm{cl}}^m$ we can canonically associate its principal symbol[9] $\mathrm{Sym}_{\mathrm{p}}^{\mathrm{cl}}(a) = \{\sigma_\psi^{m_1}(a); \sigma_{\mathrm{e}}^{m_2}(a), \sigma_{\psi,\mathrm{e}}^m(a)\}$. $\sigma_\psi^{m_1}(a)$ is called the homogeneous principal interior symbol and the pair $\{\sigma_{\mathrm{e}}^{m_2}(a), \sigma_{\psi,\mathrm{e}}^m(a)\}$ the homogeneous principal exit symbol of a.

To our aim, the following two results, included in [ES97], are central. The first says that, just like for the general $\mathbf{SG}$ symbols, classical symbols are identified by asymptotic expansion. Through the second one, we obtain the decription of the topology of the classical symbols spaces.

Theorem 3.1. *Let $a_k \in \mathbf{SG}_{l,\mathrm{cl}}^{m-ke}$, $k \in \mathbb{N}$ be an arbitrary sequence of classical symbols and $a \sim \sum_{k=0}^{\infty} a_k$ the asymptotic sum in the sense of the general $\mathbf{SG}$ symbols, as recalled in section 2. Then, $a \in \mathbf{SG}_{l,\mathrm{cl}}^m$.*

Theorem 3.2. *Let B^n denote the unitary ball of $\mathbb{R}^n$ and let χ be a diffeomorphism from the interior of B^n to $\mathbb{R}^n$ such that $|x| > 2/3 \Rightarrow \chi(x) = \dfrac{x}{|x|(1 - |x|)}$. Choosing a C^∞ function $[x]$ on $\mathbb{R}^n$ such that $|x| > 2/3 \Rightarrow [x] = |x|$ and $\forall x \in \mathbb{R}^n\, 1 - [x] \neq 0$, for any $a \in \mathbf{SG}_{l,\mathrm{cl}}^m$ denote by $(D^m a)(y, \eta)$ the function*

$$b(y, \eta) = (1 - [\eta])^{m_1}(1 - [y])^{m_2} a(\chi(y), \chi(\eta)). \tag{3.1}$$

Then, D^m extends to a homeomorphism from $\mathbf{SG}_{l,\mathrm{cl}}^m$ to $C^\infty(B^n \times B^n)$.

---

[7] That is, $\omega(\xi) \in C^\infty(\mathbb{R}^n, [0, 1])$ is zero in a neighbourhood of the origin and is 1 for large values of $|\xi|$. The notion of asymptotic equivalence here is analogous to the $\mathbf{SG}$ one, but in a single variable.

[8] Our $\mathbf{SG}_{l,\mathrm{cl}}^{(m_1, m_2)}$ is denoted by $S_{\mathrm{cl}(x,\xi)}^{m_1, m_2}$ in the cited book. We also write $\sigma_{\psi,\mathrm{e}}^m$ with $m \in \mathbb{R}^2$ rather than $\sigma_{\psi,\mathrm{e}}^{m_1, m_2}$.

[9] See again [ES97]. Compare also with [MP99].

4 Proof of the Composition Theorem

The proof of Theorem 1.1 is obtained through different steps, dealing with compositions of classical symbols. We achieve it proving that, under the hypotheses of the Theorem, the terms of the asymptotic expansion (2.4) are classical. The following Lemma is immediate, by the natural inclusions among the classical symbols spaces and the formulae for the derivatives of the exponential in (4.1) below[10].

Lemma 4.1. *Let* $\varphi \in \mathbf{SG}^e_{l,\mathrm{cl}}$. *Then, for* $\psi(x,y,\xi)$ *defined in (2.5), we have*

$$D^y_\beta e^{i\psi(x,y,\xi)}\Big|_{y=x} \in \mathbf{SG}^{(E(|\beta|/2),E(-|\beta|/2))}_{l,\mathrm{cl}}. \tag{4.1}$$

Definition 4.1. We denote by $\mathcal{P}_{cl}$ the subset of classical phase functions. It is defined as the set of all $\varphi \in \mathcal{P}$ such that $\varphi \in \mathbf{SG}^e_{l,\mathrm{cl}}$. Moreover, by $\mathcal{P}^\epsilon_{cl}$ we denote the subset of regular classical phase functions. Using the notations above, we simply set $\mathcal{P}^\epsilon_{cl} = \{\varphi \in \mathcal{P}_{cl} \mid \varphi \in \mathcal{P}^\epsilon\}$. A classical **SG** FIO is then any **SG** FIO defined as in (1.2) such that $\varphi \in \mathcal{P}_{cl}$ and $a,b \in \mathbf{SG}^m_{l,\mathrm{cl}}$.

Lemma 4.2. *If* $p \in \mathbf{SG}^m_{l,\mathrm{cl}}$ *and* $\omega \in \mathbf{SG}^{e_1}_{l,\mathrm{cl}}$ *is a vector-valued classical symbol satisfying* $\exists C > 0 \mid C^{-1} \langle\xi\rangle \le \langle\omega(x,\xi)\rangle \le C \langle\xi\rangle$, *we have* $q(x,\xi) = p(x,\omega(x,\xi)) \in \mathbf{SG}^m_{l,\mathrm{cl}}$.

Proof (sketchy). This lemma corresponds to Lemma 2.1 in the classical case. It can be proved analysing the function $\tilde{q}$ that corresponds to q under the homeomorphism D^m introduced in Theorem 3.2. $\tilde{q}$ can be expressed through $\tilde{\omega} = D^{e_1}\omega$ and $\tilde{p} = D^m p$, which are in $C^\infty(B^n \times B^n)$ by hypothesis. Since $\tilde{q}$ turns out to be in $C^\infty(B^n \times B^n)$ itself and D^m is a homeomorphism, we can conclude $q \in \mathbf{SG}^m_{l,\mathrm{cl}}$. $\square$

Theorem 4.1. *For all* $p \in \mathbf{SG}^m_{l,\mathrm{cl}}$, $a \in \mathbf{SG}^t_{l,\mathrm{cl}}$, $\varphi \in \mathcal{P}_{cl}$, *we have* $p \circ_\varphi a \in \mathbf{SG}^{m+t}_{l,\mathrm{cl}}$.

Proof. Our assumptions imply that $\omega = d_x\varphi \in \mathbf{SG}^{e_1}_{l,\mathrm{cl}}$ satisfies all the requirements of Lemma 4.2 above, so that we have

$$\forall \alpha \in \mathbb{N}^n \ (\partial^\alpha_\xi p)(x,(d_x\varphi)(x,\xi)) \in \mathbf{SG}^{m-|\alpha|e_1}_{l,\mathrm{cl}}. \tag{4.2}$$

Then, the desired result follows by the multiplication properties of classical symbols, by (2.4), (4.1) and (4.2) and by Theorem 3.1. $\square$

---

[10] See, e.g., [Co98a] or [Co98c]. In (4.1), we denote by $E(a)$ the integer part of the real number a.

5   The Cauchy Problem in the SG Classical Environment

In this section we apply the theory of classical **SG** FIOs to the solution of the Cauchy problem for a classical **SG** ψdo of order e, similarly to what was done in [Co98b]. The differences in the present treatment amount to the proof that the involved objects are classical in the sense described above.

Consider a symbol $\lambda \in C^\infty(J; \mathbf{SG}^e_{l,\mathrm{cl}})$, $J = [-T, T]$, $T > 0$, which we take scalar-valued, for simplicity. Setting $D_t = -i\partial_t$ and choosing $u_0 \in H^s$, $s \in \mathbb{R}^2$, we know[11] that, if $\Lambda(t) = \mathrm{Op}\,(\lambda(t; ., ..))$ is **SG**-hyperbolic (i.e., if there exists a real-valued $\lambda_e \in C^\infty(J; \mathbf{SG}^e_{l,\mathrm{cl}})$ such that $\lambda_0 = \lambda - \lambda_e \in C^\infty(J; \mathbf{SG}^0_{l,\mathrm{cl}})$), the solution of

$$\begin{cases} (D_t - \Lambda(t))u(t) = 0, & t \in J \\ u(0) = u_0 \end{cases} \tag{5.1}$$

exists and is unique. We also know that $u(t) = A_{\varphi(t),a(t)}u_0$, where $A(t) = A_{\varphi(t),a(t)}$ is a **SG** FIO, with phase function and amplitude depending smoothly on $t \in J' = [-T', T'] \subseteq J$, $T' > 0$. Note that $\mathrm{Sym}^{\mathrm{cl}}_{\mathrm{p}}(\lambda)$ real-valued implies Λ hyperbolic: indeed, we can take[12]

$$\lambda_e(t; x, \xi) = \omega(\xi)\sigma^1_\psi(\lambda)(t; x, \xi) + \omega(x)\left(\sigma^1_e(\lambda)(t; x, \xi) - \omega(\xi)\sigma^{(1,1)}_{\psi,e}(\lambda)(t; x, \xi)\right),$$

where ω is an excision function.

We prove Theorem 1.2 analysing separately the solution of the eikonal and of the transport equations, as in [Co98b]. We deal first with the Hamiltonian system (5.3) associated with the principal symbol of λ, since, as it is well known, the solution of the eikonal equation

$$\begin{cases} \dfrac{\partial\varphi}{\partial t}(t; x, \xi) = \lambda_e(x, (d_x\varphi)(t; x, \xi)) \\ \varphi(0; x, \xi) = \langle x|\xi\rangle . \end{cases} \tag{5.2}$$

can be given in terms of the solution of such a system.

Proposition 5.1. *Let* $\lambda_e \in C^\infty(J; \mathbf{SG}^e_{l,\mathrm{cl}})$. *The solution* $(q(t; x, \xi), p(t; x, \xi))$ *of the Hamiltonian system*

$$\begin{cases} \dot{q}(t; x, \xi) = (\nabla_\xi\lambda_e)(t; q(t; x, \xi), p(t; x, \xi)) \\ \dot{p}(t; x, \xi) = (-d_x\lambda_e)(t; q(t; x, \xi), p(t; x, \xi)) \\ q(0; x, \xi) = x, \quad p(0; x, \xi) = \xi \end{cases} \tag{5.3}$$

satisfies $q \in C^\infty(J'; \mathbf{SG}^{e_2}_{l,\mathrm{cl}})$ *and* $p \in C^\infty(J'; \mathbf{SG}^{e_1}_{l,\mathrm{cl}})$ *with* $J' = [-T', T'] \subseteq J$, $T' > 0$.

---

[11] See [Co98b] and [CR99].
[12] See [ES97].

Proof (sketchy). The fact that the solutions of (5.3) are **SG** symbols is already known (see [C95]). Moreover, it is possible to prove that the usual recursive scheme for obtaining the local solution of first order systems of ODE gives rise to two sequences, $\{p_k\} = \{p_k(t; x, \xi)\}$ and $\{q_k\} = \{q_k(t; x, \xi)\}$, converging, respectively, in $C^\infty(J, \mathbf{SG}_l^{e_1})$ and $C^\infty(J, \mathbf{SG}_l^{e_2})$. Through the homeomorphisms D^{e_1} and D^{e_2} of Theorem 3.2, it is possible to prove that $\{p_k\}$ and $\{q_k\}$ are bounded in the corresponding classical symbol spaces. Since the classical **SG** symbols of given order are nuclear Frechét spaces, this implies that each one of the two sequences $\{p_k\}$ and $\{q_k\}$ admits a subsequence converging, respectively, to a symbol in $\mathbf{SG}_{l,\mathrm{cl}}^{e_1}$ or $\mathbf{SG}_{l,\mathrm{cl}}^{e_2}$. Finally, since the topology of the classical **SG** symbols is stronger than the one inherited as subspace of the corresponding non-classical symbols, this shows that the two subsequences converge to the solutions of (5.3), which turn out to be, indeed, classical **SG** symbols. $\qquad\square$

The following Proposition can be proved with a technique similar to that used in the proof of Lemma 4.2.

Proposition 5.2. *The symbol $q(t; y, \xi)$, **SG** diffeomorphism with $\mathbf{SG}^0$ parameter dependence[13], admits an inverse $\bar{q}(t; x, \xi) \in \mathbf{SG}_l^{e_2}$ depending smoothly on t. $\bar{q}(t; x, \xi)$ is defined on a closed subinterval of J including 0, which we continue to denote by[14] J'. Moreover, $\bar{q}(t; x, \xi) \in C^\infty(J', \mathbf{SG}_{l,\mathrm{cl}}^{e_2})$.*

Proof of Theorem 1.2. After our preparation, the proof of Theorem 1.2 follows the same lines of the analogous Theorem 4.14 of [Co98b]. We can set

$$\psi(t; y, \xi) = \langle y|\xi\rangle + \int_0^t [\lambda_e(s; q(s; y, \xi), p(s; y, \xi))$$
$$- \langle (\nabla_\xi \lambda_e)(s; q(s; y, \xi), p(s; y, \xi))|p(s; y, \xi)\rangle] \, ds \qquad (5.4)$$

and also $\varphi(t; x, \xi) = \psi(t; \bar{q}(t; x, \xi), \xi) \Leftrightarrow \varphi(t; q(t; y, \xi), \xi) = \psi(t; y, \xi)$. As is well known, the function φ is the solution of (5.2), and, as such, it is a regular phase function. Moreover, using the results obtained in Lemma 4.2 and Propositions 5.1 and 5.2, it turns out that, indeed, $\varphi \in \mathcal{P}_{cl}^\epsilon$.

The amplitude $a = a(t; x, \xi)$ can be obtained solving the so-called transport equations, each one giving a term $a^{(j)} \in C^\infty(J'; \mathbf{SG}_l^{-je})$ of its asymptotic expansion[15]. By theorem 3.1, it is enough to prove that $\forall j \in \mathbb{N}$ $a^{(j)} \in C^\infty(J'; \mathbf{SG}_{l,\mathrm{cl}}^{-je})$. We give the details of the proof of this fact for $a^{(0)}$. This term must satisfy the equation

$$\begin{cases} (D_t - \lambda_0)a^{(0)} - \sum_{|\alpha|=1} \partial_\xi^\alpha \lambda_e D_\alpha^x a^{(0)} - \left(\sum_{|\alpha|=2} \frac{\partial_\xi^\alpha \lambda_e}{\alpha!} \tilde{\sigma}_\alpha \right) a^{(0)} = 0 \\ a^{(0)}(0; .) = 1 \end{cases} \qquad (5.5)$$

---

[13] See [Co98b], Proposition 4.13 and Theorem 4.14.

[14] Actually, the present J' can be strictly included in the J' on which $q(t; y, \xi)$ is defined.

[15] See again, e.g., [Co98b] for the **SG** case of this well known construction.

with suitable $\tilde{\sigma}_\alpha \in C^\infty(J'; \mathbf{SG}_{l,\mathrm{cl}}^{e_1-e_2})$, depending polynomially on the derivatives of φ. Through the change of variables $a_\star^{(0)}(t; y, \xi) = a^{(0)}(t; q(t; y, \xi), \xi)$, we are reduced to the equation

$$\begin{cases} \dfrac{d}{dt} a_\star^{(0)} - iH a_\star^{(0)} = 0 \\ a_\star^{(0)}(0; .) = 1, \quad H \in C^\infty(J', \mathbf{SG}_{l,\mathrm{cl}}^0). \end{cases} \tag{5.6}$$

The solution of (5.6) is $a_\star^{(0)}(t; y, \xi) = \exp\left(i \displaystyle\int_0^t d\tau\, H(\tau; y, \xi)\right)$. Now, $H \in C^\infty(J', \mathbf{SG}_{l,\mathrm{cl}}^0)$ implies $\tilde{H}(t; z, \zeta) = (D^0 H)(t; z, \zeta) = H(t; \chi(z), \chi(\zeta)) \in C^\infty(J', C^\infty(B^n \times B^n))$ and also, clearly, $a_\star^{(0)} \in C^\infty(J', \mathbf{SG}_{l,\mathrm{cl}}^0)$. Again by Lemma 4.2 and Propositions 5.1 and 5.2, we obtain $a^{(0)} \in C^\infty(J', \mathbf{SG}_{l,\mathrm{cl}}^0)$, as desired. The argument can be repeated with almost no change for all the other terms $a^{(j)}$, $j \geq 1$, and this allows us to conclude. $\qquad\square$

Acknowledgements

Thanks are due to Prof. B.-W. Schulze, Prof. E. Schrohe and Dr. J. Seiler, Institute of Mathematics, University of Potsdam, and to Dr. D. Kapanadze, Mathematical Institute of the Academy of Sciences of Georgia (Tbilisi), for many helpful hints and discussions with one of the authors in Clausthal and in Potsdam, where part of this work was completed.

References

[C95]   H. O. Cordes. *The Technique of Pseudodifferential Operators*. Cambridge Univ. Press, 1995.

[Co98a]   S. Coriasco. Fourier Integral Operators in **SG** classes I. Composition Theorems and Action on **SG** Sobolev Spaces. 1998. To appear in *Rend. Sem. Mat. Univ. Pol. Torino*.

[Co98b]   S. Coriasco. Fourier Integral Operators in **SG** classes II. Application to **SG** Hyperbolic Cauchy Problems. *Ann. Univ. Ferrara*, 47:81–122, 1998.

[Co98c]   S. Coriasco. *Fourier Integral Operators in **SG** classes with Applications to Hyperbolic Cauchy Problems*. PhD thesis, Università di Torino, 1998.

[CR99]   S. Coriasco and L. Rodino. Cauchy problem for **SG**-hyperbolic equations with constant multiplicities. *Ric. di Matematica*, 48, (Suppl.):25–43, 1999.

[ES97]   Y. Egorov and B.-W. Schulze. *Pseudo-Differential Operators, Singularities, Applications*. Birkhäuser, 1997.

[Mo83]   A. Mohamed. *"Etude Spectrale des Operateurs Pseudodifferentiels Hypoelliptiques"*. PhD thesis, Institut de Mathématiques et d'Informatiques de l'Université de Nantes, 1983.

[MP99]   L. Maniccia and P. Panarese. Eigenvalues Asymptotics for a Class of Elliptic Pseudodifferential Operators on Manifolds with Cylindrical Ends. 1999. To appear.

[Pa72]   C. Parenti. Operatori pseudodifferenziali in $\mathbb{R}^n$ e applicazioni. *Ann. Mat. Pura Appl.*, 93:359–389, 1972.

[Sc86]   E. Schrohe. Spaces of Weighted Symbols and Weighted Sobolev Spaces on Manifolds. In H. O. Cordes, B. Gramsch, and H. Widom, editors, *Proceedings, Oberwolfach*, number 1256 in Springer LNM, New York, pages 360–377, 1986.

[Wi98]   I. Witt. A Calculus for Classical Pseudo-Differential Operators with Non-Smooth Symbols. *Math. Nachr.*, 194:239–284, 1998.

Addresses

SANDRO CORIASCO, Dipartimento di Matematica, Università di Torino, V. C. Alberto, n. 10, I-10123, Torino, Italy

E-MAIL: Coriasco@dm.unito.it

PAOLO PANARESE, Dipartimento di Matematica, Università di Torino, V. C. Alberto, n. 10, I-10123, Torino, Italy

E-MAIL: Panarese@dm.unito.it

2000 Mathematics Subject Classification. Primary 35S30; Secondary 35L45

Operator Theory:
Advances and Applications, Vol. 126
© 2001 Birkhäuser Verlag Basel/Switzerland

A Semigroup Criterion for the Completeness of Scattering Systems

M. Demuth[*], E. Giere, K.B. Sinha

Abstract. Let H_0 be a radially symmetric function of the momentum $-i\nabla$ in $L^2(\mathbb{R}^n)$ and let H be a selfadjoint operator bounded from below. It is given a criterion for the asymptotic completeness of the scattering system $\{H_0, H\}$ in terms of the semigroup difference $e^{-sH} - e^{-sH_0}$, where $s > 0$ is an arbitrary fixed value. The result is applied to $H_0 = (-\Delta)^{\alpha/2}$.

1   Introduction

The purpose of this paper is to give a criterion for the asymptotic completeness of scattering systems in terms of semigroup differences. The scattering systems consists of a pair of selfadjoint operators (H_0, H). For H_0 we allow the following class of operators.

Definition 1.1. H_0 is a function of the quantum mechanical momentum $-i\nabla$ in $L^2(\mathbb{R}^n)$, i.e.

$$H_0 := P(-i\nabla) . \tag{1.1}$$

For the function $P : \mathbb{R}^n \to \mathbb{R}_+$ we require the following conditions:

- P is a radially symmetric C^∞-function with values in the non-negative real numbers, that is, with $h : \mathbb{R}_+ \to \mathbb{R}_+$:

$$P(p) = h(|p|) \qquad (p \in \mathbb{R}^n)$$

- $h' \geq 0$ and $h(|p|) \to \infty$ as $|p| \to \infty$.

Here $|p| = \sqrt{p_1^2 + \ldots p_n^2}$ is the Euclidian norm in $\mathbb{R}^n$. For convenience, we set $h(0) = 0$. At first we define H_0 on $C_0^\infty(\mathbb{R}^n)$. Its selfadjoint extension in $L^2(\mathbb{R}^n)$ is again denoted with H_0 with the domain $\mathcal{D}(H_0)$. The spectrum of H_0 is purely absolutely continuous; see Remark 2.1 below.

Next we introduce two sets of critical points that are important for the application of the stationary phase theorem.

---

[*] The first author would like to dedicate this article to **Jean Michel Combes** on the occasion of his sixtieth birthday

Definition 1.2. Let h be the function defining H_0 in Definition 1.1. Set

$$C_1 := \{k \in \mathbb{R}_+ \mid h'(k) = 0\} \tag{1.2}$$

and

$$C_2 := \{k \in \mathbb{R}_+ \mid h''(k) = (h'(k))^2\} . \tag{1.3}$$

C_1 is the set where the gradient of P vanishes. Since we apply in particular the stationary phase theorem we need to control the situation where a phase vanishes. In order to control this case we have to assume that C_2 is a small set in the following sense which classifies the admissible operators H_0.

Assumption 1.1. The sets C_1 and C_2 are discrete in $\mathbb{R}_+$ and have an accumulation point at most at infinity.

To specify H and H_0 we consider only selfadjoint operators in $L^2(\mathbb{R}^n)$ and bounded from below and set

$$S_0(t) := e^{-tH_0} \quad \text{and} \quad S(t) := e^{-tH} \quad (t \geq 0) . \tag{1.4}$$

Occasionally we shall write $S_0 = e^{-H_0} = \varphi(H_0)$ if we use that S_0 is a function of the unperturbed operator H_0 and we set $S = e^{-H} = \varphi(H)$.

Assumption 1.2. Let $S_0(t)$ and $S(t)$ as above and suppose

(i) $S(t) - S_0(t)$ is compact $\forall t \geq 0$
(ii) $(S - S_0)w_\gamma$ is a bounded operator for $\gamma > 1$.

We use the following notation, w_γ denotes the multiplication operator $f \mapsto w_\gamma f$ with $w_\gamma(x) = (1 + x^2)^{\gamma/2}$ and $x^2 := |x|^2$.

To formulate our results we recall the definition of the wave operators

$$\Omega^\pm(H, H_0) := s - \lim_{t \to \pm\infty} e^{itH} e^{-itH_0} P_{ac}(H_0)$$

Under these conditions we get the following completeness criterion.

Theorem 1.1. *Let $H_0 = P(-i\nabla)$ with P satisfying the conditions in Definitions 1.1, 1.2 and Assumption 1.1. Let H be another selfadjoint operator, bounded below such that the Assumption 1.2 holds true. Then the wave operators $\Omega^\pm(H, H_0)$ exist and the scattering system (H, H_0) is asymptotically complete that is*

$$\mathrm{Ran}(\Omega^\pm) = \mathcal{H}_{pp}^\perp(H)$$

Consequently, the singular continuous spectrum of H is empty.

Remark 1.1. The conditions given in Assumption 1.2 allows one to consider not only perturbations of H_0 by potentials such that $H = H_0 + V$ for some suitable function V, it is particular convenient for the study of perturbations by obstacles where the latter means that we introduce Dirichlet boundary conditions on a subset of $\mathbb{R}^n$.

In the proof of our result we make use of the invariance principle allowing us to investigate semigroup differences of the corresponding operators and use the Feynman-Kac formula. We shall verify the conditions given in Assumption 1.2 to prove existence and asymptotic completeness for scattering systems where the unperturbed operator H_0 is the polyharmonic operator $(-\Delta)^{\alpha/2}$ with $\alpha \in (0, 2)$ or the relativistic operator $H_0 = \sqrt{-\Delta + m^2} - m$.

As for the proof of the asymptotic completeness we follow the lines of standard time-dependent techniques, see e.g. [RS79]. Moreover, we make use of the stationary phase theorem as outlined in Appendix 1 in [RS79, Sec. XI.3] to prove the existence of the wave operators. Finally, we mention that the paper is motivated by the work of Demuth and Sinha [DS99] who showed analogous results for the Laplacian on $\mathbb{R}^n$. Concerning the scattering theoretical part analogous results have been shown by Simon [Sim79] and Yafaev [Yaf85]. By different methods Muthuramalingam [Mut85b,Mut85a] show also asymptotic completeness for similar scattering systems.

2 Existence of the Wave Operators

We start with the proof of a technical lemma which is essentially an adaption of standard assumptions in this time-dependent setting to the special situation of radially symmetric functions P that are used to construct the unperturbed operator H_0. For a two times differentiable function $f : \mathbb{R}^n \to \mathbb{R}$ we denote with Hess f the Hessian matrix of f.

$$\operatorname{Hess} f(p) := \left(\frac{\partial^2 f}{\partial p_i \partial p_j}(p) \right)_{i,j}$$

Lemma 2.1. *Let* $\Phi(p) = e^{-P(p)}$ *with* $P(p) = h(|p|)$ *and* h *given as in Definition 1.1, let Assumption 1.1 hold true. Moreover, we define*

$$M := \{p \in \mathbb{R}^n \mid \nabla \Phi(p) = 0 \ or \ \det[\operatorname{Hess} \Phi(p)] = 0\}$$

Then we have that

$$M = \bigcup_{j=1}^{N} S_j \quad with \ N \in \mathbb{N} \cup \{\infty\} \tag{2.1}$$

is the union of countably many hypersurfaces S_j where $S_j = \{p \in \mathbb{R}^n \mid |p| = k_j\}$ and the set $\{k_j\}_{j \in \mathbb{N}}$ is discrete in $\mathbb{R}_+$. In case of $N = \infty$ the points $k_j \in \mathbb{R}_+$ accumulate at infinity.

Proof: For $\Phi(p)$ one easily calculates

$$\nabla \Phi(p) = -\Phi(p)\nabla P(p) = -\Phi(p)h'(|p|)\frac{p}{|p|}$$

and for the Hessian matrix we get

$$\mathrm{Hess}\,\Phi(p) = -\Phi(p)\left\{\frac{h'(|p|)}{|p|}\mathrm{Id} + \left(h''(|p|) - (h'(|p|))^2 - \frac{h'(|p|)}{|p|}\right)\mathrm{Pr}_p\right\}$$

where $\mathrm{Pr}_p = \frac{1}{|p|^2}\langle p, \cdot \rangle p$ is the projection operator onto the one-dimensional subspace $\mathrm{Ran}(\mathrm{Pr}_p)$. Since Pr_p is symmetric we can diagonalise $\mathrm{Hess}\,\Phi(p)$ by an orthogonal transformation. This implies for $|p| \notin C_1 \cup \{0\}$

$$\det[\mathrm{Hess}\,\Phi(p)] = \left(-\Phi(p)\frac{h'(|p|)}{|p|}\right)^n |p| \left(\frac{h''(|p|)}{h'(|p|)} - h'(|p|)\right) .$$

Hence we get the criterion for $|p| \notin C_1 \cup \{0\}$

$$\det[\mathrm{Hess}\,\Phi(p)] = 0 \iff h''(|p|) = (h'(|p|))^2 . \tag{2.2}$$

$\square$

Remark 2.1. We mention that by the construction of H_0 the spectrum is purely absolutely continuous. In fact, P is continuously differentiable and $\nabla P(p) = 0$ only on a set of Lebesgue measure zero which follows from Assumption 1.1. Hence, the multiplication operator $\mathrm{M}_P : \psi \mapsto P\psi$ has purely absolutely continuous spectrum (for an exposition of the easy argument see e.g. Example 1.9 in subsection X.1.2 in [Kat80]). The Fourier transform relates H_0 unitarily with M_P, hence $\sigma(H_0) = \sigma_{ac}(H_0)$.

The same argument applies for S_0, therefore $\mathcal{H}_{ac}(S_0) = L^2(\mathbb{R}^n)$ where $\mathcal{H}_{ac}(S_0)$ is the absolutely continuous subspace with respect to S_0. Consequently we have $\sigma(S_0) = \sigma_{ac}(S_0)$ and $\sigma_{sing}(S_0) = \emptyset$, as well.

Now, we are going to state the existence of the wave operators

$$\Omega^{\pm}(S, S_0) := s - \lim_{t \to \pm\infty} e^{itS}e^{-itS_0}P_{ac}(S_0)$$

Proposition 2.1. *Let $H_0 = P(-i\nabla)$ with P satisfying the conditions given in Definition 1.1 and suppose that the assumptions 1.1 and 1.2 are satisfied. Then the wave operators $\Omega^{\pm}(S, S_0)$ exist.*

Proof: In order to use Cook's criterion we consider

$$\|(S - S_0)e^{-itS_0}u\| \leq \|(S - S_0)w_\gamma\| \, \|w_{-\gamma}e^{-itS_0}u\|$$
$$\leq c\|w_{-\gamma}e^{-itS_0}u\| \tag{2.3}$$

for u in an appropriate dense set in $L^2(\mathbb{R}^n)$. Using Theorems XI.14 and XI.15 in [RS79] one shows that

$$\|w_{-\gamma}u_t\| \leq c\,|t|^{-\gamma} \qquad (t > 1) \tag{2.4}$$

which implies the needed integrability $\|w_{-\gamma}e^{-itS_0}u\| \in L^1(\mathbb{R}_+, dt)$ since $\gamma > 1$. $\square$

Having proved the existence of the wave operators $\Omega^\pm(S, S_0)$, we are going to conclude the existence of $\Omega^\pm(H, H_0)$ with the help of the invariance principle.

Corollary 2.1. *Let H and H_0 be selfadjoint operators bounded from below such that Assumption 1.1 and 1.2 hold. Then the wave operators $\Omega^\pm(H, H_0)$ exist and*

$$\Omega^\pm(H, H_0) = \Omega^\mp(S, S_0). \tag{2.5}$$

Proof: The result follows from the invariance principle in its general form given in Theorem XI.23 in [RS79]. Setting $w(t) := e^{itS}e^{-tS_0}u$ one can prove with (2.3) and (2.4) that

$$|t^\beta| \, \|w'(t)\| \in L^1(I_\pm) \quad \text{and} \quad \|w'(t)\| \in L^1(I_\pm) \cap L^2(I_\pm) \tag{2.6}$$

for some $\beta > 0$ with $I_+ = [1, \infty)$ and $I_- = (-\infty, -1]$. Here u is again taken from an appropriate dense subset of $L^2(\mathbb{R}^n)$ such that the invariance principle holds. $\square$

3 Asymptotic Completeness

For the proof of Theorem 1.1 a central point is Perry's estimate (see [DS99]). Let $A = \frac{1}{4i}(x \cdot \nabla + \nabla \cdot x)$ be the generator of the dilation group and let T be the unitary operator diagonalising the operator A:

$$Af = T^*M_aTf, \text{ for } f \in \mathcal{D}(A) \text{ where}$$
$$M_ag(a, \omega) = ag(a, \omega) \quad \text{with } a \in \mathbb{R}, \ \omega \in S^{n-1} \ . \tag{3.1}$$

Define for $\mathbb{R}_+^* = (0, \infty)$ and $\mathbb{R}_-^* = (-\infty, 0)$

$$P_\pm := T^*\chi_{\mathbb{R}_\pm^*}T \qquad P_{at} := T^*\chi_{(-\alpha t, \alpha t)}T \ . \tag{3.2}$$

For $\psi \in C_0^\infty(\mathbb{R} \setminus P(M))$ there is an $\alpha > 0$ such that

$$\|P_{at}e^{\mp iS_0t}\psi(H_0)P_\pm\| \leq c_l(1 + t)^{-l} \tag{3.3}$$

for any $l \in \mathbb{N}$ and $t \geq 0$. For the proof of Perry's estimate (3.3) one needs essentially the conditions in Definition 1.1 and Definition 1.2.

Following the proof in [DS99], Perry's estimate implies finally

$$\|f\|^2 = \lim_{T \to \infty} \frac{1}{T} \int_0^T \langle e^{itS} f, \Omega^+ \psi(H_0) P_- e^{-itS} f \rangle \tag{3.4}$$

for all $f \in \mathcal{H}_c(H)$ - the continuous subspace of H. The right hand side is zero for $f_0 \in \mathcal{H}_c(H)$ but $f_0 \perp \mathrm{Ran}(\Omega^+)$. Hence $\mathrm{Ran}(\Omega^+) = \mathrm{Ran}(\Omega^-) = \mathcal{H}_c(H)$.

By the invariance principle $\Omega^\pm(S, S_0) = \Omega^\mp(H, H_0)$ thus $\mathcal{H}_{ac}(S) = \mathcal{H}_c(S) = \mathrm{Ran}(\Omega^\pm(S, S_0)) = \mathrm{Ran}(\Omega^\pm(H, H_0))$. Furthermore, as $\varphi(\lambda) = e^{-\lambda}$ satisfies $\varphi'(\lambda) < 0$ for all $\lambda \in \mathbb{R}$, we get by a straightforward calculation with the spectral measures $\langle f, E_H(d\lambda)f \rangle$ and $\langle f, E_S(d\lambda)f \rangle$ that $\mathcal{H}_{ac}(H) = \mathcal{H}_{ac}(S)$ and $\mathcal{H}_c(H) = \mathcal{H}_c(S)$. So, we arrive at

$$\mathcal{H}_{ac}(H) = \mathcal{H}_{ac}(S) = \mathcal{H}_c(S) = \mathcal{H}_c(H)$$

and therefore $\mathcal{H}_{sc}(H) = \{0\}$ implying $\sigma_{sc}(H) = \emptyset$.

4 Application to $H_0 = (-\Delta)^{\alpha/2}$

In a rather general framework, see [DvC00] for an exposition in the context of Feller operators, the semigroup $S_0(t) = e^{-tH_0}$ is related to a stochastic process $(X_t)_{t \geq 0}$ via

$$S_0(t)f(x) = \mathbb{E}_x[f(X_t)] \qquad t \geq 0, \ x \in \mathbb{R}$$

on an appropriate probability space $(\Omega, \mathcal{F}, \mathbb{P}_x)$. The operator $H_0 = (-\Delta)^{\alpha/2}$ for $\alpha \in (0, 2)$ is even a Lévy-type operator - and thus a Feller operator - and the transition function has a density p_t

$$S_0(t)f(x) = \int_{\mathbb{R}^n} p_t(x, y)f(y)dy$$

with the following properties

$$
\begin{aligned}
p_t(x, y) &= p_t(|x - y|) & &\text{(symmetry)} \\
p_t(x) &= t^{-n/\alpha} p_1\left(\tfrac{x}{t^{1/\alpha}}\right) & &\text{(scaling)} \\
p_1(x) &\leq c|x|^{-n-\alpha} \text{ for } |x| \geq 1 & &\text{(asymptotics)} .
\end{aligned}
$$

In this setting the potential perturbing H_0 has to satisfy

$$\lim_{t \searrow 0} \sup_{x \in \mathbb{R}^n} \mathbb{E}_x\left[\int_0^t |V(X_s)|ds\right] = 0 .$$

This kind of potentials, known as the Kato-class with respect to H_0, allow to get the semigroup $S(H)$ generated by $H = H_0 + V$ via the Feynman-Kac formula:

$$S(t)f(x) = \mathbb{E}_x[e^{-\int_0^t V(X_s)ds} f(X_t)] \qquad f \in L^2(\mathbb{R}^n) . \tag{4.1}$$

Again we have a density $p_t^V(x, y)$ for the transition function which can be constructed explicitly (see [FPY93] or [DvC00]). We have

$$S(t)g(x) = \int_{\mathbb{R}^n} \mathbb{E}_{x,y}^t[e^{-\int_0^t V(X_s)ds}]p_t(x, y)g(y)dy \tag{4.2}$$

where $\mathbb{E}_{x,y}^t$ denotes the expectation with respect to the law of the bridge process constructed from $(X_t)_{t\geq 0}$ starting in x by conditioning that process to arrive at y at time t.

Proceeding analogously to [DvC00, p. 65] or [Szn98] we obtain

$$p_t^V(x, y) \leq p_t^{1/q}(x, y)\Big(\sup_{u,v} p_t(u, v)\Big)^{1/p} C_1 e^{C_2 t}$$

where $1/p + 1/q = 1$ and $q > 1$.

The heart of the matter are the following weighted L^p-estimates:

Proposition 4.1. *Let $S_0(t) = e^{-tH_0}$ where $H_0 = (-\Delta)^{\alpha/2}$. Then, for $\gamma \geq 0$, $p \in [1, \infty)$ and $0 \leq t \leq 1$ the map*

$$w_{-\gamma}S_0(t)w_\gamma : L^p(\mathbb{R}^n) \to L^p(\mathbb{R}^n) \tag{4.3}$$

is bounded uniformly in t, given $n + \alpha > p\gamma$.

As a consequence of the Feynman-Kac formula (4.1) and (4.2) we get with Hölder's inequality

Corollary 4.1. *Let $S(t) = e^{-tH}$ with $H = H_0 + V$, $V^- \in K(H_0)$ and $V^+ \in K_{loc}(H_0)$. Then the map*

$$w_{-\gamma}S(t)w_\gamma : L^p(\mathbb{R}^n) \to L^p(\mathbb{R}^n) \tag{4.4}$$

is bounded for $n + \alpha > \gamma p$ and $p \in (1, \infty)$

We use Proposition 4.1 and Corollary 4.1 to prove the conditions necessary for the existence of the wave operators and the asymptotic completeness of the corresponding scattering system.

Proposition 4.2. *Let $H_0 = (-\Delta)^{\alpha/2}$, $H = H_0 + V$ with $V^- \in K(H_0)$ and $V^+ \in K_{loc}(H_0)$. Moreover, V has to be short-range, i.e. for some $R_0 > 0$ and $\gamma > 1$ we have $\sup_{|x| \geq R_0} |V(x)w_\gamma(x)| < \infty$ and $n + \alpha > 2\gamma$. Then the map*

$$(S - S_0)w_\gamma : L^2(\mathbb{R}^n) \to L^2(\mathbb{R}^n) \tag{4.5}$$

is bounded.

Proof: It holds

$$(S - S_0)w_\gamma f = \chi_{R_0}(S - S_0)w_\gamma f + \chi_{R_0}^c(S - S_0)w_\gamma f \ . \tag{4.6}$$

With the weighted L^p-estimates one finds immediately a bound for the norm of the first term in (4.6). For the second summand one uses Duhamel's formula

$$\chi_{R_0}^c(S - S_0)w_\gamma f = -\int_0^1 \chi_{R_0}^c S_0(t)VS(1-t)w_\gamma f \, dt$$

and then again the estimates obtained in Proposition 4.1 and Corollary 4.1 together with asymptotic behaviour of V on $B_{R_0}^c$. For V on the ball B_{R_0} one uses additionally that Kato-class potentials are locally integrable. $\square$

To proceed, note that $D_R(t) := \chi_R(S(t) - S_0(t))$ is a Hilbert-Schmidt operator. With the knowledge of the asymptotical behaviour of the the transition densities p_t and p_t^V the following can be shown.

Proposition 4.3. *Let H_0, H and V as above and suppose that for some $R > 0$ we have $\sup_{|x| \geq R} |Vw_\gamma|(x) < \infty$ where $\gamma > 1$ and $n + \alpha > 2\gamma$. Then, for each $t > 0$*

$$\lim_{R \to \infty} \|D_R(t) - D(t)\| = 0 \ . \tag{4.7}$$

Hence the difference $D(t) = S(t) - S_0(t)$ is compact for each $t > 0$.

Remark 4.1. Note that we need $\gamma > 1$ to apply the scattering theoretical techniques whereas the weighted L^p-estimates are valid for arbitrary $\gamma \geq 0$. By duality one even has that $\gamma \in \mathbb{R}$ is possible for those estimates. Thus one gets a restriction on the parameter α in one dimension since we need $n + \alpha > 2\gamma$ for some $\gamma > 1$, hence we have to assume $\alpha > 1$ in one dimension.

Eventually we note that the relativistic case $H_0 = \sqrt{-\Delta + m^2} - m$ is simpler since the density for the transition function $p_t(x - y)$ decays even exponentially with the distance $|x - y| \to \infty$. This can be seen from the representation, see [DvC00, p.290]

$$p_t(x) = \frac{1}{(2\pi)^n} \frac{t}{\sqrt{|x|^2 + t^2}} \int_{\mathbb{R}^n} e^{mt} e^{-\sqrt{(|x|^2+t^2)(p^2+m^2)}} \, dp$$

In [LY88] one finds the even more explicite expression for the kernel of the relativistic semigroup e^{-tH_0}

$$e^{-tH_0}(x,y) = \frac{m^2 t}{2\pi^2} \frac{1}{(|x-y|^2 + t^2)} K_2(m(|x-y|^2 + t^2)^{1/2})$$

where K_2 is the modified Bessel function whose asymptotical behaviour yields the exponential decay of the density of the transition function.

Acknowledgements

K.B. Sinha acknowledges support by the Jawaharlal Nehru Centre for Advanced Scientific Research. The work of E. Giere was supported by the DFG.

References

[DS99] M. Demuth and K. B. Sinha, *Schrödinger operators with empty singular continuous spectra*, Math. Phys., Analysis and Geometry **2** (1999), 223–244.

[DvC00] M. Demuth and J. van Casteren, *Stochastic Spectral Theory for Selfadjoint Feller Operators*, Birkäuser, Basel, 2000.

[FPY93] P. Fitzsimmons, J. Pitman, and M. Yor, *Markovian bridges: construction, Palm interpretation, and splicing*, Seminar on Stochastic Processes, 1992 (Seattle, WA, 1992), Birkhäuser Boston, Boston, MA, 1993, pp. 101–134.

[Kat80] T. Kato, *Perturbation theory for linear operators*, 2nd ed., Springer, Berlin, 1980.

[LY88] E. Lieb and H. T. Yau, *The stability and instability of relativistic matter*, Commun. Math. Phys. **118** (1988), 177–213.

[Mut85a] P. Muthuramalingam, *A note on time dependent scattering theory for $P_1^2 - P_2^2 + (1+|Q|)^{-1-\varepsilon}$ and $P_1 P_2 + (1+|Q|)^{-1-\varepsilon}$ on $L^2(\mathbf{r}^2)$*, Math. Z. **188** (1985), no. 3, 339–348.

[Mut85b] P. Muthuramalingam, *A time dependent scattering theory for a class of simply characteristic operators with short range local potentials*, J. London Math. Soc. (2) **32** (1985), no. 2, 259–264.

[RS79] M. Reed and B. Simon, *Methods of Modern Mathematical Physics III, Scattering Theory*, Academic Press, San Diego, 1979.

[Sim79] B. Simon, *Phase space analysis of simple scattering systems: extensions of some work of Enss*, Duke Math. J. **46** (1979), 119–168.

[Szn98] A.-S. Sznitman, *Brownian motion, obstacles and random media*, Springer, Berlin, 1998.

[Yaf85] D. Yafaev, *Nonstationary scattering theory for elliptic differential operators*, J. Soviet. Math. **28** (1985), 814–824.

Addresses

MICHAEL DEMUTH, TU Clausthal, Institut für Mathematik, Erzstraße 1,
38678 Clausthal-Zellerfeld, Germany

E-MAIL: demuth@math.tu-clausthal.de

ECKHARD GIERE, TU Clausthal, Institut für Mathematik, Erzstraße 1,
38678 Clausthal-Zellerfeld, Germany

E-MAIL: giere@math.tu-clausthal.de

KALYAN B. SINHA, Indian Statistical Institute, New Delhi, India

E-MAIL: kbs@isid.isid.ac.in

2000 Mathematics Subject Classification. Primary 47A40, 81U05; Secondary 47A10,
47D08, 60G52

Operator Theory:
Advances and Applications, Vol. 126
© 2001 Birkhäuser Verlag Basel/Switzerland

Parameter-Elliptic Boundary Value Problems and their Formal Asymptotic Solutions

ROBERT DENK AND LEONID VOLEVICH*

Abstract. We consider boundary value problems for mixed-order systems of partial differential operators which depend on a complex parameter but which are not parameter-elliptic in the sense of Agmon and Agranovich–Vishik. Such systems are closely related to the theory of singularly perturbed problems. Under the condition of so-called weak parameter-ellipticity it is possible to construct the formal asymptotic solution which shows, in particular, the existence of boundary layers.

1 Introduction

Let $A(D) = \big(A_{ij}(D)\big)_{i,j=1,\dots,N}$ be a matrix of partial differential operators and suppose that this matrix is elliptic in the sense of Douglis and Nirenberg. In this case there exist $2N$ integers $s_1,\dots,s_N,t_1,\dots,t_N$ such that $\operatorname{ord} A_{ij} \le s_i + t_j$. We will assume in the following that s_i and t_i are nonnegative. Without loss of generality we can suppose that the sequence $r_i := s_i + t_i$ is nonincreasing (in the opposite case we change the indexing of lines and rows). Let $A^0_{ij}(D)$ denote the principal part of the operator $A_{ij}(D)$ in the sense of Douglis–Nirenberg (we have $A^0_{ij} = 0$ if $\operatorname{ord} A_{ij} < s_i + t_j$). Ellipticity then means that $\det A^0(\xi) \ne 0$ for all $\xi \in \mathbb{R}^n \setminus \{0\}$ where $A^0(\xi) := \big(A^0_{ij}(\xi)\big)_{i,j}$ stands for the principal symbol of $A(D)$.

The aim of the present paper is to investigate boundary value problems for the parameter-dependent operator matrix given by

$$
A(D,\lambda) := \begin{pmatrix} A_{11}(D) & \cdots & A_{1N}(D) \\ \vdots & & \vdots \\ A_{N1}(D) & \cdots & A_{NN}(D) - \lambda \end{pmatrix} = A(D) - \lambda E_N \tag{1.1}
$$

and supplemented with general mixed-order boundary conditions. Here E_N stands for the $N \times N$ matrix which differs from the zero matrix only in the element at position (N,N) which equals 1.

There are several reasons for studying the parameter-dependent matrix 1.1 (see [DV00], Section 1). We only want to point out one reason. In the case of constant order matrices, the theory of ellipticity with parameter as it was developed in the

---

* Supported by the Russian Foundation of Fundamental Research, grant 00-01-00387

sixties by Agmon [A62], Agranovich–Vishik [AV64] and others can be applied to the matrix $A(D) - \lambda I_N$ (where I_N denotes the N-dimensional identity matrix). It was also remarked by Agranovich in [A90] that in the case of mixed order systems for which all numbers r_i are equal we can adjust a definite weight to the parameter λ. This again makes it possible to apply the theory of parameter-ellipticity (see also the book of Roitberg [R96] for parameter-elliptic Douglis–Nirenberg systems).

The case where some of the numbers r_j are different is much more complicated and was treated by Kozhevnikov [K96] and by the authors [DMV98]. In these papers one can find several (equivalent) definitions of ellipticity with parameter for the matrix $A(D) - \lambda I_N$ which lead to solvability results and to a priori estimates. Roughly speaking, $A(D) - \lambda I_N$ is elliptic with parameter if all submatrices of the form $A_{(k)}(D) - \lambda E_k$ are weakly parameter-elliptic in the sense defined below, where we have set $A_{(k)}(D) := \left(A_{ij}(D)\right)_{i,j=1,\ldots,k}$. This definition (also called the condition of elliptic principal minors) is essentially due to Kozhevnikov; for other descriptions we refer to [DMV98].

In [DV00] boundary value problems for systems with structure very close to (1.1) were investigated. Here the concept of weak parameter-ellipticity for such operator matrices and corresponding boundary value problems was introduced and a priori estimates in certain parameter-dependent Sobolev spaces were obtained. In the present paper we want to show that the conditions appearing in the definition of weak parameter-ellipticity are very natural from the point of view of singular perturbation theory (here λ is replaced by ε^{-1} with a small positive parameter ε), see , e.g., [VL57], [N81], [I89]. In particular, these conditions allow us to construct the so-called formal asymptotic solutions. For simplicity, we will only consider operators with constant coefficients and without lower order terms acting in the whole space or in the half-space $\mathbb{R}^n_+ := \{x = (x', x_n) \in \mathbb{R}^n : x_n > 0\}$. The same definitions and results hold for operators with variable coefficients acting on a bounded domain or on a compact manifold with boundary.

2 Weakly Parameter-Elliptic Boundary Value Problems

We start with the definition of weak parameter-ellipticity for the matrix (1.1). As above, we assume that r_i are nonincreasing and that $r_{N-1} > r_N$. We fix a closed sector $\mathcal{L} \subset \mathbb{C}$ with vertex at the origin.

Definition 2.1. Let $A(D, \lambda)$ be of the form (1.1). Then $A(D, \lambda)$ is called weakly parameter-elliptic in $\mathcal{L}$ if the inequality

$$\left| \det(A^0(\xi) - \lambda E_N) \right| \geq C |\xi|^{r_1 + \cdots + r_{N-1}} (|\xi| + |\lambda|^{1/r_N})^{r_N} \quad (\xi \in \mathbb{R}^n, \ \lambda \in \mathcal{L}) \quad (2.1)$$

holds, where here and in the following the letter C denotes an unspecified constant independent of ξ and λ.

Scalar polynomials in ξ and λ satisfying an estimate of the form (2.1) were treated in [DMV00a], [DMV00b]. It is not difficult to see that (2.1) holds if and only if $A(\xi)$ is elliptic in the sense of Douglis–Nirenberg, the same holds for $A_{(N-1)}(\xi)$, and $\det(A^0(\xi) - \lambda E_N)$ does not vanish for all $\xi \in \mathbb{R}^n \setminus \{0\}$ and all $\lambda \in \mathcal{L} \setminus \{0\}$. From this we obtain in the case $n \geq 3$ that the numbers r_j are even, for $n = 2$ we will assume this in the following.

Now let us assume that $A(D, \lambda)$ acts on the half-space $\mathbb{R}^n_+$. The partial Fourier transform with respect to the first $n-1$ variables reduces this operator to the ordinary differential operator $A(\xi', D_n, \lambda)$ (with $\xi' = (\xi_1, \ldots, \xi_{n-1})$ and $D_n = -i\partial/\partial x_n$).

Lemma 2.1. *Let $A(D, \lambda)$ be weakly parameter-elliptic in $\mathcal{L}$.*
a) *For $\xi' \neq 0$ and $\lambda \in \mathcal{L}$ the ordinary differential equation on the half-line*

$$A^0(\xi', D_n, \lambda)w(x_n) = 0 \quad (x_n > 0)$$

has exactly R_N solutions which tend to zero for $x_n \to \infty$. Here we have set $R_j := (r_1 + \ldots + r_j)/2$ for $j = 1, \ldots, N$.

b) *For $\lambda \in \mathcal{L} \setminus \{0\}$ the ordinary differential equation in $\mathbb{R}_+$*

$$A^0(0, D_n, \lambda)w(x_n) = 0 \quad (x_n > 0) \tag{2.2}$$

has exactly $r_N/2$ solutions which tend to zero for $x_n \to +\infty$.

Proof. The dimension of the space of asymptotically stable solutions (i.e. solutions which tend to zero for $x_n \to \infty$) in the case a) and b), respectively, coincides with the number of zeros of $\det A^0(\xi', \cdot, \lambda)$ and $\det A^0(0, \cdot, \lambda)$, respectively, with positive imaginary part, counted with multiplicities. As the first determinant has $2R_N$ zeros in $\mathbb{C}$ and no real zeros, it follows by standard homotopy argument that it has half of its zeros in $\mathbb{C}_+ := \{z \in \mathbb{C} : \operatorname{Imi} z > 0\}$. For the second determinant we use

$$\det A^0(0, \tau, \lambda) = \det A^0(0, \tau) - \lambda \det A^0_{(N-1)}(0, \tau).$$

As both determinants are homogeneous and elliptic polynomials there are constants $a_1, a_2 \in \mathbb{C}$ such that $\det A^0(0, \tau) = a_1 \tau^{r_1 + \cdots + r_N}$ and $\det A^0_{(N-1)}(0, \tau) = a_2 \tau^{r_1 + \cdots + r_{N-1}}$. Therefore the zeros in $\mathbb{C}_+$ of $\det A^0(0, \tau, \lambda)$ are the zeros of the equation

$$a_1 \tau^{r_N} - a_2 = 0 \quad \text{in } \mathbb{C}_+.$$

As the last equation has no real zeros due to the condition of weak parameter-ellitpicity and as r_N is even, it has exactly $r_N/2$ zeros in $\mathbb{C}_+$. $\qquad\square$

Let us assume that we have a matrix of boundary operators of the form

$$B(D) = \left(B_{jk}(D)\right)_{\substack{j=1,\ldots,R_N \\ k=1,\ldots,N}}$$

where for the boundary conditions the inequality ord $B_{jk} \leq m_j + t_k$ holds, where $m_1, \ldots, m_{R_N}$ are integer numbers satisfying $m_1 \leq \ldots \leq m_{R_N}$ and

$$m_{R_k} < m_{R_k+1} \quad (k = 1, \ldots, N-1).$$

The principal part B^0 of B is defined in the same way as for A. We also set

$$B_{(N-1)}(D) := \left(B_{jk}(D) \right)_{\substack{j=1,\ldots,R_{N-1} \\ k=1,\ldots,N-1}}.$$

The following definition is essentially taken from [DV00].

Definition 2.2. The boundary value problem $(A(D, \lambda), B(D))$ is called weakly parameter-elliptic in $\mathcal{L}$ if the following conditions are satisfied:

(i) $A(D, \lambda)$ is weakly parameter-elliptic in $\mathcal{L}$ in the sense of Definition 2.1.

(ii) For every $\xi' \in \mathbb{R}^{n-1} \setminus \{0\}$, $\lambda \in \mathcal{L}$ and every $g = (g_1, \ldots, g_{R_N}) \in \mathbb{C}^{R_N}$ the boundary value problem in $\mathbb{R}_+$

$$\begin{aligned}
A^0(\xi', D_n, \lambda)w(x_n) &= 0 \quad (x_n > 0), & (2.3) \\
B^0(\xi', D_n)w(x_n)\big|_{x_n=0} &= g, & (2.4) \\
w(x_n) \to 0 &\quad \text{for } x_n \to \infty &
\end{aligned}$$

has a unique solution.

(iii) For every $\xi' \in \mathbb{R}^{n-1}\setminus\{0\}$ and every $h \in \mathbb{C}^{R_{N-1}}$ the problem

$$\begin{aligned}
A^0_{(N-1)}(\xi', D_n)v(x_n) &= 0 \quad (x_n > 0), & (2.5) \\
B^0_{(N-1)}(\xi', D_n)v(x_n)\big|_{x_n=0} &= h, & (2.6) \\
v(x_n) \to 0 &\quad \text{for } x_n \to \infty &
\end{aligned}$$

has a unique solution.

(iv) For every every vector $h \in \mathbb{C}^{r_N/2}$ and every $\lambda \in \mathcal{L}$ with $|\lambda| = 1$ the problem

$$\begin{aligned}
A^0(0, D_n, \lambda)v(x_n) &= 0 \quad (x_n > 0), & (2.7) \\
B^0_{(N,1..N)}(0, D_n)v(x_n)\big|_{x_n=0} &= h, & (2.8) \\
v(x_n) \to 0 &\quad \text{for } x_n \to \infty &
\end{aligned}$$

has a unique solution. Here we have set

$$B_{(N,1..N)}(D) := \left(B_{jk}(D) \right)_{\substack{j=R_{N-1}+1,\ldots,R_N \\ k=1,\ldots,N}}.$$

Note that conditions (i) and (ii) in the above definition are very natural and correspond to similar conditions in traditional elliptic theory. Conditions (iii) and (iv) are connected with the case $\xi' = 0$ where the analog of (ii) does not hold. In the next section the meaning of these two conditions will become clear in the context of singular perturbation theory.

3   Formal Asymptotic Solutions

Consider the boundary value problem

$$A(D, \lambda)u = 0, \tag{3.1}$$
$$B(D)u = g \tag{3.2}$$

in the half-space $\mathbb{R}^n_+$ where we assume throughout this section that (A, B) are weakly parameter-elliptic in the sense of Definition 2.2 with $\mathcal{L} = [0, \infty)$. Setting $\lambda = \varepsilon^{-r_N}$ and multiplying the last equation of the system (3.1) by ε^{r_N}, we obtain the system

$$A_\varepsilon(D)u(x', x_n) = 0 \quad (x_n > 0) \tag{3.3}$$

where $A_\varepsilon(D) := \mathrm{diag}\,(1, \ldots, 1, \varepsilon^{r_N})A(D) - E_N$. For simplicity, let us now assume that all operators coincide with their principal parts. We are interested in the case of $\varepsilon \to 0$; more precisely, our goal is to find the formal asymptotic solution (FAS)

$$\sum_{l=0}^{\infty} \varepsilon^l u^{(l)}(x, \varepsilon),$$

i.e. the formal power series in ε for which the partial sums satisfy (3.3),(3.2) up to an arbitrary power of ε. The construction of the FAS will show the boundary layer structure of the solution of (3.3),(3.2) and give a deeper insight to the conditions of Definition 2.2. It seems to us that this construction cannot be found in literature for the boundary value problem considered here.

Following the Lyusternik-Vishik method, we will construct the FAS as the sum of the so-called exterior expansion

$$u(x, \varepsilon) = \sum_{l=0}^{\infty} \varepsilon^l u^{(l)}(x) \tag{3.4}$$

and the so-called interior expansion or boundary layer

$$v(x', x_n/\varepsilon, \varepsilon) = \sum_{l=0}^{\infty} \varepsilon^{l_0+l}\mathrm{diag}\,(\varepsilon^{t_1}, \ldots, \varepsilon^{t_N})v^{(l)}(x', x_n/\varepsilon.) \tag{3.5}$$

The number l_0 will be chosen later.

We will now show that, due to the conditions of weak parameter-ellipticity, it is possible to describe $u^{(l)}$ and $v^{(l)}$ as the solutions of boundary value problems which appear in the definition of weak parameter-ellipticity and for which the right-hand sides can be computed recursively.

(i) *Differential equations for the exterior expansion.* Substituting (3.4) into (3.3) and posing $u'^{(l)} = (u_1^{(l)}, \ldots, u_{N-1}^{(l)})$ we obtain

$$\sum_{l=0}^{\infty} \varepsilon^l \left[A_{(N-1)}(D)u'^{(l)} + \begin{pmatrix} A_{1N}(D) \\ \vdots \\ A_{N-1,N}(D) \end{pmatrix} u_N^{(l)} \right] = 0,$$

$$\sum_{l=0}^{\infty} \varepsilon^l \left(\varepsilon^{r_N} \sum_{j=1}^{N} A_{Nj}(D)u_j^{(l)} - u_N^{(l)} \right) = 0.$$

Equate to zero all terms corresponding to the same power of ε we obtain the relations

$$u_N^{(l)} = -\sum A_{Nj}(D)u_j^{(l-r_N)}, \tag{3.6}$$

$$A_{(N-1)}(D)u'^{(l)} = - \begin{pmatrix} A_{1N}(D) \\ \vdots \\ A_{N-1,N}(D) \end{pmatrix} u_N^{(l)} = \mathcal{F}(u^{(0)}, \ldots, u^{(l-r_N)}). \tag{3.7}$$

(ii) *Differential equations for the interior expansion.* Pose $t = x_n/\varepsilon$. Then

$$A_\varepsilon(D)v(x', x_n/\varepsilon, \varepsilon) = A_\varepsilon(D', \tfrac{1}{\varepsilon}D_t)v(x', t, \varepsilon)$$
$$= \left[\mathrm{diag}(\varepsilon^{-s_1}, \ldots, \varepsilon^{-s_{N-1}}, \varepsilon^{t_N})A(\varepsilon D', D_t)\mathrm{diag}(\varepsilon^{-t_1}, \ldots, \varepsilon^{-t_N}) - E_N\right]v(x', t, \varepsilon).$$

Multiplying this by $\mathrm{diag}(\varepsilon^{s_1}, \ldots, \varepsilon^{s_{N-1}}, \varepsilon^{-t_N})$ from the left and replacing $v(x, t, \varepsilon)$ by the expansion (3.5), we obtain from (3.3) the equation

$$\sum_{l=0}^{\infty} \varepsilon^{l_0+l} v^{(l)}(A(\varepsilon D', D_t) - E_N)v^{(l)} = 0. \tag{3.8}$$

Now we use the Taylor expansion of $A(\varepsilon D', D_t)$ with respect to $\varepsilon D'$ which is of the form

$$A(\varepsilon D', D_t) = A(0, D_t) + \sum_{|\alpha| \geq 1} \varepsilon^{|\alpha|} A^{(\alpha)}(0, D_t)D'^{\alpha}/\alpha! = A(0, D_t) + \sum_{k \geq 1} \varepsilon^k C_k(D)$$

and substitute this into (3.8). We get the recurrence relations

$$A(0, D_t)v^{(l)}(x', t) = -\sum_{k \geq 1} C_k(D)v^{(l-k)}. \tag{3.9}$$

Note that the left-hand sides of (3.7) and (3.9) coincide with the operators appearing in conditions (iii) and (iv), respectively, of the definition of weak parameter-ellipticity. So we see that the vector functions $u'^{(l)}, u^{(l)}$ and $v^{(l)}$ can be found recursively, provided that we know the boundary values

$$g'_{lj} := B_j(D)u'^{(l)}(x', 0), \quad j = 1, \ldots, R_{N-1}, \quad l = 0, 1, \ldots$$

and

$$g_{lj}'' := B_j(0, D_t)v^{(l)}(x', 0), \quad j = R_{N-1}+1, \ldots, R_N, \quad l = 0, 1, \ldots.$$

(Note that due to Definition 2.2 in the case of constant coefficients and no lower order terms the boundary value problems for $u^{(l)}$ and $v^{(l)}$ are uniquely solvable. In the case of variable coefficients the question of unique solvability is nontrivial; we hope to discuss this in a future paper.)

(iii) *Boundary conditions.* First of all note that

$$B_j(D)u(x', 0, \varepsilon) = \sum_{l=0}^{\infty} \varepsilon^l B_j(D)u^{(l)}(x', 0). \tag{3.10}$$

For the inner expansion we argue as before and get

$$B_j(D)v(x', 0, \varepsilon) = \sum_{l=0}^{\infty} \varepsilon^{l_0+l} B_j(D', \frac{1}{\varepsilon}D_t)\mathrm{diag}(\varepsilon^{t_1}, \ldots, \varepsilon^{t_N})v^{(l)}(x', 0)$$

$$= \sum_{l=0}^{\infty} \varepsilon^{l+l_0-m_j} B_j(\varepsilon D', D_t)v^{(l)}(x', 0).$$

Replacing $B_j(\varepsilon D', D_t)$ by

$$B_j(0, D_t) + \sum_{k=1}^{\infty} \varepsilon^k C_k(D)$$

and gathering terms with the same power of ε we finally obtain

$$B_j(D)v(x', 0, \varepsilon) = \sum_{l=l_0-m_j}^{\infty} \varepsilon^l \Big[B_j(0, D_t)v^{(l-l_0+m_j)}(x', 0)$$
$$+ C_1(D)v^{(l-l_0+m_j-1)}(x', 0) + \ldots \Big]. \tag{3.11}$$

Now we pose $l_0 = m_{R_{N-1}+1}$. According to our assumption $l_0 > m_j$ holds for $j = 1, \ldots, R_{N-1}$ and the first R_{N-1} boundary conditions are of the form

$$B_j(D)u^{(l)}(x', 0) = \delta_{0l}g_j(x') + B_j(0, D_t)v^{(l-l_0+m_j)}(x', 0)$$
$$+ C_1(D)v^{(l-l_0+m_j-1)}(x', 0) + \ldots. \tag{3.12}$$

If we already know $u^{(k)}$ and $v^{(k)}$ for $k = 1, \ldots, l-1$ this gives us the value of

$$B_j(D)u'^{(l)}(x', 0), \quad j = 1, \ldots, R_{N-1}.$$

Using the system (3.7) and these boundary conditions we can define $u'^{(l)}$ and, consequently, $u^{(l)}$.

For $j = R_{N-1} + 1$ equation (3.12) gives

$$B_j(0, D_t)v^{(l)}(x', 0) = \delta_{0l}g_l - B_j(D)u^{(l)}(x', 0) - \sum_{k \geq 1} C_k(D)v^{(l-k)}(x', 0).$$

To find the boundary conditions for $j > R_{N-1} + 1$, we apply the operator $B_j(D)$ to the term obtained from equating to zero the coefficient before $\varepsilon^{l+R_{N-1}+1-j}$. In this way we get for $j = R_{N-1} + 2, \ldots, R_N$

$$B_j(0, D_t)v^{(l)} = \delta_{0,l+R_{N-1}+1-j}g_l - B_j(D)u^{(l+R_{N-1}+1-j)} - \sum_{k \geq 1} C_k(D)v^{(l-k)}.$$

Now we can find $v^{(l)}$ and continue our process.

References

[A62]   Agmon, S.: On the eigenfunctions and on the eigenvalues of general elliptic boundary value problems. *Comm. Pure Appl. Math.* **15** (1962), 119-147.

[A90]   Agranovich, M. S.: Nonselfadjoint problems with a parameter that are Agmon-Douglis-Nirenberg elliptic (Russian). *Funktsional. Anal. i Prilozhen.* **24** (1990), No. 1, 59–61. English transl. in *Functional Anal. Appl.* **24** (1990), No. 1, 50–53.

[AV64]   Agranovich, M. S., Vishik, M. I.: Elliptic problems with parameter and parabolic problems of general form (Russian). *Uspekhi Mat. Nauk* **19** (1964), No. 3, 53-161. English transl. in *Russian Math. Surv.* **19** (1964), No. 3, 53-157.

[DMV98]   Denk, R., Mennicken, R., Volevich, L.: The Newton polygon and elliptic problems with parameter. *Math. Nachr.* **192** (1998), 125-157.

[DMV00a] Denk, R., Mennicken, R., Volevich, L.: Boundary value problems for a class of elliptic operator pencils. To appear in *Integral Equations Operator Theory*.

[DMV00b] Denk, R., Mennicken, R., Volevich, L.: On elliptic operator pencils with general boundary conditions. To appear in *Integral Equations Operator Theory*.

[DV00]   Denk, R., Volevich, L.: A priori estimate for a singularly perturbed mixed order boundary value problem. *Russian J. Math. Phys.* **7** (2000), 288–318.

[F79]   Frank, L.: Coercive singular perturbations. I. A priori estimates. *Ann. Mat. Pura Appl.* (4) **119** (1979), 41–113.

[F97]   Frank, L.: *Singular perturbations in elasticity theory.* Analysis and its Applications, 1. IOS Press, Amsterdam, 1997.

[G96]   Grubb, G.: *Functional Calculus of Pseudodifferential Boundary Problems.* Second edition. Progress in Mathematics, 65, Birkhäuser, Boston, 1996.

[I89]   Il'in, A. M.: *Matching of asymptotic expansions of solutions of boundary value problems* (Russian). "Nauka", Moscow, 1989. English transl. as Transl. Math. Monogr. **102**, Amer. Math. Soc., Providence, RI, 1992.

[K96]   Kozhevnikov, A.: Asymptotics of the spectrum of Douglis–Nirenberg elliptic operators on a closed manifold. *Math. Nachr.* **182** (1996), 261–293.

[N81]   Nazarov, S. A.: The Vishik–Lyusternik method for elliptic boundary value problems in regions with conic points. I. The problem in a cone (Russian). *Sibirsk. Mat. Zh.* **22** (1981), No. 4, 142-163. English transl. in *Siberian Math. J.* **22** (1982), 594-611.

[R96]   Roitberg, Y.: *Elliptic Boundary Value Problems in the Spaces of Distribu-tions*. Mathematics and its Applications, 384. Kluwer Academic Publishers, Dordrecht, 1996.

[VL57]   Vishik, M. I., Lyusternik, L. A.: Regular degeneration and boundary layer for linear differential equations with small parameter (Russian). *Uspehi Mat. Nauk (N.S.)* **12** (1957), No. 5 (77), 3-122. English transl. in *Amer. Math. Soc. Transl.* (2) **20** (1962), 239-364.

Addresses

ROBERT DENK, NWF I-Mathematik, Universität Regensburg, D-93040 Regensburg, Germany

E-MAIL: robert.denk@mathematik.uni-regensburg.de

LEONID VOLEVICH, Keldysh Institute of Applied Mathematics, Russian Acad. Sci., Miusskaya sqr. 4, 125047 Moscow, Russia

E-MAIL: volevich@spp.keldysh.ru

2000 Mathematics Subject Classification. Primary 35J40 ; Secondary 35B25

Operator Theory:
Advances and Applications, Vol. 126
© 2001 Birkhäuser Verlag Basel/Switzerland

Solutions of q-Deformed Equations with Quantum Conformal Symmetry

V.K. DOBREV, S.T. PETROV, AND B.S. ZLATEV

Abstract. We construct explicit solutions of a hierarchy of q-deformed equations which are quantum conformal invariant. The solutions are given in terms of two new q-deformations of the plane wave written in conjugated bases.

1   Introduction

One of the purposes of quantum deformations is to provide an alternative of the regularization procedures of quantum field theory. Applied to Minkowski space-time the quantum deformations approach is also an alternative to Connes' noncommutative geometry [Co94]. The first step in such an approach is to construct a noncommutative quantum deformation of Minkowski space-time There are several possible such deformations, cf. [CSSW90,SWZ91,Ma91,Do95a]. We follow the deformation of [Do95a] which is different from the others, the most important aspect being that it is related to a deformation of the conformal group.

The first problem to tackle in a noncommutative deformed setting is to analyze the behavior of the wave equation analogues. Here we continue the study of a hierarchy of deformed equations derived in [Do95b] with the use of quantum conformal symmetry. The hierarchy involves the massless representations of the conformal group and is parametrized by a nonnegative integer r. The case $r = 0$ corresponds to the q-d'Alembert equation, while for each $r > 0$ there are two couples of equations involving fields of conjugated Lorentz representations of dimension $r+1$. The construction of solutions of the hierarchy was started in [DK98] with the q-d'Alembert equation. One of the solutions given was a deformation of the plane wave as a formal power series in the noncommutative coordinates of q-Minkowski space-time and four-momenta. This q-plane wave has some properties analogous to the classical one but is not an exponent or q-exponent, cf. [GR90]. Thus, it differs conceptually from the classical plane wave and may serve as a regularization of the latter. For the equations labelled by $r > 0$ it turned out that one needs a second q-deformation of the plane wave in a conjugated basis [DGPZ98]. The solutions of the hierarchy in terms of the two q-plane waves were given in [DGPZ98] for $r = 1$ and in [DZ00] for $r > 1$.

In the present paper we extend these results by introducing more general q-deformations of the plane waves. We give also the solutions of the hierarchy in terms of these new q-plane waves.

2 Preliminaries

First we introduce new Minkowski variables:

$$x_\pm \equiv x_0 \pm x_3 , \qquad v \equiv x_1 - ix_2 , \qquad \bar{v} \equiv x_1 + ix_2 , \qquad (2.1)$$

which, (unlike the x_μ), have definite group–theoretical interpretation as part of a six-dimensional coset of the conformal group $SU(2,2)$ (as explained in [Do95a]). The d'Alembert equation in terms of these variables is:

$$\Box \varphi = (\partial_- \partial_+ - \partial_v \, \partial_{\bar{v}}) \, \varphi = 0 . \qquad (2.2)$$

In the q-deformed case we use the noncommutative q-Minkowski space-time of [Do95a] which is given by the following commutation relations (with $\lambda \equiv q - q^{-1}$):

$$x_\pm v = q^{\pm 1} v x_\pm , \quad x_\pm \bar{v} = q^{\pm 1} \bar{v} x_\pm , \quad x_+ x_- - x_- x_+ = \lambda v \bar{v} , \quad \bar{v} v = v \bar{v}, \qquad (2.3)$$

with the deformation parameter being a phase: $|q| = 1$. Relations (2.3) are preserved by the anti-linear anti-involution ω :

$$\omega(x_\pm) = x_\pm , \quad \omega(v) = \bar{v} , \quad \omega(q) = \bar{q} = q^{-1} , \quad (\omega(\lambda) = -\lambda) . \qquad (2.4)$$

The solution spaces consist of formal power series in the q-Minkowski coordinates (which we give in two conjugate bases):

$$\varphi = \sum_{j,n,\ell,m \in \mathbb{Z}_+} \mu_{jn\ell m} \, \varphi_{jn\ell m} , \qquad \varphi_{jn\ell m} = \hat{\varphi}_{jn\ell m}, \, \tilde{\varphi}_{jn\ell m} , \qquad (2.5a)$$

$$\hat{\varphi}_{jn\ell m} = v^j \, x_-^n \, x_+^\ell \, \bar{v}^m , \qquad (2.5b)$$

$$\tilde{\varphi}_{jn\ell m} = \bar{v}^m \, x_+^\ell \, x_-^n \, v^j = \omega(\hat{\varphi}_{jn\ell m}) . \qquad (2.5c)$$

The solution spaces (2.5) are representation spaces of the quantum algebra $U_q(sl(4))$. For the latter we use the rational basis of Jimbo [Ji85]. The action of $U_q(sl(4))$ on $\hat{\varphi}_{jn\ell m}$ was given in [Do94], (cf. also [DK98]), and on $\tilde{\varphi}_{jn\ell m}$ in [DGPZ98]. Further we suppose that q is not a nontrivial root of unity.

In order to write our q-deformed equations in compact form it is necessary to introduce some additional operators. We first define the operators:

$$\hat{M}_\kappa^\pm \, \varphi = \sum_{j,n,\ell,m \in \mathbb{Z}_+} \mu_{jn\ell m} \, \hat{M}_\kappa^\pm \, \varphi_{jn\ell m} , \qquad \kappa = \pm, v, \bar{v} , \qquad (2.6a)$$

$$T_\kappa^\pm \, \varphi = \sum_{j,n,\ell,m \in \mathbb{Z}_+} \mu_{jn\ell m} \, T_\kappa^\pm \, \varphi_{jn\ell m} , \qquad \kappa = \pm, v, \bar{v} , \qquad (2.6b)$$

and $\hat{M}_+^\pm$, $\hat{M}_-^\pm$, $\hat{M}_v^\pm$, $\hat{M}_{\bar{v}}^\pm$, resp., act on $\varphi_{jn\ell m}$ by changing by ± 1 the value of j, n, ℓ, m, resp., while $T_+^\pm$, $T_-^\pm$, $T_v^\pm$, $T_{\bar{v}}^\pm$, resp., act on $\varphi_{jn\ell m}$ by multiplication by $q^{\pm j}, q^{\pm n}, q^{\pm \ell}, q^{\pm m}$, resp. Now we can define the q-difference operators:

$$\hat{\mathcal{D}}_\kappa \, \varphi \; = \; \frac{1}{\lambda} \, \hat{M}_\kappa^{-1} \, \left(T_\kappa - T_\kappa^{-1} \right) \, \varphi \,. \tag{2.7}$$

Note that when $q \to 1$ then $\hat{\mathcal{D}}_\kappa \to \partial_k$. Using (2.6) and (2.7) then the q-d'Alembert equation may be written as ([Do95b,DK98]):

$$\left(q \, \hat{\mathcal{D}}_- \, \hat{\mathcal{D}}_+ \, T_v \, T_{\bar{v}} \; - \; \hat{\mathcal{D}}_v \, \hat{\mathcal{D}}_{\bar{v}} \right) \, T_v \, T_- \, T_+ \, T_{\bar{v}} \, \hat{\varphi} \; = \; 0 \,, \tag{2.8a}$$

$$\left(\hat{\mathcal{D}}_- \, \hat{\mathcal{D}}_+ \; - \; q \, \hat{\mathcal{D}}_v \, \hat{\mathcal{D}}_{\bar{v}} \, T_v \, T_{\bar{v}} \right) \, T_- \, T_+ \, \tilde{\varphi} \; = \; 0 \,. \tag{2.8b}$$

Note that when $q \to 1$ both equations in (2.8) go to (2.2). Note that the operators in (2.6), (2.7), (2.8) for different variables commute, i.e., we have passed to commuting variables. However, keeping the normal ordering it is straightforward to pass back to noncommuting variables.

3 Solutions of the Hierarchy Equations

As we mentioned, for $r \in \mathbb{N}$ there are two couples of equations involving fields of conjugated Lorentz representations of dimension $r + 1$. One of the couples is [Do95b]:

$$\left([r - N_{\bar{z}}]_c \hat{\mathcal{D}}_+ T_{\bar{v}} T_v^{-1} - q^{r+1} \hat{\mathcal{D}}_v \hat{\mathcal{D}}_{\bar{z}} T_- \right) \, T_- T_{\bar{v}} \, \hat{\varphi} \; = \; 0 \,, \tag{3.1a}$$

$$\left([r - N_{\bar{z}}]_c \hat{\mathcal{D}}_{\bar{v}} - q^{r+1} \hat{\mathcal{D}}_- \hat{\mathcal{D}}_{\bar{z}} T_v^2 T_- \right) \, T_{\bar{v}} \, \hat{\varphi} \; = \; 0 \,. \tag{3.1b}$$

The spin dependence is encoded in the spin variable $\bar{z}$ in which the solutions depend polynomially of degree r. As it was shown in [Do95b], if a function satisfies (3.1) then it satisfies also the q-d'Alembert equation (2.8a). Thus, it is justified to look for solutions in terms of q-deformations of the plane wave:

$$\widehat{\exp}_q(k \cdot x) \; = \; \sum_{s=0}^{\infty} \frac{1}{[s]_q!} \, \hat{h}_s \,, \tag{3.2}$$

$$[s]_q! \equiv [s]_q [s-1]_q \cdots [1]_q \,, \quad [0]_q! \equiv 1 \,, \quad [n]_q \equiv \frac{q^n - q^{-n}}{q - q^{-1}} \,,$$

$$\hat{h}_s \; = \; \beta^s \sum_{a,b,n \in \mathbb{Z}_+} \frac{(-1)^{s-a-b} \, q^{n(s-2a-2b+2n) \, + \, a(s-a-1) \, + \, b(-s+a+b+1)} \, q^{P_s(a,b)}}{\Gamma_q(a - n + 1)\Gamma_q(b - n + 1)\Gamma_q(s - a - b + n + 1)[n]_q!} \times$$

$$\times \, k_v^{s-a-b+n} k_-^{b-n} k_+^{a-n} k_{\bar{v}}^n v^n x_-^{a-n} x_+^{b-n} \bar{v}^{s-a-b+n} \,, \tag{3.3}$$

$$(\beta^s)^{-1} \; = \; \sum_{p=0}^{s} \frac{q^{(\varepsilon-p)(p-1)+p}}{[p]_q! \, [s-p]_q!} \,,$$

where the momentum components $(k_v, k_-, k_+, k_{\bar{v}})$ are supposed to be non-commutative between themselves (obeying the same rules (2.3) as the q-Minkowski coordinates), and commutative with the coordinates. Further, Γ_q is q-deformation of the Γ-function, of which here we use only the properties: $\Gamma_q(p) = [p-1]_q!$ for $p \in I\!N$, $1/\Gamma_q(p) = 0$ for $p \in Z\!\!\!Z_-$; $P_s(a,b)$ is a polynomial in a, b. This deformation of the plane wave generalizes the original one from [DK98]. To obtain the latter one has to replace $P_s(a,b)$ by 0. Each $\hat{h}_s$ satisfies the q-d'Alembert equation (2.8a) on the momentum q-cone:

$$\mathcal{L}_q^k \;\; \equiv \;\; k_- k_+ \; - \; q^{-1}\, k_v k_{\bar{v}} \; = \; 0 \; . \tag{3.4}$$

Note that the polynomials $P_s(a,b)$ by which we generalize the q-plane wave of [DK98] cannot be extended to involve dependence on the summation variable n since then $\hat{h}_s$ would not satisfy the q-d'Alembert equation.

We look for the solutions of (3.1) in a form analogous to (3.2):

$$\hat{\varphi} \; = \; \sum_{s=0}^{\infty} \frac{1}{[s]_q!} \, \hat{\varphi}_s \; . \tag{3.5}$$

The solutions are constructed component-wise, i.e., we solve (3.1) separately for each $\hat{\varphi}_s$ and we find that

$$\hat{\varphi}_s \; = \; \sum_{m=0}^{r} \hat{\gamma}_m^{rs} \left(\prod_{i=-r+1}^{-m} (k_+ - q^{i-A_s} k_v \bar{z}) \right) \left(\prod_{j=-m+1}^{0} (k_{\bar{v}} - q^{j-A_s} k_- \bar{z}) \right) \hat{h}_s \tag{3.6a}$$

$$P_s(a,b) \; = \; A_s a + P_s(b), \tag{3.6b}$$

where $\hat{\gamma}_m^{rs}$ are $r+1$ independent constants, A_s is an arbitrary constant, and $P_s(b)$ is an arbitrary polynomial in b. The check that this is a solution is done for commutative Minkowski coordinates and noncommutative momenta on the q-cone. Note that the factors preceding $\hat{h}_s$ depend on the constant A_s but not on the polynomial $P_s(b)$. In order to be able to write the general solution of the system (3.1) in terms of the deformed plane wave we have to suppose that the $\hat{\gamma}_m^{rs}$, A_s for different s coincide: $\hat{\gamma}_m^{rs} = \hat{\gamma}_m^{r}$, $A_s = A$. Then we have:

$$\hat{\varphi} \; = \; \sum_{m=0}^{r} \hat{\gamma}_m^{r} \left(\prod_{i=-r+1}^{-m} (k_+ - q^{i-A} k_v \bar{z}) \right) \left(\prod_{j=-m+1}^{0} (k_{\bar{v}} - q^{j-A} k_- \bar{z}) \right) \widehat{\exp}_q(k \cdot x) \; . \tag{3.7}$$

When $A = A_s = 0 = P_s(b)$ these solutions were obtained in [DZ00].

We pass now to the other couple of equations written for the conjugate basis:

$$\left([r - N_z]_q D_v - q^r D_z D_- T_- \right) T_v \, \tilde{\varphi} = \; 0, \tag{3.8a}$$

$$\left([r - N_z]_q D_+ T_{\bar{v}}^{-1} - q^r D_z D_{\bar{v}} T_- T_v \right) T_- \, \tilde{\varphi} \; = \; 0. \tag{3.8b}$$

Analogous to the first case we use the expansion

$$\tilde{\varphi} \;=\; \sum_{s=0}^{\infty} \frac{1}{[s]_q!}\,\tilde{\varphi}_s\;,\tag{3.9}$$

and we solve it again component-wise. Here we shall use another deformation of the plane wave:

$$\widetilde{\exp}_q(k\cdot x) \;=\; \sum_{s=0}^{\infty} \frac{1}{[s]_q!}\,\tilde{h}_s\;,\tag{3.10}$$

$$\tilde{h}_s \;=\; \tilde{\beta}^s \sum_{a,b,n} \frac{(-1)^{s-a-b}\; q^{n(2a+2b-2n-s)\,+\,a(a-s-1)\,+\,b(s-a-b+1)}\; q^{Q_s(a,b)}}{\Gamma_q(a-n+1)\,\Gamma_q(b-n+1)\,\Gamma_q(s-a-b+n+1)\,[n]_q!} \;\times$$

$$\times\; k_{\bar{v}}^{n}\,k_{+}^{a-n}\,k_{-}^{b-n}\,k_{v}^{s-a-b+n}\,\bar{v}^{s-a-b+n}\,x_{+}^{b-n}\,x_{-}^{a-n}\,v^{n}\;,\tag{3.11}$$

$$\left(\tilde{\beta}^s\right)^{-1} \;=\; \sum_{p=0}^{s} \frac{q^{(p-s)(p-1)+p}}{[p]_q!\,[s-p]_q!}\;,$$

where $Q_s(a,b)$ are arbitrary polynomials. If the latter are zero then $\widetilde{\exp}_q(k\cdot x)$ becomes the q-plane wave deformation found in [DGPZ98]. The $\tilde{h}_s$ have the same properties as the $\hat{h}_s$ but the conjugated basis is used; in particular, they satisfy the q-d'Alembert equation (2.8b) on the momentum q-cone (3.4). The solutions of (3.8) are polynomials of degree r in the other spin variable z. Explicitly they are given by:

$$\tilde{\varphi}_s \;=\; \sum_{m=0}^{r} \tilde{\gamma}_m^{rs} \left(\prod_{i=-r+2}^{-m+1} (k_+ - q^{i+B_s}k_{\bar{v}}z)\right) \left(\prod_{j=-m+2}^{1} (k_v - q^{j+B_s}k_-z)\right) \tilde{h}_s\tag{3.12a}$$

$$Q_s(a,b) \;=\; Q_s(a) + B_s b,\tag{3.12b}$$

where $\tilde{\gamma}_m^{rs}$ are $r+1$ independent constants, $Q_s(a)$ is an arbitrary polynomial in a, B_s is an arbitrary constant. Note that the factors preceding $\tilde{h}_s$ depend on the constant B_s but not on the polynomial $Q_s(a)$. In order to be able to write the general solution of the system (3.8) in terms of the deformed plane wave we have to suppose that the $\tilde{\gamma}_m^{rs}$, A_s for different s coincide: $\tilde{\gamma}_m^{rs} = \hat{\gamma}_m^r$, $B_s = B$. Then we have:

$$\tilde{\varphi} \;=\; \sum_{m=0}^{r} \tilde{\gamma}_m^{r} \left(\prod_{i=-r+2}^{-m+1} (k_+ - q^{i+B}k_{\bar{v}}z)\right) \left(\prod_{j=-m+2}^{1} (k_v - q^{j+B}k_-z)\right) \widetilde{\exp}_q(k\cdot x)\;.$$

$$\tag{3.13}$$

When $B = B_s = 0 = Q_s(a)$ these solutions were obtained in [DGPZ98] for $r=1$ and in [DZ00] for $r>1$.

Acknowledgements

V.K.D. is thankful to the organizers of the conference for the invitation to give a talk and for the financial support.

References

[CSSW90] U. Carow-Watamura et al, Zeit. f. Phys. **C48** (1990) 159.
[Co94] A. Connes, *Noncommutative Geometry*, (Academic Press, 1994).
[Do94] V.K. Dobrev, J. Phys. **A27** (1994) 4841 & 6633.
[Do95a] V.K. Dobrev, Phys. Lett. **341B** (1994) 133 & **346B** (1995) 427.
[Do95b] V.K. Dobrev, J. Phys. **A28** (1995) 7135.
[DK98] V.K. Dobrev and B.S. Kostadinov, Phys. Lett. **439B** (1998) 337.
[DGPZ98] V.K. Dobrev, R.I. Gushterski, S.T. Petrov and B.S. Zlatev, Invited talk by V.K.D. at III International Workshop on "Classical and Quantum Integrable Systems" (Yerevan, 1998), Proceedings, eds. L.G. Mardoyan, G.S. Pogosyan and A.N. Sissakian, (Publishing Department of JINR, Dubna, 1998) p. 39.
[DZ00] V.K. Dobrev and B.S. Zlatev, Czech. J. Phys. **50** (2000) 53.
[GR90] G. Gaspar and M. Rahman, *Basic Hypergeometric Series* (Cambridge U. Press, 1990).
[Ji85] M. Jimbo, Lett. Math. Phys. **10** (1985) 63; Lett. Math. Phys. **11** (1986) 247.
[Ma91] S. Majid, J. Math. Phys. **32** (1991) 3246.
[SWZ91] W.B. Schmidke, J. Wess and B. Zumino, Zeit. f. Phys. **C52** (1991) 471.

Addresses

V.K. DOBREV, School of Computing and Mathematics, University of Northumbria, Ellison Place, Newcastle upon Tyne, NE1 8ST, UK;
permanent address: Institute of Nuclear Research and Nuclear Energy, Bulgarian Academy of Sciences, 72 Tsarigradsko Chaussee, 1784 Sofia, Bulgaria

E-MAIL: vladimir.dobrev@unn.ac.uk,dobrev@inrne.bas.bg

S.T. PETROV, Institute of Nuclear Research and Nuclear Energy, Bulgarian Academy of Sciences, 72 Tsarigradsko Chaussee, 1784 Sofia, Bulgaria

E-MAIL: spetrov@inrne.bas.bg

B.S. ZLATEV, Institute of Nuclear Research and Nuclear Energy, Bulgarian Academy of Sciences, 72 Tsarigradsko Chaussee, 1784 Sofia, Bulgaria

E-MAIL: zlatev@inrne.bas.bg

2000 Mathematics Subject Classification. Primary 81R50; Secondary 17B37, 20G42

Operator Theory:
Advances and Applications, Vol. 126
© 2001 Birkhäuser Verlag Basel/Switzerland

On the Regularization and Stabilization of Approximation Schemes for C_0-Semigroups

SIMONE FLORY, FRANK NEUBRANDER, AND YU ZHUANG

Abstract. Many temporal discretization methods for linear evolution equations converge uniformly on compact time intervals at the rate $\frac{1}{n^\alpha}$ only for sufficiently smooth initial data. It is shown that these methods can be regularized such that the new schemes converge 'in the average' at the rate $\frac{1}{n^\alpha}$ for all initial data. Examples given include the Crank-Nicholson scheme and the alternating direction implicit method.

1   Introduction

In this paper, we will consider approximation schemes for abstract initial value problems

$$u'(t) = Au(t), \ t \geq 0, \ u(0) = f \in D(A), \tag{1.1}$$

where A is the generator of a strongly continuous semigroup $T(\cdot)$ on a Banach space X with $\|T(t)f\| \leq Me^{wt}\|f\|$ for all $t \geq 0$ and $f \in X$. In particular, the domain $D(A)$ is a dense subset of X, there exist constants M, w such that $R(\lambda, A) := (\lambda I - A)^{-1}$ is bounded and

$$\|R(\lambda, A)^n\| \leq \frac{M}{(\lambda - w)^n} \text{ for all } \lambda > w.$$

For simplicity, we will often assume that the spectrum $\sigma(A)$ of A is strictly contained in the left half-plane; i.e., $s(A) := \sup\{Re\lambda : \lambda \in \sigma(A)\} < 0$. For any initial value $f \in D(A)$, the function $u(\cdot) := T(\cdot)f \in C^1([0, \infty), X)$ is the unique 'classical' solution of (1.1) and, for $f \in X$, the function $u(\cdot) := T(\cdot)f \in C([0, \infty), X)$ is the unique 'mild' solution of (1.1); i.e., for any sequence $f_n \in D(A)$ with $f_n \to f$, the corresponding 'classical' solutions $T(\cdot)f_n$ converge uniformly on compact time intervals to $T(\cdot)f$, which is then a solution of the integral equation

$$u(t) = f + A \int_0^t u(s) \, ds, \ t \geq 0.$$

Since we will consider approximation schemes which do not necessarily converge to $T(t)f$ for all $f \in X$ but only on $D(A)$ or on a subset of $D(A)$, the following remark might be useful. In practice, the physically meaningful initial values f and instantaneous states $u(t)$ of the system described by the initial value problem (1.1) are always contained in $D(A)$, and often even in a true subset (cone) $D \subset$

$D(A)$. However, for the mathematical discussion of (1.1), it is convenient to enlarge D to a linear space (which is then the domain of a linear operator A) and to think of the states as points in a (complex) Banach space X, even if the imposed linearity, completeness and complexity forces us to accept as elements of X some points (functions, vectors) not having direct significance as 'classical' states of the physical system. Thus, the consideration of approximation schemes that converge for $D \subset D(A)$ and not necessarily on X is a meaningful enterprise as long as the schemes are numerically stable.

In the study of approximations of the solutions of (1.1), the Chernoff Product Formula and the Lax Equivalence Theorem are outstanding results on which much of the theory is based. The basic idea behind these results is the Euler formula; i.e., let $\{V(t), t \in [0, \delta]\}$ be a family of linear operators on X with $V(0) = I$ such that the consistency condition

$$\frac{V(t)f - f}{t} \to Af, t \to 0, f \in D \subset D(A) \tag{1.2}$$

holds. Then $V(t)f \sim (I + tA)f$ (t small), or $V(\frac{t}{n})^n f \sim (I + \frac{t}{n}A)^n f \sim e^{tA}f$ as $n \to \infty$. It was shown by Lax and Richtmyer in 1956 [LR56] and Chernoff in 1974 [Ch74] that these formal considerations are justifiable if the scheme is stable; i.e.,

$$\|V(t)^n\| \le M e^{\omega n t} \tag{1.3}$$

for some $\omega \ge 0$ and all $n \in \mathbb{N}_0, t \in [0, \delta]$. The purpose of this paper is to discuss linear (not necessarily bounded) approximation schemes that are consistent (satisfy (1.2)) but do not necessarily satisfy (1.3). Examples which will be discussed below are the Crank-Nicholson scheme,[1] the alternating direction implicit method (ADI method) of Peaceman and Rachford, the Lie-Trotter product formula, and the Strang scheme. In many cases one can show that schemes which do not necessarily satisfy (1.3) are regularizable and/or stabilizable in the following sense. An approximation scheme $\{V(t), t \in [0, \delta]\}$ is called regularizable if there exists $k \in \mathbb{N}_0$ and a convergence rate $g : \mathbb{N} \to \mathbb{R}$ with $g(n) \to 0$ as $n \to \infty$ such that

$$\left\|V(\frac{t}{n})^n f - T(t)f\right\| \le g(n)\|A^k f\| \tag{1.4}$$

for all $f \in D(A^k)$ and $t \in [0, T]$.[2] If (1.4) holds, then we will show in Section 3 that there exists a convolution norm $\||| \cdot \|||$ on $[0, T]$ and a regularized approximation scheme $V_{reg}(\cdot)$ such that

$$\left\|\left\|V_{reg}(\frac{\cdot}{n})^n f - T(\cdot)f\right\|\right\| \le g(n)\|f\| \tag{1.5}$$

---

[1] For a discussion of higher order rational approximations, see [FNZ00].

[2] Recall that we assume that $s(A) < 0$; in this case $\|A^k f\|$ is equivalent to the graph norm $\|f\| + \|A^k f\|$.

for all $f \in X$, and $n \in \mathbb{N}$. One should notice that the concept of regularization is meaningful even for stable approximation schemes. For stable schemes, (1.4) and the density of $D(A^k)$ imply that

$$\|V(\frac{t}{n})^n f - T(t)f\| \to 0 \tag{1.6}$$

as $n \to \infty$ for all $f \in X$, uniformly for $t \in [0,T]$. However, in the step from (1.4) to (1.6) one loses the rate of convergence $g(n)$ for initial data $f \in X \backslash D(A^k)$. In contrast, the rate of convergence is retained in (1.5) for all initial values $f \in X$.

In order to obtain estimates of the form (1.4) it is sometimes necessary to stabilize the approximation scheme first. A consistent approximation scheme $\{V(t), t \in [0,\delta]\} \subset \mathcal{L}(X)$ is called stabilizable if there exists $\{W(t), t \in [0,\delta\} \subset \mathcal{L}(X)$ and $j, k \in \mathbb{N}$ such that

$$\|W(\frac{t}{n})^j V(\frac{t}{n})^{n-j} f - T(t)f\| \leq g(n)\|A^k f\| \tag{1.7}$$

for all $n \geq j$, $f \in D(A^k)$, and $t \in [0,T]$, where $g(\cdot)$ is as above. The outline of the paper is as follows. In Section 2 we review the relation between stability, consistency and convergence and collect some of the known results about the Crank-Nicholson scheme, the ADI method, and the Lie-Trotter product formula. In Section 3 we discuss the regularization principle and stabilization technique outlined above and present some applications.

2 Remarks on Consistency and Stability

When an approximate solution of (1.1) is obtained by finite difference methods, the time variable t assumes discrete values $t_n := n\Delta t$ $(n \in \mathbb{N}_0)$, and correspondingly, one obtains a discrete sequence u_n of approximate states of the physical system at time $n\Delta t$. Thus, single step temporal approximation methods

$$u_n = V(\Delta t)u_{n-1}$$

approximate the solution at time $n\Delta t$ by applying an operator $V(\Delta t)$ to the approximate solution u_{n-1} at time $(n-1)\Delta t$; i.e., $u_n = V(\Delta t)^n u_0$ is an approximation of the instantaneous solution $u(n\Delta t)$ of (1.1), where u_0 is an approximation of the initial value $u(0) = f$. In particular, if one chooses $\Delta t := t/n$ and $u_0 := f \in D(A)$, then a central topic in approximation theory is to show that

$$V(\frac{t}{n})^n f \to T(t)f \tag{2.1}$$

as $n \to \infty$. Clearly, as important as convergence is the question of the rate of convergence and the stability of the scheme; i.e., instead of (2.1) one would prefer the statement

$$\|V(\frac{t}{n})^n f - T(t)f\| \leq g(n)\|f\| \tag{2.2}$$

for $f \in D(A)$ and $t \in [0,T]$, where $g : \mathbb{N} \to \mathbb{R}$ is a given function with $g(n) \to 0$ as $n \to \infty$. Unfortunately, for generators of arbitrary strongly continuous semigroups on Banach spaces, an estimate of the form (2.2) rarely exists.[3] As we will see below, one must either stabilize the schemes to obtain (2.2), or one must change one of the norms. To see this, we first recall the notions of 'consistency' and 'stability' which play an important role in the discussion of this convergence issue.

Definition 2.1. An approximation scheme $\{V(t), t \in [0,\delta]\}$ of linear operators defined on a dense subset $D \subset D(A)$ is called D-consistent if, for all $f \in D$, $V(0)f = f$ and $\frac{V(t)f - f}{t} \to Af$ as $t \to 0$.

It follows from $\frac{1}{t}[V(t)f - T(t)f] = [\frac{V(t)f - f}{t} - Af] - [\frac{T(t)f - f}{t} - Af]$ that an approximation scheme $V(t)$ is D-consistent if and only if

$$\|V(t)f - T(t)f\| = o(t) \text{ as } t \to 0 \text{ for all } f \in D. \tag{2.3}$$

An approximation scheme $\{V(t), t \in [0,\delta]\}$ is called D-*consistent of order* $\alpha > 0$ if $V(0) = I$ on D and, for all $f \in D$ and $r_0 > 0$, there exists a constant $M(r_0, f)$ such that

$$\|V(t)f - T(t)f\| \le M(f)t^{\alpha+1} \text{ for all } t \in [0,\delta], \tag{2.4}$$

In practice, the order of consistency plays an important role since it allows an estimate on the rate of convergence. This is shown in the following well known proposition (see, e.g., [BW77]), whose elementary proof we recall for the sake of completeness.

Proposition 2.1. *Let D be a $T(t)$-invariant subset and let $\{V(t), t \in [0,\delta]\}$ be a D-consistent approximation scheme of order $\alpha > 0$ that commutes with $T(t)$ on D. If there exist constants M, w such that $\|V(t)^n\| \le Me^{wnt}$ for all $t \in [0,\delta]$ and $n \in \mathbb{N}_0$, then, for all $f \in D$, there exists a constant $M(f)$ such that*

$$\|V(\frac{t}{n})^n f - T(t)f\| \le M(f)\frac{1}{n^\alpha}$$

for all $n > T/\delta$ and $t \in [0,T]$. Moreover, if the linear span of D is dense in X, then $V(\frac{t}{n})^n f_n \to T(t)f$ as $n \to \infty$ uniformly for $t \in [0,T]$ for all sequences $f_n \to f \in X$.

Proof. We may assume that $\|T(t)\| \le M$ and $\|V(t)^n\| \le Me^{wnt}$ for all $t \in [0,T]$ and $n \in \mathbb{N}_0$, where $V(t) := 0$ for $t > \delta$. Let $f \in X$. By assumption, there exists

---

[3] Exceptions are Trotter-Kato and other factorized product formulas for nonnegative selfadjoint generators on Hilbert spaces, see [IT00] and [NZ99], or A-stable rational approximation schemes vanishing at ∞ for generators of analytic semigroups, see [Ha99] and [LTW91].

a constant $M(\delta, f)$ such that $\|V(t)f - T(t)f\| \le M(\delta, f)t^{\alpha+1}$ for all $t \in [0, \delta]$. It follows from

$$V(\tfrac{t}{n})^n f - T(t)f = \sum_{i=0}^{n-1} V(\tfrac{t}{n})^{n-i-1}[V(\tfrac{t}{n})f - T(\tfrac{t}{n})]T(\tfrac{it}{n})f$$

that $\|V(\tfrac{t}{n})^n f - T(t)f\| \le M^2 M(\delta, f)\frac{n-1}{n}e^{wt}t^{\alpha+1}\frac{1}{n^\alpha} \le M(f)\frac{1}{n^\alpha}$. If the linear span of D is dense in X, then the uniform boundedness of the operators $V(\tfrac{t}{n})^n - T(t)$ for $t \in [0, T]$ and $n > T/\delta$ implies the uniform convergence of $V(\tfrac{t}{n})^n f_n$ to $T(t)f$ for all $f_n \to f \in X$. $\qquad\square$

Corollary 2.1. *[Backward Euler Scheme]. Let A be the generator of a strongly continuous semigroup $T(\cdot)$ with $\|T(t)\| \le Me^{wt}$ $(t \ge 0)$ for some $M, w \ge 0$, and let $0 < \delta < 1/w$ be such that $e^{-(w+1)t} \le 1 - wt$ for all $t \in [0, \delta]$. Then the backward Euler scheme (implicit scheme)*

$$V_{be}(t) := (I - tA)^{-1}, \ t \in [0, \delta],$$

is consistent of order 1 on $D(A^2)$, commutes with $T(t)$ on $D(A^2)$ and satisfies $\|V_{be}(t)^n\| \le Me^{(w+1)nt}$ for all $t \in [0, \delta]$ and $n \in \mathbb{N}_0$. In particular

(i) for all $T > 0$ there exists a constant $M(T)$ such that, for all $f \in D(A^2)$ and $n > T/\delta$, $\|V_{be}(\tfrac{t}{n})^n f - T(t)f\| \le M(T)\frac{1}{n}\|A^2 f\|$ for $t \in [0, T]$, and
(ii) $V_{bs}(\tfrac{t}{n})^n f_n = (I - \tfrac{t}{n}A)^{-n} f_n \to T(t)f$ as $n \to \infty$ uniformly for $t \in [0, T]$ for all sequences $f_n \to f \in X$ and all $T > 0$.

Proof. If A generates a strongly continuous semigroup with $\|T(t)\| \le Me^{wt}$, then $(I - tA)^{-1}$ exists for $t \in [0, 1/w)$ and $\|(I - tA)^{-n}\| \le \frac{M}{(1-tw)^n} \le Me^{n(w+1)t}$ for all $t \in [0, \delta]$ and $n \in \mathbb{N}_0$. Moreover, $D(A^2)$ is dense in X and $T(t)$-invariant. For $f \in D(A^{n+1})$, easy inductions yield the power series developments

$$T(t)f = \sum_{i=0}^{n} \frac{t^i}{i!}A^i f + \int_0^t \frac{(t-s)^n}{n!}T(s)A^{n+1}f\,ds, \quad \text{for all } t \ge 0 \text{ and} \tag{2.5}$$

$$(I - tA)^{-1}f = \sum_{i=0}^{n} t^i A^i f + t^{n+1}(I - tA)^{-1}A^{n+1}f \quad \text{for all } t \in [0, \delta]. \tag{2.6}$$

Thus, there exists a constant $M(T)$ such that, for $f \in D(A^2)$ and $t \in [0, T]$,
$\|V_{bs}(t)f - T(t)f\| = \|t^2(I - tA)^{-1}A^2 f - \int_0^t (t-s)T(s)A^2 f\,ds\| \le t^2 M(T)\|A^2 f\|$.
Now, the statement follows from Proposition 2.1 $\qquad\square$

Corollary 2.2. *[Crank-Nicolson Scheme]. Let A be the generator of a strongly continuous semigroup $T(\cdot)$ on a Banach space X with $\|T(t)\| \le Me^{wt}$ $(t \ge 0)$ for some $M, w \ge 0$, and let $0 < \delta < 1/w$. Then the Crank-Nicolson scheme*

$$V_{cn}(t) := (I + \tfrac{t}{2}A)(I - \tfrac{t}{2}A)^{-1}, \ t \in [0, \delta]$$

is consistent of order 2 on $D(A^3)$ and commutes with $T(t)$ on $D(A^3)$. If X is a Hilbert space and $T(\cdot)$ is a contraction semigroup,[4] then $\|V_{cn}(t)^n\| \leq 1$ for all $t \in [0,\infty)$ and $n \in \mathbb{N}_0$. Thus, for contraction semigroups on Hilbert spaces the following statements hold.

(i)   For all $T > 0$ there exists a constant $M(T)$ such that, for all $f \in D(A^3)$ and $n \in \mathbb{N}$, $\|V_{cn}(\frac{t}{n})^n f - T(t)f\| \leq M(T)\frac{1}{n^2}\|A^3 f\|$ for $t \in [0,T]$,

(ii) $V_{cn}(\frac{t}{n})^n f_n \to T(t)f$ as $n \to \infty$ uniformly for $t \in [0,T]$ for all sequences $f_n \to f \in X$ and all $T > 0$.

If X is a Banach space, A generates an analytic[5] semigroup $T(\cdot)$, and $s(A) < 0$, then

(iii) $\|V_{cn}(\frac{t}{n})^n f - T(t)f\| \leq M\frac{t^2}{n^2}\|A^2 f\|$ for all $t \geq 0, n \in \mathbb{N}$ and $f \in D(A^2)$.

(iv) $\|V_{be}(\frac{t}{n})^2 V_{cn}(\frac{t}{n})^{n-2} f - T(t)f\| \leq M\frac{1}{n^2}\|f\|$ for all $t \geq 0, 2 < n \in \mathbb{N}$ and $f \in X$.

Proof. It follows from (2.6) that $(I + \frac{t}{2}A)(I - \frac{t}{2}A)^{-1} = f + tAf + \frac{t^2}{2}A^2 f + \frac{t^3}{4}(I - \frac{t}{2}A)^{-1}A^3 f$ for all $f \in D(A^3)$. Thus, by (2.5),

$$V_{cn}(t)f - T(t)f = \frac{t^3}{4}(I - \frac{t}{2}A)^{-1}A^3 f - \int_0^t \frac{(t-s)^2}{2}T(s)A^3 f \, ds$$

for $f \in D(A^3)$. As in the proof of Corollary 2.1 we conclude that the Crank-Nicolson scheme is consistent of order 2 on $D(A^3)$. If A generates a contraction semigroup on a Hilbert space, then $Re\langle Af, f\rangle \leq 0$ for all $f \in D(A)$. Let $t \geq 0$. Clearly, $\|V_{cn}(t)^n\| \leq 1$ if $\|V_{cn}(t)\| = \|(I + \frac{t}{2}A)(I - \frac{t}{2}A)^{-1}\| \leq 1$ or, equivalently, if

$$\|(I + \frac{t}{2}A)f\| \leq \|(I - \frac{t}{2}A)f\| \text{ for all } f \in D(A).$$

This inequality holds if $\|f\|^2 + tRe\langle Af, f\rangle + \frac{t^2}{4}\|Af\|^2 = \|(I + \frac{t}{2}A)f\|^2 \leq \|(I - \frac{t}{2}A)f\|^2 = \|f\|^2 - tRe\langle Af, f\rangle + \frac{t^2}{4}\|Af\|^2$ or, equivalently, if $Re\langle Af, f\rangle \leq 0$. The proofs of the statements for generators A of analytic semigroups on Banach spaces X with $s(A) < 0$ as well as further references can be found in [LTW91] and [Ha99]. Notice that statement (iv) contains the stabilization procedure outlined in (1.7) above. $\qquad\square$

The backwards Euler scheme and the Crank-Nicholson scheme are special cases of rational approximation schemes of the exponential function due to M.H. Padé [Pa92] (see also [Tw84], 4.5.1, or [FNZ00]). However, as in the case of the Crank-Nicholson scheme, these higher order approximation schemes might become unstable for arbitrary semigroup generators in Banach spaces.

---

[4] The proof extends to quasi-contractive semigroups in Hilbert spaces (i.e., $\|T(t)\| \leq e^{wt}$ for $t \geq 0$).

[5] For the discussion of the Crank-Nicholson scheme for generators of arbitrary semigroups in Banach spaces we refer to [FNZ00].

The proofs of statements (ii) of Corollary 2.1 and (ii) of Corollary 2.2 show that the convergence of $V(\frac{t}{n})^n f$ to $T(t)f$ for all $f \in X$ depends very much on the stability of the method; i.e., $\|V(t)^n\| \le Me^{\omega nt}$ for all $n \in \mathbb{N}, t \in [0, \delta]$.

This dependence was first recognized in 1928 by Courant, Friedrichs and Lewy [CFL28] and later characterized by Lax, with the 'uniform consistency' require-ment, in a seminar talk at New York University in 1954 (see [LR56]). With con-sistency defined as in Definition 1.1, the implication $(i) \to (ii)$ is a well-known consequence of the celebrated *Chernoff product formula* (see, e.g., Section III.5 in [EN99]). We include here the proof of the implication $(ii) \to (i)$ of the following theorem since we could not find it in the literature known to us.

Theorem 2.1. *[Lax-Chernoff Equivalence Theorem]. Let A be the generator of a strongly continuous semigroup $T(\cdot)$ and let $\{V(t), t \in [0, \delta]\}$ be a D-consistent approximation scheme of bounded linear operators with $\sup_{t \in [0, \delta]} \|V(t)\| < \infty$. The following are equivalent.*

(i) *The approximation scheme is stable; i.e., there exist constants $M \ge 1$, $w \ge 0$ such that $\|V(t)^n\| \le Me^{\omega nt}$ for all $t \in [0, \delta]$ and $n \in \mathbb{N}_0$.*

(ii) *The approximation scheme is uniformly convergent for each $f \in X$; i.e., $V(\frac{t}{n})^n f \to T(t)f$ $(n \to \infty)$ uniformly for $t \in [0, T]$ for each $T > 0$ and $f \in X$.*

(iii) *The approximation scheme is uniformly convergent on compacts; i.e., for any $T > 0$ and compact subset $X_C \subset X$, $V(\frac{t}{n})^n f \to T(t)f$ $(n \to \infty)$ uniformly for $(t, f) \in [0, T] \times X_C$.*

Proof. To prove the implication $(i) \to (iii)$, observe first that the boundedness of the operators $V(t)$ and their consistency on the dense set D imply that $V(0) = I$. Defining $V(t) := 0$ for $t > \delta$, the family $\{V(t), t \ge 0\}$ satisfies the assumptions of the Chernoff product formula (see Corollary III.5.3 in [EN99]).

Let $X_C \subset X$ be compact, $T > 0$, and $L_n(t) := V(\frac{t}{n})^n - T(t)$. There exists $K > 0$ such that $\|L_n(t)\| \le K$ for all $n \in \mathbb{N}_0$ and $t \in [0, T]$. Moreover, by Chernoff's product formula, $L_n(t)f \to 0$ as $n \to \infty$ uniformly in $t \in [0, T]$ for all $f \in X$. Let $\epsilon > 0$. Then there exists a finite set $G := \{g_1, g_2, \ldots, g_k\} \subset X_C$ such that for any $f \in X_C$ there exists a $g_f \in G$ with $\|f - g_f\| \le \epsilon/2K$. Choose n_0 such that $\|L_n(t)g\| \le \epsilon/2$ for all $t \in [0, T]$, $g \in G$, and $n \ge n_0$. Then, for all $f \in X_C$, $t \in [0, T]$ and $n \ge n_0$, we have $\|L_n(t)f\| \le \|L_n(t)\|\|f - g_f\| + \|L_n(t)g_f\| \le \epsilon$. This finishes the proof of $(i) \to (iii)$. Clearly, $(iii) \to (ii)$. Thus, it remains to be shown that $(ii) \to (i)$.

Assume that (ii) holds. Let $t \in [0, T]$, $f \in X$, and $K > 0$ such that $\|T(t)\|$ and $\|V(t)\|$ are bounded by K for all $t \in [0, T]$, where $V(t) := 0$ for $t > \delta$. There exists $n_f \in \mathbb{N}$ such that $\|V(\frac{t}{n})^n f\| \le \|V(\frac{t}{n})^n f - T(t)f\| + \|T(t)\|\|f\| \le 1 + K\|f\|$ for all $n \ge n_f$ and $t \in [0, T]$. Since $\|V(\frac{t}{n})^n f\| \le K^{n_f}\|f\|$ for all $0 \le n \le n_f$ and $t \in [0, \delta]$, it follows that there exists $M_f > 0$ such that $\|V(\frac{t}{n})^n f\| \leq M_f$ for all $t \in [0, T]$ and $n \in \mathbb{N}_0$. By the Uniform Boundedness Principle,

$$M(T) := \sup_{t \in [0,T], n \in \mathbb{N}_0} \|V(\frac{t}{n})^n\| < \infty.$$

Clearly, the function $M : T \to M(T)$ is positive, increasing, and

$$\|V(t)^n\| = \|V(\frac{tn}{n})^n\| \leq \sup_{s \in [0,nt], n \in \mathbb{N}_0} \|V(\frac{s}{n})^n\| = M(nt) \leq M(T) \qquad (2.7)$$

for all $t \in [0, \delta], n \in \mathbb{N}_0$, and $T \geq nt$. Now, let $0 < t \leq \delta$ be fixed and $w := \frac{1}{\delta} \ln M(2\delta)$. Let $[\frac{2\delta}{t}]$ denote the smallest integer less than $\frac{2\delta}{t}$. Then, $[\frac{2\delta}{t}] \geq 2$ and $[\frac{2\delta}{t}]t \leq 2\delta$. Thus,

$$\|V(t)^{[\frac{2\delta}{t}]}\| \leq M(2\delta) = e^{w\delta} \leq e^{[\frac{2\delta}{t}] \cdot \frac{w\delta}{\frac{2\delta}{t}-1}} = e^{[\frac{2\delta}{t}] \cdot \frac{w\delta t}{2\delta-t}} \leq e^{[\frac{2\delta}{t}] \cdot \frac{w\delta t}{2\delta-\delta}} = e^{[\frac{2\delta}{t}]wt}.$$

Let $N := [\frac{2\delta}{t}]$. Then, by the estimate above, $\|V(t)^N\| \leq e^{Nwt}$. For any $n \in \mathbb{N}$ there exist $k, m \in \mathbb{N}_0$ such that $n = kN + m$ and $m < N$. Since $mt \leq Nt = [\frac{2\delta}{t}]t \leq 2\delta$, it follows from (2.7) that $\|V(t)^m\| \leq M(2\delta)$, and

$$\|V(t)^n\| = \|V(t)^{kN+m}\| \leq \|V(t)^m\|\|V(t)^N\|^k \leq M(2\delta)e^{Nkwt} \leq M(2\delta)e^{nwt}.$$

$$\square$$

In addition to the Crank-Nicholson scheme (Corollary 2.2) and other higher order rational approximation schemes, there are many other approximation schemes $\{V(t), t \in [0, \delta]\}$ of linear operators which are D-consistent, but do not necessarily satisfy the Lax-Chernoff stability condition $\|V(t)^n\| \leq Me^{wnt}$ for all $t \in [0, \delta]$ and $n \in \mathbb{N}_0$ in general Banach spaces. Typical examples are the alternating direction implicit method (ADI-method) and the Lie-Trotter product formula. The ADI-method

$$V_{adi}(t) := (I - \frac{t}{2}A_1)^{-1}(I + \frac{t}{2}A_2)(I - \frac{t}{2}A_2)^{-1}(I + \frac{t}{2}A_1)$$

was proposed by Peaceman and Rachford [PR55] for two dimensional parabolic problems, where A is split into directional components A_1, A_2; i.e., $A = \overline{A_1 + A_2}$ with $D(A) \subset D(A_i)$. It is an approximate factorization of the Crank-Nicolson method $V_{CN}(t) = (I + \frac{t}{2}A)(I - \frac{t}{2}A)^{-1}$, which in turn is an approximate factorization of the backwards Euler method $V_{BS}(t) = (I - tA)^{-1}$. Since the operators $V_{adi}(t)$ may not be bounded if the operators A_1, A_2 do not commute, the ADI-method can in general not be expected to be stable in the sense of the Lax-Chernoff Equivalence Theorem. Nevertheless, in numerical applications (see, e.g., [Zh00a,Zh00b]), the ADI-method exhibits excellent numerical stability and retains the order of convergence of the Crank-Nicolson scheme. The advantage of the ADI-method over the Crank-Nicolson scheme is that it reduces the computation of the 2D-operator $(I - tA)^{-1}$ into that of two 1D-operators $(I - \frac{t}{2}A_1)^{-1}$ and $(I - \frac{t}{2}A_2)^{-1}$, resulting in great computation reduction especially when A is not self-adjoint. Other generally unstable approximation methods for semigroups are the Lie-Trotter product formula

$$V_{lt} := T_1(t)T_2(t)$$

or the Strang scheme

$$V_{fs} := T_1(\frac{t}{2})T_2(t)T_1(\frac{t}{2})$$

where $A = \overline{A_1 + A_2}$ is the generator of a strongly continuous semigroup $T(\cdot)$, and A_1, A_2 are generating strongly continuous semigroups $T_1(\cdot), T_2(\cdot)$. The Lie-Trotter product formula is consistent of order 1 on $D(A^2)$ and the factorized scheme is consistent of order 2 on $D(A^3)$. Kühnemund and Wacker [KW00] showed that there are strongly continuous semigroups generators A_1, A_2 such that $A = \overline{A_1 + A_2}$ generates a strongly continuous semigroup and V_{lt} is unstable.

To discuss such consistent but in general unbounded and/or unstable approximation methods we employ the stabilization and regularization methods described in the following section.

3   Regularization and Stabilization

To demonstrate the basic idea of regularization, let us assume that we are given a $D(A)$-consistent approximation scheme of linear (not necessarily bounded) operators such that

$$\|V(\frac{t}{n})^n f - T(t)f\| \le g(n)\|Af\| \tag{3.1}$$

for all $f \in D(A)$ and $t \in [0, T]$, where $g(n) \to 0$ as $n \to \infty$ and A is the generator of the strongly continuous semigroup $T(\cdot)$ with $s(A) < 0$. We also assume that $t \mapsto V(t)^n f$ is continuous on $[0, \delta]$ for $f \in D(A)$ and $n \in \mathbb{N}$ and observe that (3.1) implies that $V(t)^n \in \mathcal{L}([D(A)], X)$ where $[D(A)]$ denotes the Banach space $D(A)$ with the graph norm. Obviously, (3.1) implies that

$$\|V(\frac{t}{n})^n A^{-1}f - T(t)A^{-1}f\| \le g(n)\|f\|$$

for all $f \in X$. Since $T(t)f - f = A \int_0^t T(s)f\,ds$ for all $f \in X$, it follows that

$$\|V(\frac{t}{n})^n A^{-1}f - A^{-1}f - \int_0^t T(s)f\,ds\| \le g(n)\|f\| \tag{3.2}$$

for all $f \in X$. Define[6] $V_{reg}^n(t)f := \frac{d}{dt}[V(\frac{t}{n})^n A^{-1}f - A^{-1}f]$ $(f \in X)$ and, for $v \in C([0,T], X)$,

$$|||v||| := \|1 * v\|_\infty := \sup_{t \in [0,T]} \| \int_0^t u(s)ds\|.$$

Then the regularized scheme $\{V_{reg}^n(t), t \in [0,\delta]\}$ consists of bounded linear operators and

$$|||V_{reg}^n(\cdot)f - T(\cdot)f||| \le g(n)\|f\| \tag{3.3}$$

for all $f \in X$. Notice that if the original scheme $V(\cdot)$ is not stable, then (3.1) does not imply that $\|V(\frac{t}{n})^n f - T(t)f\| \to 0$ as $n \to \infty$ for all $f \in X$. Thus, in the absence of stability, (3.2) and (3.3) are the best conclusions for the convergence of $V(\frac{t}{n})^n f$ when either $f \in X \backslash D(A)$, or when only $\|f\|$ but not $\|Af\|$ can be estimated properly. Clearly, (3.2) and (3.3) allow estimates on the time averages of $T(\cdot)$; i.e., let $0 \le a < b \le T$ and define

$$V_{av}^n[a,b]f := \frac{1}{b-a}[V(\frac{b}{n})^n A^{-1}f - V(\frac{a}{n})^n A^{-1}f].$$

Then (3.1) implies that

$$\|V_{av}^n[a,b]f - \frac{1}{b-a} \int_a^b T(s)f ds\| \le \frac{g(n)}{b-a}\|f\| \tag{3.4}$$

for all $f \in X$.

Since the convolution operator $C : u \mapsto 1 * u$ is an injective operator on $C([0,T], X)$ with dense range in $C_0([0,T], X)$, it extends to an isometric isomorphism between $R([0,T], X)$, the completion of $C([0,T], X)$ under the norm $|||\cdot|||$, and $C_0([0,T], X)$. Thus, the Banach space $R([0,T], X)$ coincides with the generalized (distributional) derivatives of functions in $C_0([0,T]; X)$. This space contains the Bochner integrable functions as well as the Riemann (or Henstock) integrable functions[7] ; if $f \in R([0,T], X)$ and $f_n \in C([0,T], X)$ with $f_n \to f$ (with respect to the norm $||| \cdot |||$), then $\int_a^b f(t)dt := \lim_{n \to \infty} \int_a^b f_n(t)dt$ is well-defined. Moreover, the continuous functions f_n are Cauchy with respect to $||| \cdot |||$ if and only if for all $\epsilon > 0$ and $\mu > 0$ there exists n_0 such that $|\frac{1}{b-a} \int_a^b f_m(s) - f_n(s)ds| \le \epsilon$ for all $m, n \ge n_0$ and all $0 \le a < b \le T$ with $b - a \le \mu$. Thus, we call $R([0,T], X)$ the Banach space of (generalized) Riemann integrable functions with the topology of uniform convergence in the average (see also [BLN98]).

The following generalization of the Chernoff product formula contains an application of the concept mentioned above (see statement (ii)).

---

[6] In many cases, $t \to V(\frac{t}{n})^n A^{-1}f - A^{-1}f$ is continuously differentiable; e.g., if $V = V_{be}$, then $V_{reg}^n(t)f = V_{be}(\frac{t}{n})^{n+1}$. In any case, $t \to V(\frac{t}{n})^n A^{-1}f - A^{-1}f \in C_0([0,T]; X)$ is differentiable in the sense of generalized functions (distributions), see [BLN98] or the remarks on the space $R([0,T]; X)$ below.

[7] Notice, if $\dim(X) = \infty$, then Riemann integrability does not imply measurability.

Theorem 3.1. *Let A be the generator of a strongly continuous semigroup $T(\cdot)$ with $s(A) < 0$ and let $\{V(t), t \in [0, \delta]\}$ be a strongly continuous[8] approximation scheme of linear operators with $D(A) \subset D(V(t)^\infty)$ for $t \in [0, \delta]$. If there exist $M, w \geq 0$ such that*

(a) $\|V(t)^{n+1}f - V(t)^n f\| \leq Mte^{wnt}\|Af\|$ *($n \in \mathbb{N}_0, t \in [0, \delta], f \in D(A)$), and*
(b) there exists a family of linear operators $W(t)$ defined on $D(A^2)$ such that the map $t \to W(t)f$ is continuous on $f \in D(A^2)$, $W(0)f = 0$, and
$$\|V(t)^{n+1}f - V(t)^n f - tV(t)^n Af\| \leq te^{wnt}\|W(t)f\| \text{ for all } n \in \mathbb{N}_0, t \in [0, \delta],$$

then

(i) for each $T > 0, f, f_n \in D(A)$ with $\|Af_n - Af\| \to 0$ we have that $V(\frac{t}{n})^n f_n \to T(\cdot)f$ in $C([0, T], X)$
(ii) for each $T > 0, f, f_n \in X$ with $\|f_n - f\| \to 0$ we have that $V_{reg}^n(\cdot)f_n \to T(\cdot)f$ in $R([0, T], X)$.

For a proof we refer to [Zh00a] and [Zh00b] where it is also shown that the ADI-method
$$V_{adi}(t) := (I - \frac{t}{2}A_1)^{-1}(I + \frac{t}{2}A_2)(I - \frac{t}{2}A_2)^{-1}(I + \frac{t}{2}A_1)$$

satisfies the assumptions of Theorem 3.1 if X is a Hilbert space $A = A_1 + A_2$ with $D(A) \subset D(A_1) \cap D(A_2)$ generates a C_0-semigroup $T(\cdot)$, and A_1, A_2 generate quasi-contractive semigroups.

Proposition 3.1. *Let A be the generator of a strongly continuous semigroup $T(\cdot)$ on a Banach space X with $s(A) < 0$. Let $\{V(t), t \in [0, \delta]\}$ be an approximation scheme such that $D(A^k) \subset D(V(t)^\infty)$ for some $k \geq 1$, $t \mapsto V(t)^n f \in C([0, T], X)$ and $V(0)f = f$ for all $f \in D(A^k)$ and $n \geq 1$. If*

$$\|V(\frac{t}{n})^n f - T(t)f\| \leq g(n)\|A^k f\|$$

for all $f \in D(A^k)$ and $t \in [0, T]$, then

$$\||V_{reg}^n(\cdot)f - T(\cdot)f\||_k \leq g(n)\|f\|,$$

where $V_{reg}^n(t) := \frac{d^n}{dt^n}[V(\frac{t}{n})^n A^{-k}f - \sum_{i=0}^{k-1} \frac{t^i}{i!} A^{-(k-i)}f]$ and, for $v \in C([0, T], X)$, $\||v\||_k := \sup_{t \in [0,T]} \| \int_0^t \frac{(t-s)^{k-1}}{(k-1)!} v(s)ds\|$.

Proof. Using (2.5) one proceeds as in the proof of the implication (3.1) $\to$ (3.3). $\square$

An immediate consequence of Proposition 3.1 and Corollary 2.2 *(ii)* is the following

---

[8] That is, $t \mapsto V(t)f \in C([0, \delta], X)$ for all $f \in D(A)$.

Corollary 3.1. *[Crank-Nicholson]. Let A be the generator of an analytic semigroup $T(\cdot)$ on a Banach space X with $s(A) < 0$. Then the regularized Crank-Nicholson scheme $V_{reg}^n(t) := \frac{d^2}{dt^2}[V_{cn}(\frac{t}{n})^n A^{-2}f - A^{-2}f - tA^{-1}f] = (I + \frac{t}{2n^2}A)(I - \frac{t}{2n}A)^{-4}V_{cn}(\frac{t}{n})^{n-2}$ satisfies*

$$|||V_{reg}^n(\cdot)f - T(\cdot)f|||_1 \le M\frac{T^2}{n^2}\|f\|,$$

where $|||v|||_1 := \sup_{t\in[0,T]}\|\int_0^t(t-s)v(s)ds\|$.

We end this section with the following remark. It can be seen from statement (iv) of Corollary 2.2 that the backwards Euler scheme stabilizes the Crank-Nicholson scheme if A generates an analytic semigroup $T(\cdot)$ on a Banach space X with $s(A) < 0$; i.e.,

$$\|V_{be}(\frac{t}{n})^2 V_{cn}(\frac{t}{n})^{n-2}f - T(t)f\| \le M\frac{1}{n^2}\|f\| \tag{3.5}$$

for all $t \ge 0, 2 \le n \in \mathbb{N}$, and $f \in X$. In particular, $\|V_{be}(\frac{t}{n})^2 V_{cn}(\frac{t}{n})^{n-2}\| \le M'$ for some $M' > 0$ and all $t \ge 0, 2 \le n \in \mathbb{N}$. An abstract version of the convergence statement in (3.5) is the following stabilized version of the Chernoff product formula. The proof of Theorem 3.2 can be found in [FNZ00], where we will also discuss the stabilization and regularization of the Crank-Nicholson and other higher order rational approximation schemes for generators of arbitrary strongly continuous semigroups, distribution and integral semigroups, and other types of regularized semigroups like those discussed in [LN99].

Theorem 3.2. *Let A be the generator of a strongly continuous semigroup $T(\cdot)$ on a Banach space X. Let $\{V(t), t \in [0,\delta]\}$ and $\{W(t), t \in [0,\delta]\}$ be two strongly continuous families of bounded linear operators on X satisfying $V(0) = W(0) = I$, $\lim_{t\to 0}\frac{V(t)f-f}{t} = Af$ for all $f \in D(A^\infty)$, and*

$$\|W(t)^j V(t)^{n-j}\| \le Me^{\omega tn}$$

for all $t \in [0,\delta]$ and $n \ge j$. Then, for all $f \in X$,

$$W(\frac{t}{n})^j V(\frac{t}{n})^{n-j}f \to T(t)f$$

uniformly for t in compact intervals.

References

[ABHN00]   W. Arendt, C. Batty, M. Hieber and F. Neubrander, *Vector-Valued Laplace Transforms and Cauchy Problems*. Monographs in Mathematics, Birkhäuser Verlag, to appear.

[BLN98]   B. Bäumer, G. Lumer and F. Neubrander, Convolution kernels and gener-
          alized functions, Generalized Functions, Operator Theory, and Dynamical
          Systems, *C.R.C. Res. Notes Math.*, Chapman & Hall, 1998

[BW77]   P. Butzer and R. Weis, On the Lax equivalence theorem equipped with or-
          ders, *J. Approx. Theory* 19 (1977), 239-252.

[Ch74]   P. R. Chernoff, *Product Formulas, Nonlinear Semigroups, and Addition of
          Unbounded Operators*, Memoirs of the American Math. Soc. 140, American
          Mathematical Society, Providence, Rhode Island, 1974.

[CFL28]   R. Courant, K. Friedrichs and H. Lewy, Über die partiellen Differenzialgle-
          ichungen der mathematischen Physik, *Math. Ann.* 100 (1928), 32-74.

[EN99]   K.-J. Engel and R. Nagel, *One-Parameter Semigroups for Linear Evolution
          Equations*, Graduate Texts in Mathematics 194, Springer, 1999.

[FNZ00]   S. Flory, F. Neubrander and Y. Zhuang, On the regularization and stabiliza-
          tion of rational approximation methods for semigroups in Banach spaces, in
          preparation.

[Ha99]   A. Hansbo, Nonsmooth data error estimates for damped single step methods
          for parabolic equations in Banach spaces, *Calcolo* 36 (1999), 75-101.

[IT00]   T. Ichinose and H. Tamura, On the norm convergence of the Trotter-Kato
          product formula, preprint.

[KW00]   F. Kühnemund, M. Wacker, The Lie-Trotter product formula does not hold
          for arbitrary sums of generators, *Semigroup Forum* 60 (2000), 478-485.

[LN99]   G. Lumer and F. Neubrander, Asymptotic Laplace transforms and evolution
          equations, *Advances in Partial Differential Equations*, Vol. 16, Wiley-VCH
          (1999).

[LR56]   P. Lax and R. Richtmyer, Survey of stability of linear finite difference equa-
          tions, *Comm. Pure Appl. Math.* 9 (1956), 267-293.

[LTW91]   S. Larsson, V. Thomée and L. B. Wahlbin, Finite-element methods for a
          strongly damped wave equation, *IMA J. Numer. Anal.* II (1991), 115-142.

[NZ99]   H. Neidhardt and V. Zagrebenov, Trotter-Kato product formula and
          operator-norm convergence, *Comm. Math. Physics* 205 (1999), 129-159.

[Pa92]   M. H. Padé, Sur représentation approchée d'une fonction par des frac-
          tionelles, *Ann. de l'Ecole Normale Superieure* 9 (1892), p.16.

[PR55]   D. W. Peaceman and J. H. H. Rachford, The numerical solution of parabolic
          and elliptic differential equations, *J. Soc. Industrial and Applied Math.* 3
          (1955), 28-41.

[Ri57]   R. D. Richtmyer, *Difference Methods for Initial-Value Problems*, Interscience
          Publishers, 1957.

[Tw84]   E. H. Twizell, *Computational Methods for Partial Differential Equations*,
          Ellis Horwood Limited, Chichester, 1984.

[Zh00a]   Y. Zhuang, *Classically Unstable Approximations for Linear Evolution Equa-
          tions and Applications*, Dissertation, Department of Mathematics, Louisiana
          State University, Baton Rouge, 2000.

[Zh00b]   Y. Zhuang, Toward understanding the von Neumann stability condition,
          preprint.

Addresses

SIMONE FLORY, Department of Mathematics,
Louisiana State University, Baton Rouge, LA 70803, USA

E-MAIL: flory@math.lsu.edu

FRANK NEUBRANDER, Department of Mathematics,
Louisiana State University, Baton Rouge, LA 70803, USA

E-MAIL: neubrand@math.lsu.edu

YU ZHUANG, Department of Computer Science,
Louisiana State University, Baton Rouge, LA 70803, USA

E-MAIL: zhuang@math.lsu.edu

2000 Mathematics Subject Classification. Primary 47D06, 65M06 ; Secondary 34K07, 65M12

Operator Theory:
Advances and Applications, Vol. 126
© 2001 Birkhäuser Verlag Basel/Switzerland

Pseudodifferential Operators with Symbols in Weighted Function Spaces of Quasi-Subadditive Type

GIANLUCA GARELLO

Abstract. A continuity result is stated, also from the microlocal point of view, for pseudodifferential operators with symbol in a suitable class of weighted function spaces, defined by means of quasi- subadditive weight functions. Applications to semilinear partial differential equations are also given.

1 Introduction

In order to improve the study of solvability and regularity of linear partial differential equations, many generalizations of the pseudodifferential calculus appeared in the by now long history of the microlocal analysis. Let us recall here the main contributions of Lascar [L77], Beals [B75] and finally the Weyl-Hörmander pseudodifferential calculus [H79], within the related Sobolev spaces introduced by Bony-Chemin [BC94].

Let now $F(x, \{\partial^\alpha u\}_{|\alpha| \leq m}) = 0$ be a nonlinear partial differential equation, where $F(x, \zeta)$ is a smooth function, which in the rest of the paper shall be assumed analytic in the ζ variable, for suitable reasons.

It is very direct to argue about the equation:

$$\sum_{|\alpha| \leq m} \frac{\partial F}{\partial \zeta^\alpha}(x, \{\partial^\beta u\}_{|\beta| \leq m}) \partial^\alpha \partial_{x_j} u = -\frac{\partial F}{\partial x_j}(x, \{\partial^\beta u\}_{|\beta| \leq m}) \tag{1.1}$$

which is the result of differentiating $F(x, \{\partial^\alpha u\}_{|\alpha| \leq m}) = 0$ with respect to x_j.

Two new problems immediately arise. At first we hope that the coefficients $a_\alpha(x) = \frac{\partial F}{\partial \zeta^\alpha}(x, \partial^\beta u)$ belong to some useful Sobolev spaces. Secondly we need to a theory of partial differential operators, better of pseudodifferential operators whose symbols have only such a limited regularity.

As an attempt to give some answers, in what follows we will consider symbols $a(\cdot, \xi)$ in the distribution spaces $\mathcal{B}_{p,\lambda}$ of Hörmander type [H63], with a suitable quasi-subadditive condition on the weight function $\lambda(\xi)$. Then we state the continuity of the pseudodifferential operators $a(x, D)$ between suitable function spaces of type $\mathcal{B}_{p,\lambda}$, also from the microlocal point of view. Direct applications to semilinear equations are given in the conclusion.

2 Symbols with microlocal regularity in weighted spaces of quasi-subadditive type.

Any positive map $\lambda(\xi)$ on $\mathbf{R}^n$ with polynomial growth:

$$\frac{1}{C}\left(1+|\xi|\right)^{\mu_0} < \lambda(\xi) < C\left(1+|\xi|\right)^{\mu_1}, \quad C, \mu_0, \mu_1 \text{ positive}, \quad \xi \in \mathbf{R}^n \qquad (2.1)$$

is for us a *weight function*. Moreover $\lambda(\xi)$ is *quasi-subadditive* if for some positive constant C:

$$\lambda(\eta) < C\left(\lambda(\eta - \xi) + \lambda(\xi)\right), \qquad \eta, \xi \in \mathbf{R}^n; \qquad (2.2)$$
$$\lambda(t\xi) < C\lambda(\xi), \qquad\qquad |t| \leq 1, \quad \xi \in \mathbf{R}^n. \qquad (2.3)$$

Let $\hat{u}(\xi)$ be the Fourier transform of the Schwartz distribution $u \in \mathcal{S}'(\mathbf{R}^n)$

Definition 2.1. For any $1 \leq p \leq \infty$ and $\lambda(\xi)$ quasi-subadditive weight function we define the distribution space $\mathcal{B}_{p,\lambda}$ as the class of all $u \in \mathcal{S}'(\mathbf{R}^n)$ such that

$$\lambda(\xi)\hat{u}(\xi) \in L^p,$$

equipped with the natural norm: $\|u\|_{p,\lambda} = (2\pi)^{-\frac{n}{p}}\|\lambda(\xi)\hat{u}(\xi)\|_{L^p}$.

Using (2.2) we obtain rightly from (2.1), for some positive constants C and N, the *temperance* condition:

$$\lambda(\eta) < C(1+|\eta - \xi|)^N \lambda(\xi), \qquad \xi, \eta \in \mathbf{R}^n. \qquad (2.4)$$

Then by referring to Hörmander [H63] we can assure that, for $1 \leq p \leq \infty$, $\mathcal{B}_{p\lambda}$ is a Banach space and for $p \neq \infty$ contains C_0^∞ as dense subspace. Moreover $\mathcal{S}(\mathbf{R}^n) \hookrightarrow \mathcal{B}_{p,\lambda} \hookrightarrow \mathcal{S}'(\mathbf{R}^n)$ and $\mathcal{B}_{p,\lambda_1} \hookrightarrow \mathcal{B}_{p,\lambda_2}$, when $\lambda_2(\xi) \leq C\lambda_1(\xi)$, $C > 0$.
We say that $u \in \mathcal{D}'(\mathbf{R}^n)$ belongs to the local space $\mathcal{B}_{p,\lambda}^{\mathrm{loc}}(x_0)$, $x_0 \in \mathbf{R}^n$, if for some cut-off function $\phi \in C_0^\infty(\mathbf{R}^n)$, such that $\phi(x_0) \neq 0$, we have $\phi u \in \mathcal{B}_{p,\lambda}$.
In order to microlocalize the above arguments, instead of the usual conic neighborhood, let us consider for $X \subset \mathbf{R}_\xi^n$ and $\varepsilon > 0$ the following:

$$X_{\varepsilon,\lambda} = \{\xi \in \mathbf{R}^n \, ; \, \lambda(\xi - \xi_0) < \varepsilon\lambda(\xi), \quad \text{for some } \xi_0 \in X\}. \qquad (2.5)$$

For some $1 \leq p \leq \infty$, $x_0 \in \mathbf{R}_x^n$ and $X \subset \mathbf{R}_\xi^n$, u belongs to the microlocal space $\mathcal{B}_{p,\lambda}^{\mathrm{mcl}}(x_0 \times X)$ if for some cut-off function ϕ such that $\phi(x_0) \neq 0$:

$$\chi_{\varepsilon,\lambda}(\xi)\lambda(\xi)\widehat{\phi u}(\xi) \in L^p, \qquad (2.6)$$

where $\chi_{\varepsilon,\lambda}$ is the characteristic function of $X_{\varepsilon,\lambda}$.
With the only assumption that the symbol $a(x,\xi) \in \mathcal{S}'(\mathbf{R}^{2n})$, let us recall that the pseudodifferential operator defined by

$$a(x,D)f = (2\pi)^{-n}\int e^{ix\xi}a(x,\xi)\hat{f}(\xi)\,d\xi, \qquad (2.7)$$

maps continuously $\mathcal{S}(\mathbf{R}^n)$ to $\mathcal{S}'(\mathbf{R}^n)$.

Definition 2.2. Let $\lambda(\xi)$, $\Lambda(\xi)$, $\gamma(\xi)$ be weight functions, $\lambda(\xi)$ and $\Lambda(\xi)$ temperate. For $x_0 \in \mathbf{R}^n$, $X \subset \mathbf{R}^n$, $1 \leq p \leq \infty$, we say that the symbol $a(x,\xi)$ belongs to $\mathcal{B}_{p,\lambda,\Lambda}S^\gamma(x_0 \times X)$ if for any $\xi \in \mathbf{R}^n$, $a(\cdot,\xi) \in \mathcal{B}_{p,\lambda}^{\mathrm{loc}}(x_0) \cap \mathcal{B}_{p,\Lambda}^{\mathrm{mcl}}(x_0 \times X)$ and moreover for some $\phi(x)$ cut-off function, $\phi(x_0) \neq 0$:

$$\sup_\xi \left\| \frac{\lambda(\eta)\widehat{\phi a}(\eta,\xi)}{\gamma(\xi)} \right\|_{L^p_{(\eta)}} < C_\lambda; \qquad \sup_\xi \left\| \frac{\Lambda(\eta)\chi_{\varepsilon\Lambda}(\eta)\widehat{\phi a}(\eta,\xi)}{\gamma(\xi)} \right\|_{L^p_{(\tau)}} < C_\Lambda \qquad (2.8)$$

For $1 \leq p \leq \infty$ in the following q will be the conjugate index: $\frac{1}{p} + \frac{1}{q} = 1$.
The next one is a straightforward generalization of Lemma 1.4 in [BR84], for the details of the proof see [G00].

Lemma 2.1. *For $1 \leq p \leq \infty$ let us consider $\Gamma(\xi) \in L^q$ and a function $g(\eta,\xi)$, measurable in the first variable, which satisfies for some positive constant C_g:*
$\sup_{\xi \in \mathbf{R}^n} \|g(\eta,\xi)\|_{L^p_{(\eta)}} < C_g$.
Then the linear operator defined by: $Tv(\eta) = \int \Gamma(\xi)g(\eta - \xi,\xi)v(\xi)\,d\xi$ is bounded from L^p to itself, with operator norm $\|T\|_{\mathcal{L}(L^p)}$ bounded by the L^p norm of $\Gamma(\xi)$ and the constant C_g.

Theorem 2.1. *For $1 \leq p \leq \infty$, let us consider $x_0 \in \mathbf{R}^n$, $X \subset \mathbf{R}^n$, $\sigma(\xi)$, $\gamma(\xi)$ weight functions, $\lambda(\xi)$, $\Lambda(\xi)$ quasi-subadditive weight functions, which satisfy*

$$\sigma(\xi)^{-1} \in L^q \quad \text{and} \quad \sigma(\xi) \leq \lambda(\xi) \leq \Lambda(\xi) \leq \frac{\lambda(\xi)^2}{\sigma(\xi)}. \qquad (2.9)$$

Then for $a(x,\xi) \in \mathcal{B}_{p,\lambda,\Lambda}S^\gamma(x_0 \times X)$ the following map is continuous:

$$a(x,D) : \mathcal{B}_{p,\lambda\gamma}^{\mathrm{loc}}(x_0) \cap \mathcal{B}_{p,\Lambda\gamma}^{\mathrm{mcl}}(x_0 \times X) \to \mathcal{B}_{p,\lambda}^{\mathrm{loc}}(x_0) \cap \mathcal{B}_{p,\Lambda}^{\mathrm{mcl}}(x_0 \times X). \qquad (2.10)$$

Proof. Without loss of generality let us assume that $a(x,\xi)$ is compactly supported with respect to x. For $u \in \mathcal{B}_{p,\lambda\gamma}^{\mathrm{loc}}(x_0)$ set now :

$$g(\eta,\xi) = \frac{\lambda(\eta-\xi)}{\gamma(\xi)}|\hat{a}(\eta-\xi,\xi)| \in L^p_{(\eta)}, \quad v(\xi) = \gamma(\xi)\sigma(\xi)|\hat{u}(\xi)| \in L^p; \qquad (2.11)$$

$$\tilde{g}(\eta,\xi) = \frac{\sigma(\eta-\xi)}{\gamma(\xi)}|\hat{a}(\eta-\xi,\xi)| \in L^p_{(\eta)}, \quad \tilde{v}(\xi) = \gamma(\xi)\lambda(\xi)|\hat{u}(\xi)| \in L^p. \qquad (2.12)$$

The $L^p_{(\eta)}$ norms of g and $\tilde{g}$ are uniformly bounded in ξ. In view of (2.2) we have:

$$|\lambda(\eta)(a(x,D)u)\hat{}(\eta)| \leq (2\pi)^{-n} \int \lambda(\eta)|\hat{a}(\eta-\xi,\xi)||\hat{u}(\xi)|d\xi \leq$$

$$\leq C(2\tau)^{-n}\left(\int \frac{1}{\sigma(\xi)}g(\eta,\xi)v(\xi)\,d\xi + \int \frac{1}{\sigma(\eta-\xi)}\tilde{g}(\eta,\xi)\tilde{v}(\xi)\,d\xi \right). \qquad (2.13)$$

Then rightly from Lemma 2.1 we obtain $a(x, D)u \in \mathcal{B}_{p,\lambda}^{\mathrm{loc}}(x_0)$.

Let us now take $X \subset \mathbf{R}^n$ and $\varepsilon > 0$ such that $\gamma(\zeta)\Lambda(\zeta)\chi_{\varepsilon,\Lambda}(\zeta)\hat{u}(\zeta) \in L^p$ and $\frac{\Lambda(\eta)\chi_{\varepsilon,\Lambda}(\eta)\hat{a}(\eta,\xi)}{\gamma(\xi)} \in L^p_{(\eta)}$ and observe that there exists a positive constant $\varepsilon' < \varepsilon$ such that $(X_{\varepsilon'\Lambda})_{\varepsilon'\Lambda} \subset X_{\varepsilon\Lambda}$. Thus it is enough to show that:

$$\chi_{\varepsilon'\Lambda}(\eta)\Lambda(\eta)(a(x,D)\hat{u})(\eta) \in L^p. \tag{2.14}$$

Let us set $\chi(\zeta) = \chi_{\varepsilon'\Lambda}(\zeta)$, $\chi_1(\zeta) = \chi_{\varepsilon\Lambda}(\zeta)$, $\chi_2(\zeta) = \chi_{\mathbf{R}^n\setminus X_{\varepsilon\Lambda}}(\zeta)$ and write: $\hat{a}(\zeta,\xi)\hat{u}(\xi) = \sum_{i,j=1}^2 \chi_i(\zeta)\hat{a}(\zeta,\xi)\chi_j(\xi)\hat{u}(\xi)$. Then using also (2.2) we have:

$$|\chi(\eta)\Lambda(\eta)\widehat{au}(\eta)| \leq C(2\pi)^{-n}\int \chi(\eta)\left(\Lambda(\eta-\xi)+\Lambda(\xi)\right)|\hat{a}(\eta-\xi,\xi)||\hat{u}(\xi)|\,d\xi \leq$$

$$\leq C(2\pi)^{-n}\int \chi(\eta)\sum_{i,j=1}^2 \chi_i(\eta-\xi)\Lambda(\eta-\xi)|\hat{a}(\eta-\xi,\xi)|\chi_j(\xi)|\hat{u}(\xi)|\,d\xi + \tag{2.15}$$

$$+C(2\pi)^{-n}\int \chi(\eta)\sum_{i,j=1}^2 \chi_i(\eta-\xi)|\hat{a}(\eta-\xi,\xi)|\chi_j(\xi)\Lambda(\xi)|\hat{u}(\xi)|\,d\xi \tag{2.16}$$

Set now:

$$g_1(\eta,\xi) = \chi_1(\eta-\xi)\Lambda(\eta-\xi)\gamma(\xi)^{-1}|\hat{a}(\eta-\xi,\xi)|;$$
$$g_2(\eta,\xi) = \chi_2(\eta-\xi)\sigma(\eta-\xi)\Lambda(\eta-\xi)\gamma(\xi)^{-1}\Lambda(\xi)^{-1}|\hat{a}(\eta-\xi,\xi)|;$$
$$\tilde{g}_2(\eta,\xi) = \chi_2(\eta-\xi)\sigma(\eta-\xi)^{1/2}\Lambda(\eta-\xi)^{1/2}\gamma(\xi)^{-1}|\hat{a}(\eta-\xi,\xi)|;$$
$$v_1(\xi) = \chi_1(\xi)\gamma(\xi)\sigma(\xi)|\hat{u}(\xi)|, \quad \tilde{v}_1(\xi) = \chi_1(\xi)\gamma(\xi)\Lambda(\xi)|\hat{u}(\xi)|;$$
$$v_2(\xi) = \chi_2(\xi)\gamma(\xi)\sigma(\xi)|\hat{u}(\xi)|, \quad \tilde{v}_2(\xi) = \chi_2(\xi)\sigma(\xi)^{1/2}\Lambda(\eta-\xi)^{1/2}\gamma(\xi)|\hat{u}(\xi)|.$$

Then the integral in (2.15) may be written:

$$\int \chi(\eta)\frac{1}{\sigma(\xi)}g_1(\eta,\xi)v_1(\xi)\,d\xi + \int \chi(\eta)\frac{1}{\sigma(\xi)}g_1(\eta,\xi)v_2(\xi)\,d\xi + \tag{2.17}$$

$$+ \int \chi(\eta)\frac{1}{\sigma(\eta-\xi)}g_2(\eta,\xi)\tilde{v}_1(\xi)\,d\xi + \int \chi(\eta)\frac{1}{\sqrt{\sigma(\xi)\sigma(\eta-\xi)}}\tilde{g}_2(\eta-\xi)\tilde{v}_2(\xi)\,d\xi.$$

Arguing as in the first part of the proof we obtain $g_1(\eta,\xi) \in L^p_{(\eta)}$, uniformly with respect to ξ. It is easy to see that $v_1(\xi)$, $\tilde{v}_1(\xi)$, $v_2(\xi)$ realize to be in L^p. Since in both the integrals in the second line of (2.17) $\eta \in X_{\varepsilon'\Lambda}$ and $\eta - \xi \in \mathbf{R}^n \setminus X_{\varepsilon\Lambda}$, it immediately follows that $\Lambda(\eta-\xi) \leq (1/\varepsilon')\Lambda(\xi)$, then $g_2(\eta,\xi) \in L^p_{(\eta)}$ uniformly with respect to ξ. Using also (2.9) we can show that $\tilde{g}_2(\eta,\xi) \in L^p_{(\eta)}$ and $\tilde{v}_2(\eta) \in L^p$. Since $1/\sigma(\xi) \in L^q$ the integral in (2.15) realizes to be in $L^p_{(\eta)}$.

Symmetrically we can manage the arguments in (2.16), and the proof is thus concluded. $\square$

For the continuity properties stated in Theorem 2.1 the symbol $a(x,\xi)$ is required to satisfy only the weak assumptions in (2.8), though if we wish something more, e.g. a suitable symbolic calculus, estimates on the derivatives up to a right order with respect to the ξ variable would be required. For simplicity we can then consider $a(x,\xi)$ smooth with respect to the frequency ξ.

With the settings in (2.11), (2.12), from the integral (2.13) we easily obtain for $a(x,D)u$ the following estimate of Moser type:

$$\|a(x,D)u\|_{p,\lambda} < C \left\| \frac{a(x,\xi)}{\gamma(\xi)} \right\|_{p,\lambda} \|u\|_{p,\sigma} + \left\| \frac{a(x,\xi)}{\gamma(\xi)} \right\|_{p,\sigma} \|u\|_{p,\lambda}. \qquad (2.18)$$

Thus the algebra properties of the function space $\mathcal{B}_{p,\lambda}$, for $\lambda \geq \sigma$, are a trivial corollary of Theorem 2.1, see also [G96].

Moreover in the particular case $p = 2$, $\lambda(\xi) = (1 + |\xi|^2)^{\frac{s}{2}}$, $\Lambda(\xi) = (1 + |\xi|^2)^{\frac{r}{2}}$, $\frac{n}{2} < s \leq r < 2s - \frac{n}{2}$, $a(x,\xi) = a(x)$ and X open conic subset of $\mathbf{R}_\xi^n$, we get the well known Rauch result about algebra properties of the classic Sobolev spaces H^s, both from the local and microlocal point of view, see [RA79].

Although, for $\sigma(\xi)^{-1} \in L^q$, $\mathcal{B}_{p,\sigma} \subset L^\infty$ it does not seem possible to replace the norms $\| \ \|_{p,\sigma}$ in (2.18) by the respective $\| \ \|_{L^\infty}$ norms.

The continuity from $\mathcal{B}^{\mathrm{loc}}_{p,\gamma\lambda}(x_0)$ to $\mathcal{B}^{\mathrm{loc}}_{p,\lambda}(x_0)$ of the pseudodifferential operators whose symbol satisfies the first estimate in (2.8) can be verified assuming only that $\lambda(\xi)$, $\gamma(\xi)$ are temperate weight functions and $\lambda(\xi)$ satisfies the q- Beurling condition: $\frac{\lambda(\eta)}{\lambda(\xi)\lambda(\eta-\xi)} \in L^q_{(\eta)}$ with the norm uniformly bounded in ξ, see [G00].

With some simple considerations on the Taylor expansions and the above arguments, any entire analytic function $F(z)$, $z \in \mathbf{C}$, which vanishes at $z = 0$, satisfies $\|F(u)\|_{p,\lambda} \leq C\|u\|_{p,\lambda}$, where the constant C depends only on F and $\|u\|_{p,\sigma}$. Moreover for any function $F(x,\zeta)$, smooth with respect to $x \in \mathbf{R}^n$, entire analytic with respect to $\zeta \in \mathbf{C}^M$ and such that $F(x,0) = 0$, we have $F(x,u_1,\ldots,u_M) \in \mathcal{B}^{\mathrm{loc}}_{p,\lambda}(x_0)$ when all the components u_j are in the same function space.

It is quite easy at the moment to give a direct application of the previous arguments to the microlocal properties of some classes of semilinear equations. Namely, for $\lambda(\xi)$ quasi-subadditive weight function, $x_0 \in \mathbf{R}^n$, $m \in \mathbf{R}$, let me consider a smooth symbol $p(x,\xi)$ in $S^m_\lambda(x_0)$, that is $|\partial_\xi^\alpha \partial_x^\beta \phi(x) p(x,\xi)| < C_{\alpha,\beta} \lambda(\xi)^{m-|\alpha|}$, for some $\phi \in C_0^\infty(\mathbf{R}^n)$, $\phi(x_0) \neq 0$ and any $\alpha, \beta \in \mathbf{Z}_+^n$.

Since $S^m_\lambda(x_0) \subset \bigcap_s \mathcal{B}_{p,\lambda^s} S^{\lambda^m}(x_0)$, the related pseudodifferential operator $p(x,D)$ is bounded from $\mathcal{B}^{\mathrm{loc}}_{p,\lambda^{t+m}}(x_0) \cap \mathcal{B}^{\mathrm{mcl}}_{p,\lambda^{T+m}}(x_0 \times X)$ to $\mathcal{B}^{\mathrm{loc}}_{p,\lambda^t}(x_0) \cap \mathcal{B}^{\mathrm{mcl}}_{p,\lambda^T}(x_0 \times X)$, for any $t, T \in \mathbf{R}$, $X \subset \mathbf{R}^n$.

Let us assume moreover that the linear operator $p(x,D)$ is microelliptic in $x_0 \times X$, that is $|p(x,\xi)| > c\lambda^m(\xi)$ in $\{x_0\} \times X$, $|\xi| > \frac{1}{c}$.

We can then use a suitable symbolic calculus, see Rodino [RO82], and state the following microlocal property:

$$p(x,D)u \in \mathcal{B}^{\mathrm{mcl}}_{p,\lambda^{s+m}}(x_0 \times X) \Rightarrow u \in \mathcal{B}^{\mathrm{mcl}}_{p,\lambda^s}(x_0 \times X). \qquad (2.19)$$

Let me consider now the semilinear equation:

$$p(x, D)u = F(x, \{D^\alpha u\}_{\alpha \in \mathcal{I}}); \tag{2.20}$$

where $F(x, \zeta)$ is smooth in the x variable and entire analytic in $\zeta \in \mathbf{C}^M$, $M = \sum_{\alpha \in \mathcal{I}} 1$. For some positive δ assume that $\xi^\alpha \in S_\lambda^{m-\delta}(x_0)$ for $\alpha \in \mathcal{I}$. Let us consider now $u \in \mathcal{B}_{p,\lambda^s}(x_0)$, $s \geq S + m - \delta$, solution of (2.20), $\lambda^{-S}(\xi) \in L^q$. We obtain $p(x, D)u = F(x, \partial^\alpha u) \in \mathcal{B}_{p,\lambda^{s-m+\delta}}(x_0)$. In view of (2.19) $u \in \mathcal{B}_{p,\lambda^{s+\delta}}^{\mathrm{mcl}}(x_0 \times X)$, thus $F(u, \partial^\alpha u) \in \mathcal{B}_{p,\lambda^{s-m+2\delta}}^{\mathrm{mcl}}(x_0 \times X)$. We obtain then by iteration that a solution u of (2.20) assumed a priori in $\mathcal{B}_{p,\lambda^s}^{\mathrm{loc}}(x_0)$, $s \geq m + S - \delta$ satisfies:

$$u \in \mathcal{B}_{p,\lambda^t}^{\mathrm{mcl}}(x_0 \times X), \qquad \text{for} \quad t \leq 2s - m - S + 2\delta.$$

References

[B75]   R. Beals, *A general calculus of pseudodifferential operators*, Duke Math. J. **42** (1975), 1-42.

[BR84]   M. Beals, M.C. Reeds, *Microlocal regularity theorems for nonsmooth pseudo-differential operators and applications to nonlinear problems*, Trans. Am. Math. Soc. **285** (1984), 159-184.

[BC94]   J. M. Bony, I. Chemin, *Espaces fonctionnels associés au calcul de Weyl-Hörmander*, Bull. Soc. Math. France **122** (1994), 77-118.

[G96]   G. Garello, *Inhomogeneous paramultiplication and microlocal singularities for semilinear equations*, Boll. Un. Mat. Italiana **10-B**(1996), 885-902.

[G00]   G. Garello, *Continuity for pseudodifferential operators with symbols in weighted function spaces*, to appear in Proceedings Workshop "Partial Differential Operators, edited by L. Rodino, Torino, May 2000.

[H63]   L. Hörmander, "Linear partial differential operators", Springer Verlag, Berlin-Heidelberg, 1963.

[H79]   L. Hörmander, *The Weyl calculus of pseudodifferential operators*, Comm. Pure Appl. Math. **vol. XXXII** (1979), 359-443.

[L77]   R. Lascar, *Propagation des singularités des solutions d'équations pseudodifferentielles quasi-homogènes*, Ann. Inst. Fourier, Grenoble **27** (1977), 79-123.

[RA79]   J. Rauch, *Singularities of solutions to semilinear wave equations*, J. Math. Pures et Appl. **58** (1979), 299-308.

[RO82]   L. Rodino, *Microlocal analysis for spatially inhomogeneous pseudodifferential operators*, Ann. Sc. Norm. Sup. Pisa Classe di Scienze **9/2** (1982), 221-253.

Address

GIANLUCA GARELLO, Dipartimento di Matematica Università di Torino, Via Carlo Alberto 10, 10123 Torino, Italy

E-MAIL: garello@dm.unito.it

2000 Mathematics Subject Classification. 35S05 ; 35H30

Operator Theory:
Advances and Applications, Vol. 126
© 2001 Birkhäuser Verlag Basel/Switzerland

Spectral Analysis of Quantum Field Models with a Particle Number Cutoff

VLADIMIR GEORGESCU

Abstract. We determine the essential spectrum and prove the absence of singular continuous spectrum for a class of boson quantum field models with cutoffs. Our main purpose, however, is to show the power of an algebraic technique based on the study of the C^*-algebra generated by the operators which are natural candidates as hamiltonians of the system in a given physical situation.

1 Algebras of Creation-Annihilation Operators

Our aim is to study a system of at most $N < \infty$ bosons which interact between themselves and possibly also with some external system, creation and annihilation of particles being allowed. For this purpose I shall use a strategy developed in a joint work with A. Iftimovici, see his contribution to these proceedings and [GeI00]. Theorems 1.1 and 1.2 are the result of a collaboration with M. Mantoiu [GeM]. I stress that my purpose here is not to get optimal results or to improve known results but rather to point out a new method of analysing the spectral properties of systems with a nontrivial algebraic structure. Although not considered here, the case $N = \infty$ can be treated by similar techniques.

Let $\mathcal{H}$ be the (complex, infinite dimensional) Hilbert space of one boson states. We denote $\mathcal{H}^{\vee n}$ the symmetric tensor power of order n of $\mathcal{H}$ and $\Gamma(\mathcal{H}) = \bigoplus_{n=0}^{\infty} \mathcal{H}^{\vee n}$ the Fock space associated to $\mathcal{H}$. Then $\Gamma_N(\mathcal{H}) = \bigoplus_{n=0}^{N} \mathcal{H}^{\vee n}$ is the state space of the system consisting of at most N bosons. The interaction between bosons is a (generalized) polynomial in $a(u)$, $a^*(u)$, the annihilation and creation operators of a boson in the state $u \in \mathcal{H}$ (the normalizations and other conventions are as in [DeG98]). Since we are interested only in situations when the number of bosons does not exceed N, we introduce the truncated annihilation and creation operators $a_N(u) = 1_N a(u) 1_N$, $a_N^*(u) = 1_N a^*(u) 1_N$, where 1_N is the orthogonal projection $\Gamma(\mathcal{H}) \to \Gamma_N(\mathcal{H})$. The set $K(\Gamma_N(\mathcal{H}))$ of compact operators on $\Gamma_N(\mathcal{H})$ will be denoted $\mathcal{K}_N$. We state a preliminary result.

Theorem 1.1. *Let $\mathcal{A}_N$ be the unital C^*-algebra of operators on $\Gamma_N(\mathcal{H})$ generated by the operators $a_N(u)$ when u runs over $\mathcal{H}$. There is a unique unital morphism $\mathcal{A}_N \to \mathcal{A}_{N-1}$ such that the image of $a_N(u)$ is $a_{N-1}(u)$ for all $u \in \mathcal{H}$. This morphism is surjective and its kernel is equal to $\mathcal{K}_N$.*

In other terms, $\mathcal{K}_N$ is a (closed, bilateral) ideal in $\mathcal{A}_N$ and we have a canonical identification of the quotient C^*-algebra: $\mathcal{A}_N/\mathcal{K}_N \cong \mathcal{A}_{N-1}$. We shall denote by $T \mapsto \widehat{T}$ the quotient map, thus $\widehat{\mathcal{A}_N} = \mathcal{A}_{N-1}$.

The computation of $\widehat{T}$ is important in applications to spectral analysis of the preceding result and we give some examples now. Let $\mathcal{A}$ be the unital $*$-algebra of (unbounded) operators on $\Gamma(\mathcal{H})$ (with domain $\bigcup_{k=0}^{\infty} \Gamma_k(\mathcal{H})$) generated by the $a(u)$. Clearly $1_N \mathcal{A} 1_N$ is a linear subspace of $\mathcal{A}_N$ (it is not dense and it is not a subalgebra) and in fact it coincides with the linear subspace of $\mathcal{A}_N$ generated by the Wick products

$$W_N = \prod_i a_N^*(u_i) \prod_j a_N(v_j).$$

One also has

$$W_N = 1_N \prod_i a^*(u_i) \prod_j a(v_j) 1_N \equiv 1_N W 1_N.$$

For Wick products it is very easy to compute the quotient:

$$[W_N]^{\wedge} = W_{N-1} \quad \text{or} \quad [1_N W 1_N]^{\wedge} = 1_{N-1} W 1_{N-1}.$$

As a second example, let p_k be the orthogonal projection onto $\mathcal{H}^{\vee k}$. So $1_N = \sum_{k=0}^{N} p_k$ and the particle number operator can be decomposed as $\mathbf{N}_N = \sum_{k=0}^{N} k p_k$. Then $p_k \in \mathcal{A}_N$ and we have $\widehat{p}_0 = 0$ and $\widehat{p}_k = p_{k-1}$ if $1 \le k \le N$. Hence $\mathbf{N}_N$ belongs to $\mathcal{A}_N$ and its quotient is $\widehat{\mathbf{N}_N} = \mathbf{N}_{N-1} + 1_{N-1}$. According to the terminology of [Ger96] for example, $\mathbf{N}_N$ is the number of alive bosons and so $\mathbf{D}_N = N 1_N - \mathbf{N}_N$ is the number of dead bosons. Since $\widehat{1_N} = 1_{N-1}$ we see that $\widehat{\mathbf{D}_N} = \mathbf{D}_{N-1}$.

The class of hamiltonians which can be analysed by our methods with the help of Theorem 1.1 does not contain those which appear in the usual quantum field models. For this we introduce now a more general type of algebras. First a notation: if $S \in B(\mathcal{H})$ then $S^{\vee n} = S^{\otimes n}|\mathcal{H}^{\vee n}$ and

$$\Gamma_N(S) = \oplus_{n=0}^{N} S^{\vee n} \in B(\Gamma_N(\mathcal{H})).$$

Now if $\mathcal{C} \subset B(\mathcal{H})$ is a unital C^*-subalgebra, then $\mathcal{C}_N$ is the C^*-algebra of operators on $\Gamma_N(\mathcal{H})$ generated by the operators of the form $\Gamma_N(S)$ and $a_N(u)$ with $S \in \mathcal{C}$ and $u \in \mathcal{H}$. So $\mathcal{C}_N$ is the smallest C^*-algebra which contains $\Gamma_N(\mathcal{C}) \bigcup \mathcal{A}_N$. A rather natural notation for $\mathcal{C}_N$ is $\mathcal{A}_N \rtimes \mathcal{C}$ (to suggest crossed-product).

Theorem 1.2. *Assume that $\mathcal{C} \subset B(\mathcal{H})$ is a unital C^*-subalgebra with no nonzero finite rank projections. Then there is a unique morphism $\mathcal{P}_N : \mathcal{C}_N \to \mathcal{C} \otimes \mathcal{C}_{N-1}$ such that for all $u \in \mathcal{H}$ and $S \in \mathcal{C}$*

$$\mathcal{P}_N[a_N(u)] = 1 \otimes a_{N-1}(u)$$
$$\mathcal{P}_N[\Gamma_N(S)] = S \otimes \Gamma_{N-1}(S).$$

One has $\ker \mathcal{P}_N = \mathcal{K}_N$.

Thus we have a canonical embedding

$$\mathcal{C}_N / \mathcal{K}_N \equiv \widehat{\mathcal{C}_N} \hookrightarrow \mathcal{C} \otimes \mathcal{C}_{N-1}.$$

As before we denote by $T \mapsto \widehat{T}$ the canonical morphism of $\mathcal{C}_N$ onto $\widehat{\mathcal{C}_N}$ which will be identified with a C^*-subalgebra of $\mathcal{C} \otimes \mathcal{C}_{N-1}$. Note that this time the inclusion $\widehat{\mathcal{C}_N} \subset \mathcal{C} \otimes \mathcal{C}_{N-1}$ is strict.

2 Some General Facts

We summarize several definitions and results from [ABG96] and [DaG99]. A self-adjoint operator H on a Hilbert space $\mathscr{H}$ is called *affiliated* to a C^*-algebra $\mathscr{C}$ of operators on $\mathscr{H}$ if $(H - z)^{-1} \in \mathscr{C}$ for some $z \in \mathbb{C} \backslash \sigma(H)$. This implies $\varphi(H) \in \mathscr{C}$ for all $\varphi \in C_0(\mathbb{R})$, so one can associate to H a morphism $\varphi \mapsto \varphi(H)$ of the C^*-algebra $C_0(\mathbb{R})$ into $\mathscr{C}$ (by "morphism" between two C^*-algebras we understand $*$-morphism and $C_0(\mathbb{R})$ is the space of continuous complex valued functions on $\mathbb{R}$ that converge to zero at infinity). Such a morphism is called *observable affiliated to* $\mathscr{C}$. This notion makes sense even if $\mathscr{C}$ is an abstract C^*-algebra (i.e. is not realized on a Hilbert space) and has the advantage that one can speak of the image of an observable through a morphism $\mathcal{P} : \mathscr{C} \to \mathscr{D}$ (this is just the composition of the two morphisms). Note that if $\mathscr{C} \subset B(\mathscr{H})$ then there are observables affiliated to $\mathscr{C}$ which do not correspond to self-adjoint operators on $\mathscr{H}$ (if one requires, as usual, that such an operator is densely defined). However, if the observable is *strictly affiliated* to $\mathscr{C}$, then it is represented by a (densely defined) self-adjoint operator in each non-degenerate representation of $\mathscr{C}$ (see [DaG99]). We also refer to [DaG99] for a quite powerful and easy to use criterion which assures that an operator of the form $H = H_0 + V$ is (strictly) affiliated to an algebra if one knows that H_0 has the corresponding property (all the situations considered below and in [GeI00], [GeM] are covered by this result).

Assume now that a C^*-algebra $\mathscr{C}$ of operators on a Hilbert space $\mathscr{H}$ is given and that $\mathscr{C}$ contains the set of compact operators $K(\mathscr{H})$. From the physical point of view, this is the C^*-algebra generated by the operators which are natural candidates as hamiltonians for some system in a given physical situation; we call it C^*-*algebra of energy observables* of the system. Then $K(\mathscr{H})$ is an ideal (closed, bilateral) in $\mathscr{C}$ and one can construct the quotient C^*-algebra $\widehat{\mathscr{C}} = \mathscr{C}/K(\mathscr{H})$. If H is a self-adjoint operator on $\mathscr{H}$ affiliated to $\mathscr{C}$ then one can consider its image $\widehat{H}$ through the canonical morphism $\mathscr{C} \to \widehat{\mathscr{C}}$. Then $\widehat{H}$ is an observable affiliated to $\widehat{\mathscr{C}}$ and it is easy to show that the essential spectrum of H is equal to the spectrum of $\widehat{H}$ (the spectrum of an observable being defined in a natural way). Now assume that H is strictly affiliated to $\mathscr{C}$ and that a faithful non-degenerate realization of $\widehat{\mathscr{C}}$ on some Hilbert space $\widehat{\mathscr{H}}$ is given. Then the observable $\widehat{H}$ is realized as a self-adjoint operator (which we denote also by $\widehat{H}$) on $\widehat{\mathscr{H}}$ and clearly we have $\sigma_{ess}(H) = \sigma(\widehat{H})$. This is the method that we use for the computation of the essential spectrum of the hamiltonians considered in [GeI00], [GeM].

The absence of the singular continuous spectrum is proved with the help of the Mourre method in the version developed in [ABG96]. The main difficulty is to prove the Mourre estimate, so I shall indicate here the main steps. I think that one gets a particularly elegant presentation by working at a purely algebraic level (as in [ABG96]), but I shall avoid this in order not to introduce too many abstract objects.

Let $\mathscr{C}$ be the same C^*-algebra as above and A a self-adjoint operator on $\mathscr{H}$ such that $e^{-itA}\mathscr{C}e^{itA} = \mathscr{C}$ for each real t. Assume also that the map $t \mapsto e^{-itA}Se^{itA}$ is norm continuous for each $S \in \mathscr{C}$. Since $e^{-itA}K(\mathscr{H})e^{itA} = K(\mathscr{H})$, there is a norm continuous one-parameter group of automorphisms α_t of $\widehat{\mathscr{C}}$ such that $\left[e^{-itA}Se^{itA}\right]^{\wedge} = \alpha_t(\widehat{S})$ for all t and $S \in \mathscr{C}$. Finally, assume that the group α_t is unitarily implemented in the representation on $\widehat{\mathscr{H}}$, more explicitly: there is a self-adjoint operator $\widehat{A}$ on $\widehat{\mathscr{H}}$ such that, for all t and $S \in \mathscr{C}$,

$$\left[e^{-itA}Se^{itA}\right]^{\wedge} = e^{-it\widehat{A}}\,\widehat{S}\,e^{it\widehat{A}}. \tag{2.1}$$

We have to recall now some general facts concerning the Mourre estimate in the version of [ABG96]. A self-adjoint operator H is of class $C_u^1(A)$ if the map $t \mapsto e^{-itA}(H+i)^{-1}e^{itA}$ is norm C^1. We define two lower semi-continuous functions $\rho_H^A, \widehat{\rho}_H^A : \mathbb{R} \to (-\infty, \infty]$ by the following rules: $\widehat{\rho}_H^A(\lambda)$ is the upper bound of the numbers a for which there are a real function $\varphi \in C_c(\mathbb{R})$ with $\varphi(\lambda) \neq 0$ and a compact operator K such that

$$\varphi(H)[H, iA]\varphi(H) \geq a\varphi(H)^2 + K$$

(one can rigorously give a meaning to the sesquilinear form from the l.h.s. above as a bounded symmetric operator). In other terms, $\widehat{\rho}_H^A(\lambda)$ is the best constant in the Mourre estimate. Similarly, $\rho_H^A(\lambda)$ is the upper bound of the numbers a such that the preceding inequality holds for some φ and $K = 0$. The set $\tau_A(H)$ of real λ where $\widehat{\rho}_H^A(\lambda) \leq 0$ is closed and will be called the set of A-*thresholds* of H. The (closed) set $\kappa_A(H)$ of A-*critical points* of H is given by the condition $\rho_H^A(\lambda) \leq 0$.

An important property of the functions ρ_H^A and $\widehat{\rho}_H^A$ is the following. Say that $\lambda \in \mathbb{R}$ is an M-*eigenvalue* of H if it is an eigenvalue and $\widehat{\rho}_H^A(\lambda) > 0$. By the virial theorem, these eigenvalues are of finite multiplicity and are not accumulation points of eigenvalues of H. Theorem 7.2.13 from [ABG96] says that $\rho_H^A(\lambda) = 0$ if λ is a M-eigenvalue of H and otherwise $\rho_H^A(\lambda) = \widehat{\rho}_H^A(\lambda)$. Thus the functions $\rho_H^A, \widehat{\rho}_H^A$ differ only on a small set: the (discrete) set of M-eigenvalues. Also, $\rho_H^A(\lambda) > 0$ if and only if $\widehat{\rho}_H(\lambda) > 0$ and $\lambda \notin \sigma_p(H)$, or $\kappa_A(H) = \tau_A(H) \cup \sigma_p(H)$.

We go back now to the setting described before and consider a self-adjoint operator H on $\mathscr{H}$ strictly affiliated to $\mathscr{C}$ and of class $C^1_u(A)$. It is clear then that $\widehat{H}$ is of class $C^1_u(\widehat{A})$ and a straightforward computation gives $\widehat{\rho}^A_H = \rho^{\widehat{A}}_{\widehat{H}}$.

For a complete determination of the threshold set in the cases of interest for us here we need a last formula from [ABG96]. Let $\mathscr{H} = \mathscr{H}_1 \otimes \mathscr{H}_2$ and H_i, A_i self-adjoint operators in $\mathscr{H}_i$ such that H_i is bounded from below and of class $C^1_u(A_i)$. Consider the self-adjoint operators $H = H_1 \otimes 1 + 1 \otimes H_2$ and $A = A_1 \otimes 1 + 1 \otimes A_2$ on $\mathscr{H}$. Then H is of class $C^1_u(A)$ and

$$\rho^A_H(\lambda) = \inf_{\lambda=\lambda_1+\lambda_2} \left[\rho^{A_1}_{H_1}(\lambda_1) + \rho^{A_2}_{H_2}(\lambda_2) \right]. \tag{2.2}$$

3   Spectral Analysis of Hamiltonians Affiliated to $\mathcal{C}_N$

We apply now the results described in Section 2 to the self-adjoint operators affiliated to the C^*-algebra $\mathcal{C}_N$ constructed in Section 1. In order to make clear the connection with the quantum field models studied in the literature, we consider in detail the case of a spin zero boson with $\mathbb{R}^s$ as configuration space (however, it should be clear that the next arguments are of an abstract nature and so apply to much more general situations). Then the one boson Hilbert space is $\mathcal{H} = L^2(\mathbb{R}^s)$ and we choose $\mathcal{C} = \{\psi(P) \mid \psi \in C_0(\mathbb{R}^s)\}$ where $P = -i\nabla$ is the momentum observable. Let $\omega : \mathbb{R}^s \to \mathbb{R}$, the one boson kinetic energy, be a continuous divergent function. The field operators on $\Gamma(\mathcal{H})$ are denoted by $\varphi(x)$ and $\varphi_\chi(x)$ is a regularized version of the field given by an ultraviolet cutoff χ. Finally, let $g : \mathbb{R}^s \to \mathbb{R}$ be a spatial cutoff. Then the total hamiltonian of the boson field has the form

$$H = d\Gamma(\omega(P)) + \int_{\mathbb{R}^s} g(x) : \mathrm{Pol}[\varphi_\chi(x)] : \, dx \equiv d\Gamma(\omega(P)) + W$$

where Pol is a real polynomial and : : denotes Wick product. We also recall that if S is a self-adjoint bounded from below operator on $\mathcal{H}$ then the self-adjoint operator $d\Gamma(S)$ on $\Gamma(\mathcal{H})$ is defined by $\exp(-d\Gamma(S)) = \Gamma(\exp(-S))$ (for details on this formalism, see the papers of J. Derezinski and C. Gérard quoted in the references). From now on, in order to simplify the notations, we set $\omega = \omega(P)$.

We shall now add a particle number cutoff, so we take $\mathscr{H} = \Gamma_N(\mathcal{H})$ and $\mathscr{C} = \mathcal{C}_N$ in Section 2 and consider the operator $H_N = 1_N H 1_N$. Set $d\Gamma_N(\omega) = 1_N d\Gamma(\omega) 1_N$ and $W_N = 1_N W 1_N$, so $H_N = d\Gamma_N(\omega) + W_N$. Then, under quite general conditions on g and χ, the operator H_N is self-adjoint on $\Gamma_N(\mathcal{H})$ and strictly affiliated to $\mathcal{C}_N$. Indeed, $d\Gamma_N(\omega)$ is always strictly affiliated to $\mathcal{C}_N$ and W_N is in fact a bounded operator in $\mathcal{A}_N$; one gets $(H_N + \lambda)^{-1} \in \mathcal{C}_N$ for large positive λ by making a

Neumann expansion. An operator obtained as norm resolvent limit of operators of the form H_N is still affiliated to $\mathcal{C}_N$; this allows one to remove the ultraviolet cutoff if $s = 1$.

We have now to compute $\widehat{H_N}$. Note first that Theorem 1.2 gives an embedding $\widehat{\mathcal{C}_N} \subset \mathcal{C} \otimes \mathcal{C}_{N-1}$ hence a faithful and nondegenerate realization of $\widehat{\mathcal{C}_N}$ on the Hilbert space $\mathcal{H} \equiv \Gamma_N(\mathcal{H})^\wedge = \mathcal{H} \otimes \Gamma_{N-1}(\mathcal{H})$. This is the Hilbert space on which we shall represent $\widehat{H_N}$. From Theorem 1.2 and the examples given after Theorem 1.1 we obtain $\mathcal{P}_N(W_N) = W_{N-1}$. Then we compute the quotient of $d\Gamma_N(\omega)$ by noticing that it is uniquely defined by the property

$$\exp(-\mathcal{P}_N(d\Gamma_N(\omega))) = \mathcal{P}_N(\exp(-d\Gamma_N(\omega)) = \mathcal{P}_N(\Gamma_N(\exp(-\omega))$$

and by using Theorem 1.2 once again. This gives

$$[d\Gamma_N(\omega)]^\wedge = \omega \otimes 1_{N-1} + 1 \otimes d\Gamma_{N-1}(\omega).$$

Finally, by making a Neumann expansion of $(H_N + \lambda)^{-1}$ and observing that $\widehat{H_N}$ is characterized by the relation $\mathcal{P}_N[(H_N + \lambda)^{-1}] = (\widehat{H_N} + \lambda)^{-1}$, we obtain

$$\widehat{H_N} = \omega \otimes 1_{N-1} + 1 \otimes H_{N-1}. \tag{3.1}$$

As we explained in Section 2, this relation has important consequences. First, it allows us to compute the essential spectrum of H_N. We recall that $\sigma(S \otimes 1 + 1 \otimes T) = \sigma(S) + \sigma(T)$ if S, T are operators bounded from below. Let $m = \min \omega$ be the least value of the function ω; then $\sigma(\omega(P)) = [m, \infty)$ hence

$$\sigma_{ess}(H_N) = [m + \min \sigma(H_{N-1}), \infty). \tag{3.2}$$

Now we look for a conjugate operator in the sense of Mourre for the hamiltonian H_N. We start, as in [Ger96], with a self-adjoint operator $\mathfrak{a}$ on $\mathcal{H}$ such that $\omega(P)$ is of class $C_u^1(\mathfrak{a})$ and take $A \equiv A_N = d\Gamma_N(\mathfrak{a})$. Since $e^{itA_N} = \Gamma_N(e^{it\mathfrak{a}})$ we have $e^{-itA_N} a_N(u) e^{itA_N} = a_N(e^{-it\mathfrak{a}}u)$, so the algebra $\mathcal{A}_N$ is stable under the group of automorphisms associated to A_N. But $e^{-itA_N}\Gamma_N(S)e^{itA_N} = \Gamma_N(e^{-it\mathfrak{a}}Se^{it\mathfrak{a}})$, so we have to assume that $\mathcal{C}$ is stable under the automorphisms associated to $\mathfrak{a}$ in order that the condition $e^{-itA_N}\mathcal{C}_N e^{itA_N} = \mathcal{C}_N$ to be satisfied. For example, this holds if the function ω is of class C^1 and we take $\mathfrak{a} = F(P)Q + QF(P)$, where F is a vector field on $\mathbb{R}^s$ proportional to $\nabla\omega$ (cf. (7.6.15) and Lemma 7.6.4 from [ABG96]).

It is easy to deduce from Theorem 1.2 that in the present situation the relation (2.1) is satisfied with $\widehat{A_N} = \mathfrak{a} \otimes 1_{N-1} + 1 \otimes A_{N-1}$. We stress that the hat on A_N should not be interpreted in the sense of Section 1 (because this operator is not affiliated to the algebra $\mathcal{C}_N$). Now one has to put some conditions on the cutoffs g and χ which assure that W_N is of class $C_u^1(A_N)$; this is quite easy, see again the quoted papers of Derezinski and Gérard (our purpose here is not to get optimal results but rather to point out a strategy of proof). Assuming this, we see that H_N is of the same class, so we may associate to it the functions $\rho_N = \rho_{H_N}^{A_N}$ and $\widehat{\rho}_N = \widehat{\rho}_{H_N}^{A_N}$. As was explained in Section 2, $\widehat{\rho}_N$ coincides with the ρ function associated to the couple $\widehat{A_N, H_N}$. Let $\rho = \rho_\omega^{\mathfrak{a}}$; then (2.2) gives the recursion relation

$$\widehat{\rho}_N(\lambda) = \inf_{\lambda=\lambda_1+\lambda_2} [\rho(\lambda_1) + \rho_{N-1}(\lambda_2)]. \tag{3.3}$$

This has to be used in conjunction with the relation: $\rho_{N-1}(\lambda) = \widehat{\rho}_{N-1}(\lambda)$ if λ is not an M-eigenvalue of H_{N-1}; $\rho_{N-1}(\lambda) = 0 < \widehat{\rho}_{N-1}(\lambda)$ otherwise.

We shall compute now the threshold sets by induction on N under the assumption $\rho \geq 0$. Note that (3.3) and the last remark imply $0 \leq \rho_N \leq \widehat{\rho}_N$ for all N. The functions ρ and ρ_{N-1} are equal to ∞ below the spectra of the operators ω and H_{N-1} respectively and the function $\mu \mapsto \rho(\lambda - \mu) + \rho_{N-1}(\mu)$ is lower semi-continuous, hence it attains its lower bound on each compact set. Thus the infimum in (3.3) is attained. In particular. $\widehat{\rho}_N(\lambda) = 0$ if and only if one can write $\lambda = \lambda_1 - \lambda_2$ with $\rho(\lambda_1) = 0$ and $\rho_{N-1}(\lambda_2) = 0$. Equivalently $\tau_{A_N}(H_N) = \kappa_{\mathfrak{a}}(\omega) + \kappa_{A_{N-1}}(H_{N-1})$, or

$$\tau_{A_N}(H_N) = \kappa_{\mathfrak{a}}(\omega) + \tau_{A_{N-1}}(H_{N-1}) \cup \sigma_{\mathrm{p}}(H_{N-1}). \tag{3.4}$$

Since $\tau_{A_0}(H_0) = \emptyset$ and $\sigma_{\mathrm{p}}(H_0) = \{0\}$ we get successively $\tau_{A_1}(H_1) = \kappa_{\mathfrak{a}}(\omega)$, $\tau_{A_2}(H_2) = \kappa_{\mathfrak{a}}(\omega) + \kappa_{\mathfrak{a}}(\omega) \cup \sigma_{\mathrm{p}}(H_1)$ and so on.

The simplest situation is that when one can choose $\mathfrak{a}$ such that $\rho(\lambda) > 0$ if $\lambda \neq 0$ and $\rho(0) = 0$ (this happens in the usual models, when the function ω has only one critical point hence only one critical value). Then $\tau_{A_N}(H_N) = \bigcup_{n=0}^{N-1} \sigma_{\mathrm{p}}(H_n)$. In particular, $\tau(H_N) \equiv \tau_{A_N}(H_N)$ is a closed countable set and the eigenvalues of H_N which do not belong to $\tau(H_N)$ are M-eigenvalues, so of finite multiplicity and with no accumulation points outside $\tau(H_N)$. If H_N is of a slightly stronger regularity class with respect to A_N, then H_N will not have singularly continuous spectrum.

We indicate now the procedure which has to be used when the function ω has many critical values. Take $\mathfrak{a} = F(P)Q + QF(P)$ with $F(p) = \frac{1}{2}\theta(\omega(p)) |\nabla\omega(p)|^2$, where $\theta \geq 0$ is a C^∞ function, $\theta(\lambda) = 1$ if $\lambda < s$ and $\theta(\lambda) = 0$ if $\lambda > s+1$ for some large positive s. Then $\rho(\lambda) = \theta(\lambda) \min_{\omega(p)=\lambda} |\nabla\omega(p)|^2$. We see that $\rho \geq 0$ and on $(-\infty, s]$ one has $\rho(\lambda) > 0$ except when λ is a critical value of ω. We define the *threshold set* of H_N recursively by

$$\tau(H_N) = \kappa(\omega) + \tau(H_{N-1}) \cup \sigma_{\mathrm{p}}(H_{N-1}), \tag{3.5}$$

where $\kappa(\omega)$ is the set of critical values of the function ω. Note that this is a closed set depending only on the hamiltonian. Outside it H_N admits locally conjugate operators, so $\sigma_{\mathrm{p}}(H_N) \setminus \tau(H_N)$ consists of M-eigenvalues. If $\kappa(\omega)$ is countable and H_N satisfies a certain regularity condition, then H_N has no singularly continuous spectrum. Concerning ω, it suffices to assume that it is of polynomial growth with derivatives bounded by $C(1 + |\omega|)$.

4   Coupling of Two Systems

A system of bosons is often considered in interaction with an external system, like in the spin-boson case or the more general Pauli-Fierz models (see [Ger96] and [DeG99]. We state briefly an algebraic result which allows one to study such a situation.

Assume that two physical systems are given with state spaces $\mathcal{H}_1, \mathcal{H}_2$ and "algebras of energy observables" $\mathcal{C}_1 \subset B(\mathcal{H}_1), \mathcal{C}_2 \subset B(\mathcal{H}_2)$ such that $K(\mathcal{H}_i) \subset \mathcal{C}_i$, and that at least one of the algebras $\mathcal{C}_i$ is nuclear. We shall define the algebra of energy observables of the coupled system as being $\mathcal{C}_1 \otimes \mathcal{C}_2$. Since $K(\mathcal{H}_1) \otimes K(\mathcal{H}_2) = K(\mathcal{H}_1 \otimes \mathcal{H}_2)$ we are in a situation similar to the preceding ones: $K(\mathcal{H}_1 \otimes \mathcal{H}_2) \subset \mathcal{C}_1 \otimes \mathcal{C}_2 \subset B(\mathcal{H}_1 \otimes \mathcal{H}_2)$. We denote, as in Section 2, by a hat the quotient with respect to the ideal of compact operators. The main fact is that there is a canonical embedding

$$[\mathcal{C}_1 \otimes \mathcal{C}_2]^{\wedge} \subset \left[\widehat{\mathcal{C}_1} \otimes \mathcal{C}_2\right] \oplus \left[\mathcal{C}_1 \otimes \widehat{\mathcal{C}_2}\right].$$

For example, in order to treat a boson field coupled with a confined system (like in the spin-boson or Pauli-Fierz case) we take $\mathcal{C}_1 = \mathcal{C}_N$ and $\mathcal{C}_2 = K(\mathcal{H}_2)$. Then $\widehat{\mathcal{C}_2} = 0$ which makes things quite simple. One can also couple the field with a non-relativistic many-body system for which the algebra of energy observables is known (see [DaG99] for example) and the quotient is not trivial.

Acknowledgements I would like to thank Michael Demuth for financially supporting my participation at the conference. I am also indebted to the referee for correcting several errors in the first version of this article.

References

[ABG96]   Amrein, W., Boutet de Monvel, A., Georgescu, V., C_0-Groups, Commutator Methods and Spectral Theory of N-Body Hamiltonians, *Birkhauser Verlag*, 1996.

[DaG99]   Damak, M., Georgescu, V., C^*-Algebras Related to the N-Body Problem and the Self-adjoint Operators Affiliated to them, *available as preprint 99-482 at http://www.ma.utexas.edu/mp_arc*, 1999.

[DeG99]   Derezinski, J., Gérard, C., Asymptotic completeness in quantum field theory. Massive Pauli-Fierz Hamiltonians, *Rev. Math. Phys.* 11(4):383–450, 1999.

[DeG98]   Derezinski, J., Gérard, C., Spectral and scattering theory of spatially cut-off $P(\phi)_2$ hamiltonians, *available as preprint at http://math.polytechnique.fr/cmat/gerard/gerard.html*.

[GeI00]   Georgescu, V., Iftimovici, A., C^*-Algebras of Energy Observables: I. General theory and bumps algebras, *available as preprint 00-521 at http://www.ma.utexas.edu/mp_arc/*.

[GeM]   Georgescu, V., Mantoiu, M., C^*-Algebras of Hamiltonians on a Fock Space and Spectral Analysis of Quantum Field Models, *in preparation*.

[Ger96]   Gérard, C., Asymptotic completeness for the spin-boson model with a particle number cutoff, *Rev. Math. Phys.* 8(4):549–589, 1996.

Address

VLADIMIR GEORGESCU, CNRS and Département de Mathématiques de l'Université de Cergy-Pontoise, 2 Avenue Adolphe Chauvin, F 95302 Cergy-Pontoise Cedex, France

E-MAIL: Vladimir.Georgescu@math.u-cergy.fr

2000 Mathematics Subject Classification. Primary 81Q10, 81T10; Secondary 46L60, 47A10, 47D45

Operator Theory:
Advances and Applications, Vol. 126
© 2001 Birkhäuser Verlag Basel/Switzerland

On the Norm Convergence of the Trotter–Kato Product Formula with Error Bound

Takashi Ichinose and Hideo Tamura

Abstract. The norm convergence of the Trotter–Kato product formula with error bound is shown for the semigroup generated by that operator sum of two nonnegative selfadjoint operators A and B which is selfadjoint.

1 Introduction

It is well-known (e.g. [RS80]) that the Trotter–Kato product formula for the self-adjoint semigroup holds in strong operator topology. The aim of this note is to briefly announce our recent results on its operator-norm convergence with error bound. In [IT00] we have shown

Theorem 1.1. *If A and B are nonnegative selfadjoint operators in a Hilbert space $\mathcal{H}$ with domains $D[A]$ and $D[B]$ and if their sum $C := A + B$ is selfadjoint on $D[C] = D[A] \cap D[B]$, then the product formula in operator norm holds with error bound:*

$$\|(e^{-tB/2n}e^{-tA/n}e^{-tB/2n})^n - e^{-tC}\| = O(n^{-1/2}),$$
$$\|(e^{-tA/n}e^{-tB/n})^n - e^{-tC}\| = O(n^{-1/2}), \quad n - \infty. \qquad (1.1)$$

The convergence is uniform on each compact t–interval in $(0, \infty)$, and further, if C is strictly positive, uniform on $[T, \infty)$ for every fixed $T > 0$.

One of the typical examples of such a selfadjoint operator $C = A + B$ is the Schrödinger operator $H = -\Delta + P|x|^{-1} + D|x|^2 + E|x|^{2000}$ in $L^2(\mathbf{R}^3)$, where P, D and E are nonnegative constants.

Remark 1.1. The first result of such a norm convergence of the Trotter–Kato product formula (1.1) was proved by Rogava [R93] in the abstract case under an additional condition that B is A–bounded, with error bound $O(n^{-1/2} \log n)$. The next was by Helffer [H95] for the Schrödinger operators $H = H_0 + V \equiv -\frac{1}{2}\Delta + V(x)$ with C^∞ nonnegative potentials $V(x)$, roughly speaking, growing at most of order $O(|x|^2)$ for large $|x|$ with error bound $O(n^{-1})$. Each of these two results is independent of the other.

Then under some stronger or more general conditions, several further results are obtained. As for the abstract case, a better error bound $O(n^{-1}\log n)$ than Rogava's is obtained by Ichinose–Tamura [IT98b] when B is A^α–bounded for some $0 < \alpha < 1$, even though the $B = B(t)$ may be t–dependent, and by Neidhardt–Zagrebnov [NZ98], [NZ99a] when B is A–bounded with relative bound less than 1. As for the Schrödinger operators, more general results were proved for continuous nonnegative potentials $V(x)$, roughly speaking, growing of order $O(|x|^\rho)$ for large $|x|$ with $\rho > 0$, together with error bounds dependent on the power ρ (for instance, of order $O(n^{-2/\rho})$, if $\rho \geq 2$), by Ichinose–Takanobu [ITk97] (cf. [ITk98]), Doumeki–Ichinose–Tamura [DIT98], Ichinose–Tamura [IT98a], and others. It should be noted (see [G93], [S95/6]) that in all these cases of the Schrödinger operators the sum $H = H_0 + V$ is selfadjoint on the domain $D[H] = D[H_0] \cap D[V]$.

Thus the present theorem not only extends Rogava's result, but also can extend and contain all the results mentioned above, inclusive better error bounds in some cases.

Remark 1.2. Unless the sum $A + B$ is selfadjoint on $D[A] \cap D[B]$, the norm convergence of the Trotter–Kato product formula does not always hold, even though the sum is essentially selfadjoint there and B is A–form-bounded with relative bound less than 1. A counterexample is due to Hiroshi Tamura [Tm00].

The theorem also holds with the exponential function e^{-s} replaced by real-valued, Borel measurable functions f and g on $[0, \infty)$ satisfying that $0 \leq f(s) \leq 1$, $f(0) = 1$, $f'(0) = -1$, that for every small $\varepsilon > 0$ there exists a positive constant $\delta = \delta(\varepsilon) < 1$ such that $f(s) \leq 1 - \delta(\varepsilon)$ for $s \geq \varepsilon$, and that, for some fixed constant κ with $1 < \kappa \leq 2$, $[f]_\kappa := \sup_{s>0} s^{-\kappa}|f(s) - 1 + s| < \infty$, and the same for g. Of course, the functions $f(s) = e^{-s}$ and $f(s) = (1 + k^{-1}s)^{-k}$ with $k > 0$ are examples of functions having these properties.

Theorem 1.2. *If $3/2 \leq \kappa \leq 2$, it holds in operator norm that*

$$\|[g(tB/2n)f(tA/n)g(tB/2n)]^n - e^{-tC}\| = O(n^{-1/2}),$$
$$\|[f(tA/n)g(tB/n)]^n - e^{-tC}\| = O(n^{-1/2}), \quad n \to \infty. \qquad (1.2)$$

2 Outline of the Proof

To prove the theorem, it is crucial to establish the following operator-norm version of Chernoff's theorem with error bounds. For the proof we refer to [IT00].

Lemma 2.1. *Let C be a nonnegative selfadjoint operator in a Hilbert space $\mathcal{H}$ and let $\{F(t)\}_{t\geq 0}$ be a family of selfadjoint operators with $0 \leq F(t) \leq 1$. Define*

$S_t = t^{-1}(1 - F(t))$. *Then in the following two assertions, for $0 < \alpha \leq 1$, (a) implies (b):*

(a) $\qquad \|(1 + S_t)^{-1} - (1 + C)^{-1}\| = O(t^\alpha), \quad t \downarrow 0;$ $\qquad\qquad$ (2.1)

(b) $\qquad \|F(t/n)^n - e^{-tC}\| = \delta^{-2}t^{-1+\alpha}e^{\delta t}O(n^{-\alpha}), \quad n \to \infty,$ $\qquad$ (2.2)

for any $0 < \delta \leq 1$ and all $t > 0$.

Therefore, for $0 < \alpha < 1$ (resp. $\alpha = 1$), the convergence in (2.2) is uniform on each compact t–interval in $(0, \infty)$ (resp. $[0, \infty)$). Moreover, if C is strictly positive, i.e. $C \geq \eta$ for some constant $\eta > 0$, the error bound on the right-hand side of (2.2) can be replaced by $(1 + 2/\eta)^2 t^{-1+\alpha}O(n^{-\alpha})$, so that, for $0 < \alpha < 1$ (resp. $\alpha = 1$), the convergence in (2.2) is uniform on $[T, \infty)$ for every $T > 0$ (resp. on $[0, \infty)$).

Sketch of the Proof of Theorem 1.1

First note that since $C = A + B$ is itself a selfadjoint and so closed operator, by the closed graph theorem there exists a constant a such that $\|(1+A)u\| + \|(1+B)u\| \leq a\|(1 + C)u\|$ for all $u \in D[C] = D[A] \cap D[B]$.

The proof of the theorem is divided into two cases, (a) the symmetric product case $F(t) = e^{-tB/2}e^{-tA}e^{-tB/2}$ and (b) the non-symmetric product case $G(t) = e^{-tA}e^{-tB}$.

(a) In the symmetric case put

$$S_t = t^{-1}(1 - F(t)) = t^{-1}(1 - e^{-tB/2}e^{-tA}e^{-tB/2}).$$

By Lemma 2.1 we have only to show that

$$\|(1 + S_t)^{-1} - (1 + C)^{-1}\| = O(t^{1/2}), \quad t \downarrow 0.$$

Put $A_t = t^{-1}(1 - e^{-tA})$, $B_t = t^{-1}(1 - e^{-tB})$, $C_t = t^{-1}(1 - e^{-tC})$, and

$$K_t = 1 + A_t + B_{t/2} - 4^{-1}tB_{t/2}^2 \geq 1,$$

$$Q_t = 4^{-1}t^2 K_t^{-1/2}B_{t/2}A_t B_{t/2}K_t^{-1/2} - 2^{-1}tK_t^{-1/2}(A_t B_{t/2} + B_{t/2}A_t)K_t^{-1/2}.$$

We have

$$1 + S_t = 1 + A_t + B_{t/2} - 4^{-1}tB_{t/2}^2 + 4^{-1}t^2 B_{t/2}A_t B_{t/2} - 2^{-1}t(A_t B_{t/2} + B_{t/2}A_t)$$

$$= K_t^{1/2}(1 + Q_t)K_t^{1/2}.$$

Then we can show that $\|(1 + Q_t)^{-1}\| \leq 2/(3 - \sqrt{5})$, so that

$$\|(1 + S_t)^{-1}K_t^{1/2}\| = \|K_t^{-1/2}(1 + Q_t)^{-1}\| \leq 2/(3 - \sqrt{5}). \qquad (2.3)$$

Then we have

$$(1 + S_t)^{-1} - (1 + C)^{-1} = (1 + S_t)^{-1}(C - S_t)(1 + C)^{-1}$$
$$= (1 + S_t)^{-1}(A - A_t)(1 + C)^{-1} + (1 + S_t)^{-1}(B - B_{t/2})(1 + C)^{-1}$$
$$+ (1 + S_t)^{-1}[4^{-1}tB_{t/2}(1 - tA_t)B_{t/2} + 2^{-1}t(A_tB_{t/2} + B_{t/2}A_t)](1 + C)^{-1}$$
$$\equiv R_1(t) + R_2(t) + R_3(t). \tag{2.4}$$

We can show that, for some constant $c > 0$,

$$\|R_i(t)\| \le ct^{1/2}, \qquad i = 1, 2, 3. \tag{2.5}$$

For instance, we can get the bound for $R_1(t)$, via the expression

$$R_1(t) = [(1 + S_t)^{-1}K_t^{1/2}][K_t^{-1/2}(1 + A_t)^{1/2}]$$
$$\times [(1 + A_t)^{-1/2} - (1 + A_t)^{1/2}(1 + A)^{-1}](1 + A)(1 + C)^{-1},$$

by (2.3) and the spectral theorem

$$\|R_1(t)\| \le 2(3 - \sqrt{5})^{-1}a\|(1 + A_t)^{-1/2} - (1 + A_t)^{1/2}(1 + A)^{-1}\| \le ct^{1/2}.$$

(b) The non-symmetric case will follow from the symmetric case. We use the commutator argument to show that $\|G(t/n)^n - F(t/n)^n\| = O(1/n)$.

3   The Final Result

In a recent preprint [ITTZ00], we have shown that if $\kappa = 2$, then Theorem 1.2 holds with optimal error bound $O(n^{-1})$. However, the convergence is uniform on each compact t–interval in $[0, \infty)$, and, if C is strictly positive, on the whole closed half line $[0, \infty)$. The idea of proof is simply to iterate the resolvent equation of the first identity in (2.4) and to use the same arguments. Therefore it turns out that the product formula (1.1) in Theorem 1.1 holds, now with ultimate error bound $O(n^{-1})$, properly extending and containing all the known previous related results.

Acknowledgement. One of the authors (T.I.) wishes to thank Professor Michael Demuth for his kind invitation to this conference and hospitality in Clausthal.

References

[DIT98]   A. Doumeki, T. Ichinose and Hideo Tamura, Error bound on exponential product formulas for Schrödinger operators, *J. Math. Soc. Japan*, 50: 359–377, 1998.

[G93]   D. Guibourg, Inégalités maximales pour l'opérateur de Schrödinger. *C. R. Acad. Sci. Paris*, 316, Série I Math.: 249–252, 1993.

[H95]   B. Helffer, Around the transfer operator and the Trotter–Kato formula, *Operator Theory: Advances and Appl.*, 78: 161–174, 1995.

[ITk97] T. Ichinose and S. Takanobu, Estimate of the difference between the Kac operator and the Schrödinger semigroup, *Commun. Math. Phys.*, 186: 167–197, 1997.

[ITk98] T. Ichinose and S. Takanobu, The norm estimate of the difference between the Kac operator and the Schrödinger semigroup: A unified approach to the nonrelativistic and relativistic cases, *Nagoya Math. J.*, 149: 53–81, 1998.

[IT98a] T. Ichinose and Hideo Tamura, Error bound in trace norm for Trotter–Kato product formula of Gibbs semigroups, *Asymptotic Analysis*, 17: 239–266, 1998.

[IT98b] T. Ichinose and Hideo Tamura, Error estimates in operator norm of exponential product formulas for propagators of parabolic evolution equations, *Osaka J. Math.*, 35: 751–770, 1998.

[IT00]   T. Ichinose and Hideo Tamura, The norm convergence of the Trotter–Kato product formula with error bound, to appear in *Commun. Math. Phys.*

[ITTZ00] T. Ichinose, Hideo Tamura, Hiroshi Tamura and Valentin A. Zagrebnov, Note on the paper "The norm convergence of the Trotter–Kato product formula with error bound" by Ichinose and Tamura, to appear.

[NZ98]   H. Neidhardt and V. Zagrebnov, On error estimates for the Trotter–Kato product formula, *Lett. Math. Phys.*, 44: 169–186, 1998.

[NZ99a] H. Neidhardt and V. Zagrebnov, Fractional powers of selfadjoint operators and Trotter–Kato product formula, *Integr. Equat. Oper. Theory*, 35: 209–231, 1999.

[NZ99b] H. Neidhardt and V. Zagrebnov, Trotter–Kato product formula and operator-norm convergence, *Commun. Math. Phys.*, 205: 129–159, 1999.

[R93]   Dzh. L. Rogava, Error bounds for Trotter–type formulas for self-adjoint operators, *Functional Analysis and Its Applications*, 27: 217–219, 1993.

[RS80]   M. Reed and B. Simon, Methods of Modern Mathematical Physics I: Functional Analysis. *Academic Press*, New York, 1980.

[S95/6] Z. Shen, L^p estimates for Schrödinger operators with certain potentials, *Ann. Inst. Fourier, Grenoble*, 45: 513–546, 1995; Estimates in L^p for magnetic Schrödinger operators, *Indiana Univ. Math. J.*, 45: 817–841, 1996.

[Tm00]   Hiroshi Tamura, A remark on operator-norm convergence of Trotter–Kato product formula, *Integr. Equat. Oper. Theory*, 37: 350–356, 2000.

Addresses

TAKASHI ICHINOSE, Department of Mathematics, Faculty of Science, Kanazawa University, Kanazawa, 920–1192, Japan

E-MAIL: ichinose@kappa.s.kanazawa-u.ac.jp

HIDEO TAMURA, Department of Mathematics, Faculty of Science, Okayama University, Okayama, 700–8530, Japan

E-MAIL: tamura@math.okayama-u.ac.jp

2000 Mathematics Subject Classification. Primary 47D06, 47B25; Secondary 47D08, 81Q10

Operator Theory:
Advances and Applications, Vol. 126
© 2001 Birkhäuser Verlag Basel/Switzerland

Nonperturbative Techniques in the Investigation of the Spectral Properties of Many-Channel Systems

ANDREI IFTIMOVICI

Abstract. We show that the C^*-algebras generated by the "virtual" hamiltonians of a quantum system (with a qualitatively specified interaction) has in many cases an interesting and nontrivial structure. In particular, the quotient of this algebra with respect to the ideal of compact operators can be explicitly computed. This allows one to determine in a unified way the essential spectrum and to prove the Mourre estimate for large classes of hamiltonians, including: N-Body systems, stratified media, particles subject to Klaus type interactions (widely separated bumps) and other classes of (phase-space) anisotropic hamiltonians. The results presented here are based on a joint work with V. Georgescu.

1 Introduction. Examples of General Many-Channel Systems

We present a couple of quantum systems that have an interesting and rather complex many-channel structure. We will focus on studying the essential spectrum although the algebraic methods introduced in [GI00] fit quite well in the context of the Mourre theory and allow one to investigate refined properties of the spectrum of such systems.

In this section we describe some results that apparently have nothing to do with an algebraic context, and in the rest of the paper we show how they can be obtained by using a unifying C^*-algebraic formalism. We begin with a result which generalizes that of Klaus on "widely separated bumps" types of perturbations. Let $P = -i\nabla$ and Q be the momentum and position observables in $L^2(\mathbb{R}^n)$ and $\mathcal{H}^s = \mathcal{H}^s(\mathbb{R}^n)$ the Sobolev space for $s \in \mathbb{R}$. In a very rough version, we are placed in the following framework: consider a subset L of $\mathbb{R}^n$ which is *rarefied* (or *sparse*) in the sense that it is locally finite and $\lim_{x \in L, |x| \to \infty} \mathrm{dist}\,(x, L \setminus \{x\}) = \infty$. Let $h : \mathbb{R}^n \to \mathbb{R}$ be a continuous function such that $C^{-1}|x|^{2s} \leq |h(x)| \leq C|x|^{2s}$ if $|x| > R$, for some constants $s > 0$, $C > 0$ and $R < \infty$. Denote $H_0 = h(P)$. Let $\{V_l\}_{l \in L}$ be a family of symmetric operators in the space of bounded operators $B(\mathcal{H}^t, \mathcal{H}^{-t})$, for some $t \in [0, s)$, with the property that there is a number $a > 2n$ such that

$$\sup_{l \in L} \|\langle Q \rangle^a V_l\|_{B(\mathcal{H}^t, \mathcal{H}^{-t})} < \infty. \tag{1.1}$$

We give now the result in its first version:

Theorem 1.1. *Under the above hypotheses H_0 is a self-adjoint operator in $\mathscr{H}$ with $D(|H_0|^{1/2}) = \mathscr{H}^s$ and one has:*

(i) The series $\sum_{l \in L} \mathrm{e}^{-ilP} V_l \mathrm{e}^{ilP}$ converges in the strong topology of $B(\mathscr{H}^t, \mathscr{H}^{-t})$ toward a symmetric operator $V : \mathscr{H}^t \to \mathscr{H}^{-t}$.

(ii) For each $\varepsilon > 0$ there is a constant $c < \infty$ such that $\pm V \leq \varepsilon H_0 + c$. Thus the form sums $H = H_0 + V$, $H_l = H_0 + V_l$ are self-adjoint operators in $\mathscr{H}$ with the same form domain as H_0.

(iii) The essential spectrum of H is: $\sigma_{\mathrm{ess}}(H) = \bigcap_{\substack{F \subset L \\ F \text{finite}}} \overline{\bigcup_{l \in L \backslash F} \sigma(H_l)}$.

Remarks: (i) We consider in fact an arbitrary abelian locally compact group X instead of $\mathbb{R}^n$. **(ii)** If $s \leq n/2$ and $V_l : \mathbb{R}^n \to \mathbb{R}$ are Borel functions satisfying the condition $\int_{|y-x|<1} |V_l(y)| \cdot |y - x|^{-n+2s-\lambda} \, dy \leq c\langle x \rangle^{-a} \; \forall x \in \mathbb{R}^n$ for some constans $c, \lambda > 0$, then the operators V_l of multiplication by the functions V_l satisfy (1.1) for some $t < s$. If $s > n/2$ then the simpler condition $\int_{|y-x|<1} |V_l(y)| \, dy \leq c\langle x \rangle^{-a}$ suffices. If s is an integer and V_l is a differential operator of order less than $2s$, similar explicit conditions on its coefficients can be stated. **(iii)** The unnatural condition $a > 2n$ may actually be improved to $a > n$ when V_l are local operators. Anyway, the assumption $a > 2n$ may be relaxed in terms of the "degree of rarefaction" of L: the greatest lower bound for a is actually inversely proportional to it. **(iv)** There is no change in our formalism if we replace the Hilbert space $L^2(X)$ of physical states by $L^2(X; \mathbf{E})$ where $\mathbf{E}$ is a finite dimensional Hilbert space. This allows us to treat, for example, Dirac hamiltonians H_0 perturbed by the same class of potentials V.

We give now a second version of the result, involving limits at infinity along ultrafilters. If X is a locally compact space then $\widetilde{X}$ is the set of ultrafilters $\varkappa$ on X finer than the filter of neighbourhoods of infinity; $\widetilde{X}$ should be thought as the "boundary" at infinity of X. Denote $\lim_{l \in \varkappa}$ the limit along the ultrafilter $\varkappa$.

Theorem 1.2. *Under the hypotheses of the preceding theorem, one has:*

(i) The strong resolvent limit s-$\lim_{l \in \varkappa} \mathrm{e}^{-ilP} H \mathrm{e}^{ilP} = H_\varkappa$ exists for each $\varkappa \in \widetilde{L}$.

(ii) For $\varkappa \in \widetilde{L}$ one also has $H_\varkappa = \mathrm{u}\text{-}\lim_{l \in \varkappa} H_l$ (limit in the norm resolvent sense).

(iii) The essential spectrum of H is: $\sigma_{\mathrm{ess}}(H) = \bigcup_{\varkappa \in \widetilde{L}} \sigma(H_\varkappa)$.

Notice that the existence of the limits in (i) and (ii) above and the inclusion $\supset$ in (iii) are not hard to prove. If one prefers to avoid the notion of ultrafilter, the result can be stated (in a less intrinsic form) in terms of usual sequences, as follows:

Let $\mathcal{L}(H)$ be the set of sequences $\mathrm{l} \equiv \{l_k\}_{k \in \mathbb{N}}$ of $L \subset X$ such that s-$\lim_{k \to \infty} \mathrm{e}^{-il_k P} H \mathrm{e}^{il_k P} =: H_{\mathrm{l}}$ exists. Then: $\sigma_{\mathrm{ess}}(H) = \bigcup_{\mathrm{l} \in \mathcal{L}(H)} \sigma(H_{\mathrm{l}})$.

As a second example we consider quantum systems with a general configuration space anisotropy:

Theorem 1.3. *Let $X = \mathbb{R}^n$ and H a self-adjoint operator in $L^2(X)$ such that:*

(a) $\lim_{\substack{x \to 0 \\ x \in X}} \|(e^{i\langle x, P\rangle} - 1)(H + i)^{-1}\| = 0,$

(b) $\lim_{\substack{k \to 0 \\ k \in X^*}} \|[e^{i\langle Q, k\rangle}, (H + i)^{-1}]\| = 0,$

Then, for each $\varkappa \in \widetilde{X}$ the limit $H_\varkappa = \mathrm{s\text{-}lim}_{x \in \varkappa}\, e^{i\langle x, P\rangle} H e^{-i\langle x, P\rangle}$ exists (in strong resolvent sense), and $\sigma_{\mathrm{ess}}(H) = \bigcup_{\varkappa \in \widetilde{X}} \sigma(H_\varkappa).$

We stress that such a pure configuration space description of the essential spectrum is not valid in general (e.g. for systems with full phase-space anisotropy).

2 An Algebraic Method

Our goal is to show how these results, and many others of this type, involving even much complicated situations, may be treated in a unified manner, by an algebraic approach. The main ideea of this approach is that the admissible hamiltonians corresponding to a physical system subject to a certain type of interactions generate a C^*-algebra $\mathfrak{C}$ (the algebra of energy observables of the system) realized on a Hilbert space $\mathcal{H}$, such that:

(i) $K(\mathcal{H}) \subset \mathfrak{C}$

(ii) the quotient C^*-algebra $\mathfrak{C}/K(\mathcal{H})$ is, in a certain sense, *computable*.

The *consequences for the spectral theory* will concern both the essential spectrum and the Mourre estimate which is the key of a refined investigation of the spectrum (see [ABG96] for further information). Indeeed, if we suppose (for simplicity) $H \in \mathfrak{C}$, and if $\widehat{H}$ denotes its image in $\mathfrak{C}/K(\mathcal{H})$, these consequences may be stated as:

(i) $\sigma_{\mathrm{ess}}(H) = \sigma(\widehat{H})$

(ii) $\widehat{H}$ satisfies a strict Mourre estimate at some $\lambda \in \mathbb{R}$ if and only if H satisfies a Mourre estimate at λ (modulo some technical conditions).

2.1. We need some definitions. A self-adjoint operator H on $\mathcal{H}$ is *affiliated to a C^*-algebra $\mathfrak{C} \subset B(\mathcal{H})$* iff $(H - z)^{-1} \in \mathfrak{C}$ for some $z \in \mathbb{C} \setminus \sigma(H)$; then $\varphi(H) \in \mathfrak{C}$ for all $\varphi \in C_0(\mathbb{R})$ (this will sometimes be denoted by $H \,''\!\in\!''\, \mathfrak{C}$). The *$C^*$-algebra generated by a family of self-adjoint operators $\{H\}$* is, by definition, the C^*-algebra generated by the family of resolvents $(H - z)^{-1}$. If $\{\mathfrak{C}_i\}_{i \in I}$ is an arbitrary family of C^*-algebras, the direct product $\prod_{i \in I} \mathfrak{C}_i$ and direct sum $\bigoplus_{i \in I} \mathfrak{C}_i$ are defined as:

$$\textstyle\prod_{i \in I}\mathfrak{C}_i = \{S = (S_i)_{i \in I} \mid S_i \in \mathfrak{C}_i \text{ and } \|S\| := \sup_{i \in I} \|S_i\| < \infty\}$$

$$\textstyle\bigoplus_{i \in I}\mathfrak{C}_i = \{S = (S_i)_{i \in I} \mid S_i \in \mathfrak{C}_i \text{ and } \|S_i\| \to 0 \text{ as } i \to \infty\}.$$

These are C^*-algebras for the usual operations and $\bigoplus_{i \in I} \mathfrak{C}_i$ is an ideal in $\prod_{i \in I} \mathfrak{C}_i$.

2.2. The notion of *"computable"* for a quotient algebra $\mathfrak{C}/K(\mathcal{H})$ is crucial for the method we present here. However, we must be aware that its precise definition depends on the complexity of the problem. Here we describe two situations of increasing complexity, for which the meaning of "computable" is different:

A) Systems as in the case of the usual and generalized N-body problems (the lattice of partitions can be replaced by an arbitrary, even infinite, lattice), or various configuration and phase-space anisotropic systems (e.g. 1-dim operators, multistratified strips, constant magnetic fields). Then the algebra $\mathfrak{C}/K(\mathscr{H})$ is "computable" in the following elementary sense: one can find a family $\{\mathfrak{C}_i\}_{i\in I}$ of C^*-algebras (which all have a "much simpler" structure than that of $\mathfrak{C}$) such that:

$$\mathfrak{C}/K(\mathscr{H}) \hookrightarrow \prod_{i\in I}\mathfrak{C}_i. \tag{2.1}$$

In other terms, a family of morphisms $\mathcal{P}_i : \mathfrak{C} \to \mathfrak{C}_i$ is given such that $\bigcap_{i\in I}\ker\mathcal{P}_i = K(\mathscr{H})$. If we set $\mathcal{P}[S] = \prod_{i\in I}\mathcal{P}_i[S]$ we shall get a morphism $\mathcal{P} : \mathfrak{C} \to \prod_{i\in I}\mathfrak{C}_i$ with kernel $K(\mathscr{H})$, which induces the required embedding. Now let, for simplicity, H be a bounded admissible hamiltonian and set $H_i = \mathcal{P}_i[H]$. Then, by (2.1), we have an identification $\widehat{H} = \prod_{i\in I}H_i$ and thus:

$$\sigma_{\mathrm{ess}}(H) = \overline{\bigcup_{i\in I}\sigma(H_i)}. \tag{2.2}$$

This covers the *theorems of HVZ type* known in the usual many-channel situations. We mention that in [DG] is pointed out an abstract class of C^*-algebras graded by infinite semilattices, whose quotient is of the preceding form.

B) A worse situation, which covers systems with infinitely many, widely separated "bumps" cf. [K83]: one can find a family $\{(\mathfrak{C}_i,\mathfrak{J}_i)\}_{i\in I}$ of ("much simpler" than $\mathfrak{C}$) C^*-algebras $\mathfrak{C}_i$ equipped with ideals $\mathfrak{J}_i$ such that there is a natural embedding:

$$\mathfrak{C}\big/K(\mathscr{H}) \hookrightarrow \prod_{i\in I}\mathfrak{C}_i\big/\bigoplus_{i\in I}\mathfrak{J}_i\,. \tag{2.3}$$

Note that the preceding case (2.1) is obtained when $\mathfrak{J}_i = \{0\}$ for all $i \in I$.

Assume, for example, that each $\mathfrak{C}_i$ is realized on a Hilbert space $\mathscr{H}_i$ and that $\mathfrak{J}_i = K(\mathscr{H}_i)$. We show that if H is a self-adjoint operator affiliated to $\mathfrak{C}$ and $\widehat{H}$ is, in the representation given by the right hand side of (2.3), the quotient of some family $\{H_i\}_{i\in I}$ of operators H_i affiliated to $\mathfrak{C}_i$, then

$$\sigma_{\mathrm{ess}}(H) = \bigcap_{\substack{F\subset I\\F\,\mathrm{finite}}}\left\{\left[\bigcup_{i\in F}\sigma_{\mathrm{ess}}(H_i)\right] \cup \left[\overline{\bigcup_{i\in I\setminus F}\sigma(H_i)}\right]\right\}. \tag{2.4}$$

Versions of this formula that are more similar to a classic HVZ result can be obtained, as we shall see, by the approach based on ultrafilters.

2.3. We shall describe now some of the ways in which we may construct algebras of energy observables.

A) *Algebras of energy observables constructed as cross products:*

Let $\mathscr{A}$ be a closed $*$-subalgebra of $C^u_b(\mathbb{R}^n)$ which contains the constants and is stable under translations; put $\mathscr{A}^\infty = \{\varphi \in C^\infty(\mathbb{R}^n) \mid \partial^\alpha\varphi \in \mathscr{A},\ \forall\alpha\}$, let $\Delta = P^2$ be the (positive) Laplace operator and consider self-adjoint operators of the form $H = \Delta + V$, where V is a first order differential operator with coefficients in $\mathscr{A}^\infty$.

We shall denote by X the configuration space $\mathbb{R}^n$ and by X^* its dual. Then:

(i) *the C^*-algebra of operators on $L^2(X)$ generated by the operators H is equal to the norm closed subspace generated by operators of the form $\varphi(Q)\psi(P)$, where $\varphi \in \mathscr{A}$ and $\psi \in C_0(X)$*. This last algebra is denoted by $[\![\mathscr{A} \cdot C_0(X^*)]\!]$.

(ii) *the algebra $[\![\mathscr{A} \cdot C_0(X^*)]\!]$ is actually the crossed product $\mathscr{A} \rtimes X$ of $\mathscr{A}$ by the* natural action of the translation group on it.

This justifies the interpretation of a general class of crossed products $\mathscr{A} \rtimes X$ as algebras of energy observables. In fact, the class of "algebras of energy observables" is much larger than the class of algebras obtained as cross products, the latter corresponding only to the case of Q-anisotropic or P-anisotropic systems. Other types of algebras of energy observables, which are not crossed products, may be studied by the same method:

B) The simplest type of phase space (or Q-P) anisotropy, where the algebra is (essentially) of the form $[\![\mathscr{A} \cdot \mathscr{B}]\!]$, where $\mathscr{A}$ and $\mathscr{B}$ are C^*-subalgebras of $C^u_b(X)$ and $C^u_b(X^*)$ respectively.

C) Full phase space anisotropy: here we consider C^*-algebras naturally associated to the Heisenberg group, which is the simplest type of nilpotent group. The algebra we construct will be called *graded symplectic algebra*.

D) Algebras of creation-anihilation operators on a Fock space. These are generated by hamiltonians of quantum fields with a particle number cut-off. Their quotient with respect to the ideal of compact operators has an especially nice form (see the lecture of V. Georgescu in this volume).

2.4. The *main advantage of crossed product algebras $\mathscr{A} \rtimes X$ is that* due to $\mathbb{K}(X) = [\![C_0(X) \cdot C_0(X^*)]\!] = C_0(X) \rtimes X$ and to the good functorial properties of cross products, *it suffices to "compute" the quotient at an abelian level* for example to put in evidence an embedding $\mathscr{A}/C_0(X) \hookrightarrow \prod_{i \in I}^X \mathscr{C}_i$, (where the latter is the greatest subalgebra of $\prod_{i \in I} \mathscr{C}_i$ for which the cross product with X makes sense). Indeed, we first choose a C^*-algebra $\mathscr{A}$, viewed as the algebra of "elementary interactions" (or potentials), such that:

$$\{\, C_0(X) \subset \mathbb{C} + C_0(X) \equiv C_\infty(X) \subset \mathscr{A} \subset C^u_b(X), \mathscr{A} \text{ is stable under translations.}$$

$$(2.5)$$

If the quotient is computable at the abelian level, as above, then

$$[\mathscr{A} \rtimes X]/\mathbb{K}(X) \;=\; [\mathscr{A} \rtimes X]/[C_0(X) \rtimes X] \;\cong\; [\mathscr{A}/C_0(X)] \rtimes X$$

$$\hookrightarrow \prod_{i \in I}^X \mathscr{C}_i \rtimes X \;\hookrightarrow\; \prod_{i \in I}[\mathscr{C}_i \rtimes X].$$

The simplest situation is furnished by a two-body quantum system (or a particle in an external field); then $\mathscr{A} = C_\infty(X)$, $\mathbb{T}(X) \equiv C_\infty(X) \rtimes X$ is the "two-body energy algebra" and we get $\mathbb{T}(X)/\mathbb{K}(X) = C_0(X^*)$.

2.5. Notice that $C_\infty(X)$ is related to the simplest compactification of X, the one-point (Alexandroff) compactification: $X_\infty \equiv X \cup \{\infty\}$. Indeed, we have a natural identification $C_\infty(X) = C(X_\infty)$ and this is the simplest case of a general phenomenon: the algebras $\mathscr{A}$ with the properties (2.5) are in bijective correspodance with a certain class of compactifications of X. The spectrum $X(\mathscr{A})$ of $\mathscr{A}$ is a compact space and X is homeomorphically and densely embedded in $X(\mathscr{A})$. Also, $\mathscr{A}$ is just the set of functions in $C_{\mathrm{b}}^{\mathrm{u}}(X)$ which have continuous extensions to $X(\mathscr{A})$, i.e. $\mathscr{A} = C(X(\mathscr{A}))$. These compactifications can be described in terms of ultrafilters on X (they are quotients of the Stone-Čech compactification βX). We shall, however, not insist on this point of view, because in non trivial cases the spaces $X(\mathscr{A})$ are rather complicated.

3 "Bumps" Algebras

The class of hamiltonians considered in this section generalizes that of the Schrödinger operators with "widely separated bumps" introduced by M.Klaus in [K83] the quotient algebra being computable in the more general sense we already described in examples of type (2) from §2.2.

3.1. Let X be an abelian locally compact group and $L \subset X$. If $\varphi : X \to \mathbb{C}$, we say that *the limit L-$\lim\varphi$ exists* iff $\exists\, c \in \mathbb{C}$ such that $\forall\, \varepsilon > 0$ one can find a compact set $\Lambda \subset X$ such that $x \notin L + \Lambda \implies |\varphi(x) - c| < \varepsilon$. We denote $c = L\text{-}\lim\varphi$. Define:

$$\mathscr{C}_L = \{\varphi \in C_{\mathrm{b}}^{\mathrm{u}}(X) \mid L\text{-}\lim\varphi \text{ exists }\} \quad \text{and} \quad \mathscr{C}_{L,0} = \{\varphi \in C_{\mathrm{b}}^{\mathrm{u}}(X) \mid L\text{-}\lim\varphi = 0\}.$$

Then $\mathscr{C}_L$ and $\mathscr{C}_{L,0}$ are translation invariant C^*-subalgebras of $C_{\mathrm{b}}^{\mathrm{u}}(X)$. One has $C_\infty(X) \subset \mathscr{C}_L$ and $C_0(X) \subset \mathscr{C}_{L,0}$. Also, $\mathscr{C}_{L,0}$ is an ideal of $\mathscr{C}_L$ (and of $C_{\mathrm{b}}^{\mathrm{u}}(X)$) and $\mathscr{C}_L = \mathbb{C} + \mathscr{C}_{L,0}$. We may construct an "energy observables algebra" by taking the cross product with X; as we said before we also have the identities:

$$\mathfrak{C}_L := \mathscr{C}_L \rtimes X = [\![\mathscr{C}_L \cdot C_0(X^*)]\!] \quad \text{and} \quad \mathfrak{C}_{L,0} := \mathscr{C}_{L,0} \rtimes X = [\![\mathscr{C}_{L,0} \cdot C_0(X^*)]\!]. \quad (3.1)$$

The problem of computing the quotients $\mathfrak{C}_L / \mathbb{K}(X)$ and $\mathfrak{C}_{L,0} / \mathbb{K}(X)$ is difficult for general sets L but there are notable particular cases when this is possible, for instance the case of a "widely separated bumps" problem, when L is *rarefied*, i.e. it is locally finite and for each compact Λ of X there is a finite set $F \subset L$ such that if $l \in M = L \setminus F$ and $l' \in L \setminus \{l\}$ then $(l + \Lambda) \cap (l' + \Lambda) = \emptyset$. An explicit description of the quotient algebras will be given below. Notice that we actually can manage sets L that are subject to less restrictive assumptions than that of being rarefied (in the above sense).

3.2. It is possible to give a *new characterization of the algebras $\mathfrak{C}_L$ and $\mathfrak{C}_{L,0}$ in geometric terms in the phase-space $\Xi = X \oplus X^*$*. We begin by introducing some notations. Let $\{U_x\}_{x \in X}$ and $\{V_k\}_{k \in X^*}$ be the strongly continuous unitary representations of X and X^* in $L^2(X)$ defined by $(U_x f)(y) = f(x + y)$ and $(V_k f)(y) = \mathrm{e}^{i\langle y, k\rangle} f(y)$ respectively. Note that U_x and V_k satisfy the so called canonical commutation relations $U_x V_k = \mathrm{e}^{i\langle x, k\rangle} V_k U_x$. If X is a (real or p-adic) finite dimensional vector space, we have (with the quantum mechanical conventions): $U_x = \mathrm{e}^{i\langle x, P\rangle}$ and $V_k = \mathrm{e}^{i\langle Q, k\rangle}$. If A and T are bounded operators and A

is self-adjoint, we shall say that $\|AT^{(*)}\|$ satisfies the property (F) if and only if $\|TA\| + \|AT\|$ verifies (P). Also, $\chi_{L_\Lambda^c}$ will denote the characteristic function of the set $L_\Lambda^c \equiv X \setminus (L + \Lambda)$.

Theorem 3.1. **(1)** *A bounded operator T on $L^2(X)$ belongs to $\mathfrak{C}_{L,0}$ if and only if*
(i) $\lim_{x\to 0} \|(U_x - 1)T^{(*)}\| = 0$,
(ii) $\lim_{k\to 0} \|V_k T V_k^* - T\| = 0$,
(iii) *For each $\varepsilon > 0$ there is a compact set Λ in X such that $\|\chi_{L_\Lambda^c}(Q)T^{(*)}\| < \varepsilon$.*
(2) *$T \in \mathfrak{C}_L$ iff it satisfies (i), (ii) and there exists $\widetilde{T} \in C_0(X^*)$ such that for each $\varepsilon > 0$ there is a compact set $\Lambda \in X$ with $\|\chi_{L_\Lambda^c}(Q)(T - \widetilde{T})^{(*)}\| < \varepsilon$.*

In order to compute the quotients $\mathfrak{C}_L / \mathbb{K}(X)$ and $\mathfrak{C}_{L,0} / \mathbb{K}(X)$ it will be useful to consider the case when for each $l \in L$ is given an algebra $\mathfrak{C}_l \equiv \mathfrak{C}$ independent of $l \in L$. Then we denote $\bigoplus_{l\in L}\mathfrak{C}$ by $\mathfrak{C}^{(L)}$ and one can introduce a new C^*-subalgebra of $\prod_{l\in L}\mathfrak{C}$, namely

$$\mathfrak{C}^{[L]} := \{(S_l)_{l\in L} \in \textstyle\prod_{l\in L}\mathfrak{C} \mid \{S_l \mid l \in L\} \text{ is a relatively compact set in } \mathfrak{C}\}. \quad (3.2)$$

Then $\mathfrak{C}^{(L)}$ is an ideal in $\mathfrak{C}^{[L]}$. If L is equipped with the discrete topology, then

$$\mathfrak{C}^{(L)} = c_0(L; \mathfrak{C}) \cong c_0(L) \otimes \mathfrak{C} \quad \text{and} \quad \mathfrak{C}^{[L]} \cong l^\infty(L) \otimes \mathfrak{C} \subset l^\infty(L; \mathfrak{C}) = \textstyle\prod_{l\in L}\mathfrak{C}.$$

One more object appears naturally: the *L-asymptotic algebra* of $\mathfrak{C}$. This is the quotient algebra: $\mathfrak{C}^{\langle L\rangle} := \mathfrak{C}^{[L]} / \mathfrak{C}^{(L)}$.

Theorem 3.2. *If L is rarefied, $\mathfrak{C}_L / \mathbb{K}(X)$ and $\mathfrak{C}_{L,0} / \mathbb{K}(X)$ are "computable" in the sense that:*
(i) $\mathfrak{C}_{L,0} / \mathbb{K}(X) \cong \mathbb{K}(X)^{\langle L\rangle}$.
(ii) *One has a natural embedding: $\mathfrak{C}_L / \mathbb{K}(X) \hookrightarrow \mathbb{T}(X)^{[L]} / \mathbb{K}(X)^{\langle L\rangle}$.*

Proof. We follow the algorithm already described in §2.4. We prove first the theorem at the abelian level, i.e. we show the embeddings:

$$\mathscr{C}_{L,0} / C_0(X) \cong C_0(X)^{\langle L\rangle} \quad \text{and} \quad \mathscr{C}_L / C_0(X) \hookrightarrow C_\infty(X)^{[L]} / C_0(X)^{(L)}.$$

A precise description of these morphisms is given by the result:

There is a unique morphism $\mathscr{C}_L \to C_\infty(X)^{[L]} / C_0(X)^{(L)}$ such that the image of an element of the form $c + \sum_{l\in M} U_l^*\varphi(Q)U_l$, where $c \in \mathbb{C}$, $M \subset L$, and $\varphi \in C_0(X)$ verifies $\varphi = \chi_\Lambda\varphi$ for some compact set $\Lambda \subset X$, be the quotient of the element $\chi_M \otimes (c + \varphi(Q)) \in C_\infty(X)^{[L]}$ with respect to the ideal $C_0(X)^{(L)}$. The kernel of this morphism is $C_0(X)$ and its restriction to $\mathscr{C}_{L,0}$ induces the canonical isomorphism of $\mathscr{C}_{L,0} / C_0(X)$ with the L-asymptotic algebra $C_0(X)^{\langle L\rangle}$.

In the second step of the proof we use the functorial properties of the crossed product in order to obtain the result at the quantized level. $\qquad\square$

3.3. If H is an observable affiliated to $\mathfrak{C}_L$ and if $\widehat{H}$ is its image through the canonical morphism $\mathfrak{C}_L \to \mathfrak{C}_L / \mathbb{K}(X)$, then there is a family $(H_l)_{l\in L}$ of observables affiliated to the two-body algebra $\mathbb{T}(X)$ s.t. the quotient of $\prod_{l\in L} H_l$ with respect to the ideal $\mathbb{K}(X)^{(L)}$ be equal to the image of $\widehat{H}$ through the embedding $\mathcal{J}$ below:

$$
\begin{array}{ccc}
H\ ''\in''\ \mathfrak{C}_L & & \mathbb{T}(X)^{[L]}\ ''\ni''\ (H_l)_{l\in L} \\
\downarrow & & \downarrow \\
\widehat{H}\ ''\in''\ \mathfrak{C}_L \big/ \mathbb{K}(X) \ \overset{\mathcal{J}}{\hookrightarrow}\ \mathbb{T}(X)^{[L]} \big/ \mathbb{K}(X)^{(L)} & ''\ni'' & \mathcal{J}(\widehat{H}) = \widehat{(H_l)}_{l\in L}
\end{array}
$$

Such a family $(H_l)_{l\in L}$ will be called a *representative* of H. By the discussion above we have $\prod_{l\in L}(H_l - z)^{-1} \in \mathbb{T}(X)^{[L]}$ and the component of $(H_l - z)^{-1}$ in $C_0(X^*)$ is independent of $l \in L$, so $\sigma_{\mathrm{ess}}(H_l)$ is independent of l. Thus the next result is a consequence of (2.4).

Theorem 3.3. *If H is an observable affiliated to $\mathfrak{C}_L$ and $\{H_l\}_{l\in L}$ is a representative of H, then $\sigma_{\mathrm{ess}}(H) = \bigcap_{\substack{F\subset L \\ F\,\mathrm{finite}}} \overline{\bigcup_{l\in L\setminus F} \sigma(H_l)}$.*

4   The Examples Revisited

A) Klaus type potentials: The result, in its first version, may be restated as:

Theorem 1.1. BIS: *Under the assumptions and notations of the Theorem 1.1, $H = H_0 + V$ is strictly affiliated to $\mathfrak{C}_L$, $H_l = H_0 + V_l$ is strictly affiliated to $\mathbb{T}(X)$, and the family $\{H_l\}_{l\in L}$ is a representative of H. In particular:*

$$
\sigma_{\mathrm{ess}}(H) = \bigcap_{\substack{F\subset L \\ F\,\mathrm{finite}}} \overline{\bigcup_{l\in L\setminus F} \sigma(H_l)}.
$$

Since $H = H_0 + V$ is affiliated to $\mathfrak{C}_L$, the second version of the result (Theorem 1.2), based on ultrafilters, follows from:

Theorem 4.1. *For each $S \in \mathfrak{C}_L$ and $\varkappa \in \widetilde{L}$ the limit s-$\lim_{l\in\varkappa} U_l S U_l^* =: S_\varkappa$ exists and belongs to $\mathbb{T}(X) = C_0(X^*) + \mathbb{K}(X)$. The component of $S_\varkappa$ in $C_0(X^*)$ is equal to that of S in $C_0(X^*)$. The map $S \mapsto (S_\varkappa)_{\varkappa\in\widetilde{L}}$ is a morphism of $\mathfrak{C}_L$ into $C(\widetilde{L};\mathbb{T}(X))$ with kernel $\mathbb{K}(X)$. Its range equals the set of $(S_\varkappa)_{\varkappa\in\widetilde{L}}$ such that the component of $S_\varkappa$ in $C_0(X^*)$ is independent of $\varkappa$. Thus: $\mathfrak{C}_L / \mathbb{K}(X) \hookrightarrow C(\widetilde{L};\mathbb{T}(X))$.*

B) General configuration space anisotropy: Remember the result stated as the Theorem 1.3: apparently it has nothing to do with the preceding algebraic approach. The connection is in fact established via the following theorem.

Theorem 4.2. *Let $X = \mathbb{R}^n$ and H a self-adjoint operator in $L^2(X)$. Then the following assertions are equivalent:*
(i) H is affiliated to the crossed product algebra $C_b^u(X) \rtimes X$
(ii) H satisfies the assertions (a) and (b) of the Theorem 1.3, namely

$$\lim_{\substack{x \to 0 \\ x \in X}} \|(e^{i\langle x, P \rangle} - 1)(H + i)^{-1}\| = 0 \quad and \quad \lim_{\substack{k \to 0 \\ k \in X^*}} \|[e^{i\langle Q, k \rangle}, (H + i)^{-1}]\| = 0.$$

Hence, to prove the Theorem 1.3, it is sufficient to put in evidence the embedding:

$$C_b^u(X) \,/\, C_0(X) \;\hookrightarrow\; C(\widetilde{X}\,; C_b^u(X)).$$

Acknowledgements

I would like to thank to the organizers of the conference and especially to Professor M. Demuth for their hospitality, financial support, and for having invited me at this very interesting scientific meeting.

References

[ABG96] W. Amrein, A. Boutet de Monvel and V. Georgescu, C_0-groups, commutator methods and spectral theory of N-body hamiltonians, Birkhauser Verlag, 1996.

[DG99]   M. Damak and V. Georgescu, C^*-algebras related to the N-body problem and the self-adjoint operators affiliated to them, (available as preprint 99-482 at http://www.ma.utexas.edu/mp_arc/).

[GI00]   V. Georgescu, A. Iftimovici C^*-algebras of energy observables I. The essential spectrum (in preparation)

[K83]   M. Klaus, On $-d^2/dx^2 + V$ where V has infinitely many "bumps", Ann. Inst. H. Poincaré Sect. A (N.S.) **38** (1983), no. 1, 7–13.

Address

ANDREI IFTIMOVICI, Département de Mathématiques de l'Université de Cergy-Pontoise, 2, avenue Adolphe Chauvin, F 95302 Cergy Pontoise Cedex, France

E-MAIL: Andrei.Iftimovici@math.u-cergy.fr

2000 Mathematics Subject Classification. Primary 47D45, 81Q10 ; Secondary 81U10, 47C15, 47D25, 81R99, 47S10.

Operator Theory:
Advances and Applications, Vol. 126
© 2001 Birkhäuser Verlag Basel/Switzerland

Essential Self-Adjointness of n-Dimensional Dirac Operators with a Variable Mass Term

Hubert Kalf and Osanobu Yamada

Abstract. We give some results about the essential self-adjointness of the Dirac operator

$$H = \sum_{j=1}^{n} \alpha_j\, p_j + m(x)\, \alpha_{n+1} + V(x)\, I_N \quad \left(N = 2^{\left[\frac{n+1}{2}\right]} \right),$$

on $[C_0^\infty(\mathbf{R}^n \setminus 0)]^N$, where the α_j $(j = 1, 2, \cdots, n)$ are Dirac matrices and $m(x)$ and $V(x)$ are real-valued functions. We are mainly interested in a singularity of $V(x)$ and $m(x)$ near the origin which preserves the essential self-adjointness of H. As a result, if $m = m(r)$ is spherically symmetric or $m(x) \equiv V(x)$, then we can permit a singularity of m and V which is stronger than that of the Coulomb potential.

1 Introduction

In this note we investigate the essential self-adjointness of the Dirac operator

$$H := \sum_{j=1}^{n} \alpha_j\, p_j + m(x)\, \alpha_{n+1} + V(x)\, I_N$$

$$\left(x \in \mathbf{R}^n, \ n \geq 2, \ p_j = -i\frac{\partial}{\partial x_j}, \ N = N(n) := 2^{\left[\frac{n+1}{2}\right]} \right)$$

in the Hilbert space $\mathbf{H} := [L^2(\mathbf{R}^n)]^N$. Here I_N is the $N \times N$ unit matrix, and the α_j are $N \times N$ Dirac matrices, i.e., Hermitian matrices satisfying

$$\alpha_j\, \alpha_k + \alpha_k\, \alpha_j = 2\, \delta_{jk}\, I_N \quad (j = 1, 2, \cdots, n+1)$$

$m(x)$ and $V(x)$ are real valued functions.

In contrast to the Schrödinger case, the essential self-adjointness of H can not be destroyed by the behavior of m and V at infinity (Chernoff [Ch77]), but only by local singularities of those functions. For example, if m is constant and $V(r) = e/r$, H is essentially self-adjoint if and only if $|e| \leq \sqrt{3}/2$. Extending a seminal paper by Schmincke [Sc72], very general self-adjointness criteria were given by e.g., Levitan–Otelbaev [LO77], Arai [Ar83], Yamada [Ya86], Vogelsang [Vo87], and Kalf [Kl98] for the situation that m is a constant and V is a strongly singular potential that is not necessarily spherically symmetric.

2 Theorems

Here we employ the following abbreviations ;

$$\alpha := (\alpha_1, \alpha_2, \cdots, \alpha_n), \quad p := (p_1, p_2, \cdots, p_n), \quad \alpha \cdot p := \sum_{j=1}^{n} \alpha_j p_j,$$

$$\Omega := \mathbf{R}^n \setminus \{0\}, \quad \mathbf{R}_+ := (0, +\infty), \quad \mathbf{D} := [C_0^\infty(\Omega)]^N.$$

Theorem 2.1. *Let* $m = m(r)$ *be spherically symmetric and absolutely continuous in* $\mathbf{R}_+$ *with the derivative* $m'(r) \in L^2_{loc}(\mathbf{R}_+)$. *Assume that* $\alpha \cdot p + m\,\alpha_{n+1}$ *on* $\mathbf{D}$ *is essentially self-adjoint, and* $V(x) \in L^2_{\mathrm{loc}}(\Omega)$ *satisfies*

$$(1 + \varepsilon)^2 \left[V^2(x) + \frac{1}{4\,r^2} \right] + \left| m'(r) + \frac{m(r)}{r} \right| \leq m^2(r) + \left(\frac{n-1}{2r} \right)^2$$

$(x \in \Omega)$ *for some* $\varepsilon > 0$. *Then* H *is essentially self-adjoint. If* $n \geq 3$ *and* $rm(r)$ *is bounded, then the domain* $D(\overline{H})$ *of the closure* $\overline{H}$ *coincides with the Sobolev space* $[W^{1,2}(\mathbf{R}^n)]^N$.

The next result does not require m to be spherically symmetric and extends a theorem of Levitan–Otelbaev [LO77].

Theorem 2.2. *Let* $n \geq 3$ *and* $m, V \in L^2_{\mathrm{loc}}(\Omega)$ *be real-valued,*

$$q(r) := \sup_{|x|=r} \left[V^2(x) + m^2(x) + 2\,|m(x)|\,\sqrt{V^2(x) + \frac{1}{4\,r^2}} \right]^{1/2}$$

and

$$a := \sup_{r>0} \left(\frac{1}{r^{n-2}} \int_0^r t^{n-1} q^2(t)\,dt \right)^{1/2} < \frac{\sqrt{n}}{2}.$$

Then H *is essentially self-adjoint with* $D(\overline{H}) = [W^{1,2}(\mathbf{R^n})]^N$.

The Dirac operator is in general not essentially self-adjoint on $C_0^\infty(\mathbf{R}^n)^N$ if m, V are merely in $L^2_{loc}(\mathbf{R}^n)$. Due to a theorem of Kato [Ka72] we do however have the following result.

Theorem 2.3. *Assume that* $m(x)$ *and* $V(x)$ *are* $L^2_{loc}(\mathbf{R}^n)$ *functions satisfying* $m(x) \equiv V(x)$ *or* $m(x) \equiv -V(x)$. *Then* $H = (\alpha \cdot p) + m\,\alpha_{n+1} + V$ *on* $[C_0^\infty(\mathbf{R}^n)]^N$ *is essentially self-adjoint.*

A sketch of the proofs of these theorems for 3-dimensional case appears in Kalf–Yamada [KY00]. The detailed proofs of the general case will be given elsewhere.

Acknowledgements

The second author would like to express his gratitude to the organizers, especially to Professor Dr. M. Demuth, for their hospitalities during the PDE 2000 Conference in Clausthal.

References

[Ar83]   M. Arai, On essential selfadjointness, distinguished selfadjoint extension and essential spectrum of Dirac operators o with matrix valued potentials, *Publ. RIMS, Kyoto Univ.*, 33–57, **19** (1983).

[Ch77]   P. R. Chernoff, Schrödinger and Dirac operators with singular potentials and hyperbolic equations, *Pacific J. Math.*, 361–382, **72** (1977).

[Kl98]   H. Kalf, Essential self-adjointness of Dirac operators under an integral condition on the potential, *Letters in Math. Phys.*, 225–232, **44** (1998)

[KY00]   H. Kalf and O. Yamada, Essential self-adjointness of Dirac operators with a variable mass term, *Proc. Japan Acad. Ser. A*, 13–15, **76** (2000).

[Ka72]   T. Kato, Schrödinger operators with singular potentials, *Israel J. Math.*, 135–148, **13** (1972).

[LO77]   B. M. Levitan and M. Otelbaev, On conditions for self-adjointness of the Schrödinger and Dirac operators, *Soviet Math. Dokl.*, 1044–1048, **18** (1977).

[Sc72]   U.-W. Schmincke, Essential selfadjointness of Dirac operators with a strongly singular potential, *Math. Z.*, 71–81, **126** (1972).

[Vo87]   V. Vogelsang, Remark on essential selfadjointness of Dirac operators with Coulomb potentials, *Math. Z.*, 517–521, **196** (1987).

[Ya86]   O. Yamada, A remark on the essential self-adjointness of Dirac operators, *Proc. Japan Acad., Ser. A.*, 327–330, **62** (1986).

Addresses

HUBERT KALF, Mathematisches Institut der Universität, Theresienstr. 39, München, D-80333, Germany

E-MAIL: kalf@rz.mathematik.uni-muenchen.de

OSANOBU YAMADA, Department of Mathematics, Ritsumeikan University, Kusatsu, Shiga 525-8577, Japan

E-MAIL: yamadaos@se.ritsumei.ac.jp

2000 Mathematics Subject Classification. Primary 35P05 ;Secondary 47A55

Operator Theory:
Advances and Applications, Vol. 126
© 2001 Birkhäuser Verlag Basel/Switzerland

Towards the Spectral Analysis of Schrödinger Operator with Fractal Perturbation

VOLODYMYR KOSHMANENKO

Abstract. Let $-\Delta_{\alpha,T} = -\Delta \dotplus \alpha T$, $\alpha < 0$, denote the generalized Schrödinger operator in $L_2(\mathbf{R}^n)$, where T is a positive operator, acting in the Sobolev scale of spaces, from W_2^1 to its dual W_2^{-1}, and supported by a fractal set $\Gamma \subset \mathbf{R}^n$. The sufficient conditions ensuring finiteness of the negative spectrum for $-\Delta_{\alpha,T}$ are found.

1 Introduction

Let $-\Delta$ be the usual self-adjoint Laplace operator in the Hilbert space $L_2 \equiv L_2(\mathbf{R}^n, dx)$, $n \geq 1$. By $W_2^k \equiv W_2^k(\mathbf{R}^n)$, $k \in \mathbf{R}^1$, we denote the Sobolev space equipped with the norm,

$$\|f\|_k := (\int_{\mathbf{R}^n} | (I - \Delta)^{k/2} f(x) |^2 \, dx)^{1/2},$$

where I stands for the identity operator.

In the terminology of the references [AKK95,KK99,Ko99] an operator $-\tilde{\Delta} \neq -\Delta$, self-adjoint in L_2 is called a (pure) singular perturbation of $-\Delta$ if the set

$$\mathcal{D} := \{f \in \mathcal{D}(-\tilde{\Delta}) \cap \mathcal{D}(-\Delta) : -\Delta f = -\tilde{\Delta} f\} \text{ is dense in } L_2.$$

We say that $-\tilde{\Delta}$ belongs to the class of weak singular perturbations of $-\Delta$ and write $-\tilde{\Delta} \in \mathcal{P}_{ws}(-\Delta)$, if the following condition holds, $\mathcal{D}(-\tilde{\Delta}) \subset W_2^1$.

In this paper we deal with the negative eigenvalue problem for $-\tilde{\Delta} \in \mathcal{P}_{ws}(-\Delta)$, which is the generalized Schrödinger operator of a form $-\tilde{\Delta} \equiv -\Delta_{\alpha,T} = -\Delta \dotplus \alpha T$, $\alpha < 0$, where T is a positive L_2–singular operator acting in the Sobolev scale of spaces, $T : W_2^1 \to W_2^{-1}$, and supported by a fractal set $\Gamma \subset \mathbf{R}^n$. Here $\dotplus$ denotes the generalized operator sum (see [KaKo99]), defined in such a way that, in a sense of distributions,

$$-\tilde{\Delta}\varphi = -\Delta\varphi + T\varphi, \quad \mathcal{D}(-\tilde{\Delta}) = \{\varphi \in W_2^1 \cap \mathcal{D}(T) : -\Delta\varphi + T\varphi \in L_2\}.$$

There is a rich literature concerning estimates of the number of negative eigenvalues for the usual Schrödinger operator with a potential perturbation (see e.g. [BS91,Coj00,BEKS94,EK96,EK91,Ma85,To79,Tr97]). The main tool in this problem is the Birman-Schwinger principle. We deduce our results using this principle but without of the assumption that T is a potential or even a distribution.

2   Construction of the Generalized Schrödinger Operator

There are two ways for construction of $-\Delta_{\alpha,T}$, the form sum method and the generalized sum approach. Here we use the latter one (for details see [AK00,KaKo99] and [AKK01]), since it works even in a case not necessarily below bounded perturbations (c.f. with [PM94]).

Let $\Gamma \subset \mathbf{R}^n$ be a self-similar isotropic fractal of zero Lebesgue measure, $|\Gamma| = 0$ (for theory of fractal sets see [Fa85,Fa90,Hu81,Tr97]). Throughout this paper we assume that

$$n - 2 \;<\; d \;<\; n, \quad (0 \;<d<\; 1, \text{ if } \; n = 1), \tag{2.1}$$

where $d = \dim_H \Gamma$ denotes the Hausdorff dimension of Γ.

Let

$$\mathcal{H}_1 \equiv W_2^1 = \mathcal{M}_1 \oplus \mathcal{N}_1, \tag{2.2}$$

be the orthogonal decomposition of $\mathcal{H}_1$, where

$$\mathcal{M}_1 := W_2^1(\mathbf{R}^n \setminus \Gamma) = \{f \in W_2^1 : \mathrm{tr}_\Gamma f = 0\}, \tag{2.3}$$

and $\mathrm{tr}_\Gamma f \equiv f \mid \Gamma$ means the restriction (trace) of a function f onto Γ. We note that by the spectral synthesis theorem (see [AH96]) the trace $\mathrm{tr}_\Gamma f$ exists for any element $f \in W_2^1$. From (2.2) it follows that

$$\mathcal{N}_1 = \{\eta \in W_2^1 : \eta = f - P_{\mathcal{M}_1} f, \;\; f \in W_2^1(\mathbf{R}^n)\}, \tag{2.4}$$

where $P_{\mathcal{M}_1}$ denotes the orthogonal projection in $\mathcal{H}_1$ onto the subspace $\mathcal{M}_1$. Condition (2.1) ensures that $\mathcal{N}_1 \neq 0$, since the W_2^1-capacity of Γ is strongly positive (see [AH96]). Thus there exists a possibility to give a non-trivial sense for perturbation supported by fractal Γ in spite of the fact that the Lebesgue measure of Γ is zero and therefore the subspace $\mathcal{M}_1$ is dense in L_2 .

We observe that, for elements from the Sobolev space W_2^1, the operations of restriction onto Γ and projection onto the subspace $\mathcal{N}_1$ are in one-to-one correspondence. Moreover any $\eta \in \mathcal{N}_1$ is uniquely determined by its trace, i.e., by the function $h(\gamma) = \mathrm{tr}_\Gamma \eta(\gamma)$, $\gamma \in \Gamma$. By standard arguments one can identify the subspace

$$\mathcal{N}_1 = \mathcal{H}_1 \ominus \mathcal{M}_1$$

with the space of traces

$$\{h(\gamma) = \mathrm{tr}_\Gamma f(\gamma), \; \gamma \in \Gamma : \; f \in W_2^1(\mathbf{R}^n)\}.$$

According to [Tr97] (see Definition 20.2 and notation (25.1)) the above space of traces is denoted by $H^s(\Gamma)$ with $s = 1 - \frac{n-d}{2}$. Thus we can write (for details see again [Tr97], Sections 18, 20),

$$H^s(\Gamma) = \mathrm{tr}_\Gamma W_2^1(\mathbf{R}^n), \;\; s = 1 - \frac{n-d}{2},$$

where the space $H^s(\Gamma)$ may be defined as a complete Hilbert space of functions $\mathrm{tr}_\Gamma f(\gamma) = h(\gamma)$ with respect to the quasi-norm

$$\|h\|_{H^s} = \inf \|f\|_1,$$

where the infimum is taken over all $f \in W_2^1(\mathbf{R}^n)$ such that $\mathrm{tr}_\Gamma f = h$. Thus for all $f \in W_2^1(\mathbf{R}^n)$ with a fixed $h = \mathrm{tr}_\Gamma f$ and $\eta = P_{\mathcal{N}_1} f$ we have,

$$\|h\|_{H^s} = \|\eta\|_{W_2^1} = \|\eta\|_{\mathcal{N}_1}. \tag{2.5}$$

So, taking into account the above identification between $\eta = P_{\mathcal{N}_1} f \in \mathcal{N}_1$ and $h = \mathrm{tr}_\Gamma f \in H^s(\Gamma)$, one can write

$$\mathcal{N}_1 = H^s(\Gamma), \quad s = 1 - \frac{n-d}{2}. \tag{2.6}$$

Let $L_2(\Gamma) \equiv L_2(\Gamma, d\mu)$ denote the usual complex space, where μ is the restriction onto Γ of the d–dimensional Hausdorff measure in $\mathbf{R}^n$, i.e., $\mu = \mathcal{H}^d \mid \Gamma$. So the norm in $L^2(\Gamma, d\mu)$ is given by $\int_\Gamma \mid h(\gamma) \mid^2 d\mu(\gamma)$. The important fact which we will used say that the embedding $I_\gamma : H^s(\Gamma) \to L_2(\Gamma, d\mu)$ is compact (see Proposition 20.5 in [Tr97]).

Let τ be a positive invertible operator in $L_2(\Gamma)$ with domain $\mathcal{D}(\tau) \subset \mathcal{N}_1$. Assume that the quadratic form generated by τ is bounded in $\mathcal{N}_1$. Further let Q_T be the quadratic form in W_2^1 constructed by τ and μ,

$$Q_T[f] = \int_\Gamma (\tau h)(\gamma)\bar{h}(\gamma)d\mu(\gamma), \quad h = \mathrm{tr}_\Gamma f, \quad f \in W_2^1. \tag{2.7}$$

By construction the form $Q_T[f]$ is bounded in W_2^1 and therefore there exists the bounded self-adjoint operator associated with the closure of $Q_T[f]$ in W_2^1:

$$(\mathcal{T}f, \, f)_1 = Q_T[f], \quad f \in W_2^1. \tag{2.8}$$

We observe that

$$\mathrm{Ker}Q_T = \mathrm{Ker}\mathcal{T} = \mathcal{M}_1$$

due to (2.3) and (2.7). Recall that $\mathcal{M}_1$ is dense in L_2 since the Lebesgue measure of Γ is zero.

Lemma 2.1. *Under condition (2.1) the operator $\mathcal{T}$ is compact in W_2^1. The null-subspace of $\mathcal{T}$ is dense in L_2.*

Indeed using the identification (2.6) one can consider $\mathcal{T}$ as an operator in $H^s(\Gamma)$, where it is a product of a bounded operator τ and the compact embedding $H^s(\Gamma)$ into $L_2(\Gamma, d\mu)$.

Put $T = (I - \Delta)\mathcal{T} : W_2^1 \to W_2^{-1}$, where recall that $(I - \Delta) : W_2^1 \to W_2^{-1}$ is unitary. Clearly $\mathrm{Ker}T = \mathcal{M}_1$. This means that T is supported by fractal Γ, i.e., $Tf = 0$ for any $f \in W_2^1$ such that $\mathrm{tr}_\Gamma f = 0$. Therefore the operator T belongs to the $\mathcal{H}_1 - class$ (see [Ko99]), in particular the set $\mathrm{Ker}T \cap \mathcal{D}(-\Delta)$ is dense in the subspace $\mathcal{M}_1$. Thus for the construction of the perturbed operator we can use the generalized sum approach (for more details see [AK00,KK99,Ko00,Kosh00]). So, define the operator $-\tilde{\Delta}$ as $-\Delta_{\alpha,T} = -\Delta \tilde{+} \alpha T$, $\alpha \in \mathbf{R}$. In fact we deal with some family of operators which belong to the class $\mathcal{P}_{ws}(-\Delta)$. Every such operator we call the generalized Schrödinger operator with fractal perturbation.

Theorem 2.1. *Let* $T = (I - \Delta)\mathcal{T} : W_2^1 \to W_2^{-1}$ *with* $\mathcal{T}$ *defined by (2.8), (2.7), where* Γ *is a fractal of Hausdorff dimension d satisfying condition (2.1). Then the generalized Schrödinger operator* $-\Delta_{\alpha,T} = -\Delta \tilde{+} \alpha T$, $\alpha \in \mathbf{R}$, *is self-adjoint in* L_2. $-\Delta_{\alpha,T}$ *coincides with* $-\Delta$ *on the domain*

$$\mathcal{D} := \{f \in W_2^2 : f(x) = 0, x \in \Gamma\} = \mathrm{Ker}T \cap W_2^2, \tag{2.9}$$

which is dense in L_2. *Moreover* $-\Delta_{\alpha,T}$ *belongs to the class* $\mathcal{P}_{ws}(-\Delta)$ *and hence it is a self-adjoint extension of the symmetric operator*

$$- \overset{\circ}{\Delta} = -\Delta \upharpoonright \mathcal{D}$$

that has infinite deficiency indices,

$$\mathbf{n}^{\pm}(- \overset{\circ}{\Delta}) = \dim\mathcal{N}_1 = \infty.$$

3   The Negative Eigenvalue Problem

Let $-\tilde{\Delta} = -\Delta_{\alpha,T} = -\Delta \tilde{+} \alpha T \in \mathcal{P}_{ws}(-\Delta)$ be the generalized Schrödinger operator with fractal perturbation constructed in the previous section. Since due to Lemma 2.1 the operator $\mathcal{T} = (I - \Delta)^{-1}T$ is compact in W_2^1,

$$\sigma_{ess}(-\tilde{\Delta}) = \sigma(-\Delta) = [0, -\infty).$$

We want to study the negative eigenvalue problem:

$$-\tilde{\Delta}\psi = E\psi, \quad E < 0, \tag{3.1}$$

and to answer the question, how many solutions of (3.1) exist, counting the multiplicities. In particular, whether the counting function

$$N_-(-\tilde{\Delta}) = \#\{E < 0 : E \in \sigma_{disc}(-\tilde{\Delta})\} \tag{3.2}$$

is finite or infinite, where $\#$ means the number of elements of a set $\{\cdot\}$.

As a tool in the spectral analysis of the operator $-\tilde{\Delta}$ we will use the following theorem (cf. with Theorem 4.4 in ([AK00])).

Theorem 3.1. *A pair $E < 0$, $\psi \in \mathcal{D}(-\tilde{\Delta})$ solves the problem (3.1) if and only if ψ belongs to the deficiency subspace $\mathfrak{N}_E = \mathfrak{N}_{z=E}$ of the symmetric operator $-\overset{\circ}{\Delta}$,*

$$\psi \in \mathfrak{N}_E, \tag{3.3}$$

and satisfies the equation

$$\mathcal{T}_E \psi = -\psi, \tag{3.4}$$

where $\mathcal{T}_E := (E - \Delta)^{-1}T$. Moreover each such ψ has a form

$$\psi = (E - \Delta)^{-1}\omega,$$

with some $\omega \in \mathcal{N}_{-1} := (I - \Delta)\mathcal{N}_1$

4   The Birman-Schwinger Principle

Here we give a brief account of the main tool for investigation of the negative eigenvalue problem which is known as the Birman-Schwinger principle (for more details see [BS91]).

Let A be a bounded from below self-adjoint operator in a Hilbert space $\mathcal{H}$. Define the spectral distribution function

$$N(\lambda, A) := dim[\mathbf{E}_A(-\infty, \lambda)\mathcal{H}], \ \lambda \in \mathbf{R},$$

where $\mathbf{E}_A(\cdot)$ is the operator spectral measure of A. Obviously $N(\lambda, A) = 0$ for $\lambda \leq inf(Af, f)$, $\|f\| = 1$. If $N(\lambda, A) < \infty$ for some λ then only a finite number of eigenvalues lie on the left side from a point λ, i.e.,

$$\sigma(A) \cap (-\infty, \lambda) \subset \sigma_{disc}(A).$$

The following result is known as the Glazman lemma. Let $a[f] := (Af, f)$ be the quadratic form generated by A and $\mathcal{H}_1(A)$ denote the closure of $\mathcal{D}(A)$ with respect to the norm $\|f\|_1 = (a[f] + m_A\|f\|^2)^{1/2}$, where $m_A > |inf(Af, f)|$, $\|f\| = 1$. Then

$$N(\lambda, A) = \sup\{\dim F : F \subset \mathcal{H}_1(A), \ a[f] < \lambda\|f\|^2, 0 \neq f \in F\}, \tag{4.1}$$

where F may be taking from a dense linear set $\mathcal{F} \subset \mathcal{H}_1(A)$.

Let now the operator A be positive, $A > 0$, i.e., $a[f] = (Af, f) > 0$, $0 \neq f \in \mathcal{D}(A)$. Let H^1 denote the space obtained as the completion of $\mathcal{D}(A)$ with respect to the norm $\|f\|_{H^1} := (a[f])^{1/2}$. Consider in H^1 a bounded positive quadratic form $b[f]$. Let $\mathcal{B}$ denote the associated to $b[f]$ self-adjoint operator in H^1, $b[f] = (\mathcal{B}f, f)_{H^1}$. Define a suitable version of the spectral distribution function for $\mathcal{B}$,

$$n_+(\lambda, \mathcal{B}) = \dim[\mathbf{E}_\mathcal{B}H^1], \ \lambda > 0.$$

Using the same arguments as in Glazman's lemma one can show that

$$n_+(\lambda, \mathcal{B}) = \sup\{\dim F : F \subset H^1, \; b[f] > \lambda a[f], \; 0 \not\equiv f \in F\}. \tag{4.2}$$

Assume now that the quadratic form $a_\alpha[f] = a[f] + \alpha b[f]$, $\alpha < 0$ is bounded from below and closable in $\mathcal{H}$. Then there exists the self-adjoint operator A_α associated to the closure of $a_\alpha[f]$ in $\mathcal{H}$, $(A_\alpha f, f) = a_\alpha[f]$.

Let $N_-(A_\alpha) := N(\lambda, A_\alpha)$ with $\lambda = 0$. According to (4.1) we have

$$N_-(A_\alpha) == \sup\{\dim F : F \subset \mathcal{H}_1(A_\alpha), \; a_\alpha[f] < 0, \; 0 \not\equiv f \in F\}. \tag{4.3}$$

We observe that the condition $a_\alpha[f] = a[f] + \alpha b[f] < 0$ in (4.3) coincides with condition $b[f] > \lambda a[f]$ in (4.2) for $\lambda = -1/\alpha$ (recall that $\alpha < 0$).

Thus

$$N_-(A_\alpha) = n_+(\lambda, \mathcal{B}), \; \lambda = -\frac{1}{\alpha}.$$

Therefore if $n_+(\lambda, \mathcal{B})$ is finite for some $\alpha > 0$ then the negative spectrum of A_α is discrete for all $\alpha \in [-\frac{1}{\alpha}, 0)$. So we have

Theorem 4.1. *Let A be a positive operator in a Hilbert space $\mathcal{H}$ and let $B : \mathcal{H}_1(A) \to \mathcal{H}_{-1}(A)$ be a positive operator acting in the A-scale of spaces. Let $a[f]$ and $b[f]$ denote the corresponding to A and B quadratic forms. Assume that the quadratic form $b[f]$ is bounded in the space H^1, which is the completion of $\mathcal{D}(A)$ with respect to the norm $\|f\|_{H^1} = (a[f])^{1/2}$. Assume that the operator B associated with $b[f]$ in H^1 is compact. Then for any $\alpha < 0$ the quadratic form $a_\alpha[f] = a[f] + \alpha b[f]$ is bounded from below and closable in $\mathcal{H}$ and for the operator A_α associated to $a_\alpha[f]$ the negative spectrum is finite, $N_-(A_\alpha) < \infty$.*

5   The Poincaré Inequality

Another tool used below is the classical Poincaré inequality (see [AH96]),

$$\int_\Omega |f|^2 dx \leq c(\Omega) \int_\Omega |\nabla f|^2 dx, \tag{5.1}$$

valid for $f \in \overset{\circ}{W}{}_2^1(\Omega)$, where $\Omega \subset \mathbf{R}^n$ is a bounded open domain with C^∞ boundary, and $\overset{\circ}{W}{}_2^1(\Omega)$ stands for the completion of $C_0^\infty(\Omega)$ in the usual Sobolev norm. The important consequence of the Poincaré inequality (5.1) is,

$$\|f\|_{\overset{\circ}{W}{}_2^1(\Omega)} \leq c_1(\Omega)\|f\|_{\overset{\circ}{H}{}_2^1(\Omega)}, \tag{5.2}$$

where $\overset{\circ}{H_2^1}(\Omega)$ denotes a homogeneous version of the Sobolev space, i.e., $\overset{\circ}{H_2^1}(\Omega)$ is the completion of $C_0^\infty(\Omega)$ in the norm,

$$\|f\|_{\overset{\circ}{H_2^1}(\Omega)} = \int_\Omega |\nabla f|^2 dx)^{1/2}.$$

From (5.2) it follows that

$$\overset{\circ}{W_2^1}(\Omega) = \overset{\circ}{H_2^1}(\Omega). \tag{5.3}$$

Let $-\Delta_{\alpha,T}$ be defined in $L_2(\Omega, dx)$ by the way described above, where $-\Delta$ is the Laplace operator with Dirichlet boundary conditions. As a simple consequence of Lemma 2.1, the Birman-Schwinger principle, and equality (5.3) we derive the following result (one can also get the similar result from Theorem 30.2 in [Tr97]).

Theorem 5.1. *Let the Hausdorff dimension d of a fractal $\Gamma \subset \Omega$ satisfy condition (2.1). Then $N_-(-\Delta_{\alpha,T}) < \infty$ for all $\alpha < 0$, where $-\Delta_{\alpha,T}$ is the generalized Schrödinger operator with fractal perturbation in $L_2(\Omega, dx)$.*

6 Finiteness of Negative Spectrum

Let an operator $T : W_2^1(\mathbf{R}^n) \to W_2^{-1}(\mathbf{R}^n)$ be as above and therefore $\mathcal{T} = (I - \Delta)^{-1}T$ has a purely discrete spectrum in $W_2^1 \ominus \mathrm{Ker} T$. Assume that the range of T essentially belongs to the space H_2^{-1}, i.e., $\dim[\mathcal{R}(T) \cap H_2^{-1}] = \infty$, where H_2^{-1} is the dual space to H_2^1, which is defined as the completion of $C_0^\infty(\mathbf{R}^n)$ in the norm $\|f\|_{H_2^1} = (\int_{\mathbf{R}^n} |\nabla f|^2 dx)^{1/2}$. Assume additionally that T admits a decomposition into the sum,

$$T = T^{\mathrm{fin}} + T',$$

where T' acts from H_2^1 into H_2^{-1}, and $T^{\mathrm{fin}} : W_2^1 \to W_2^{-1}$ has a finite rank. Then by the additive property of generalized sums we have

$$-\Delta_{\alpha,T} = -\Delta \widetilde{+} \alpha T = (-\Delta \widetilde{+} \alpha T') \widetilde{+} \alpha T^{\mathrm{fin}} = -\Delta \widetilde{+} \alpha T^{\mathrm{fin}} \widetilde{+} \alpha T'.$$

Since $W_2^1 \subset H_2^1$ one can understand the quadratic form $Q_{T'}[f] = \langle T'f, f \rangle$, $f \in W_2^1$ as a positive form in H_2^1. If it is bounded, then there exists a positive bounded operator $\mathcal{T}'$ associated to the closure of $Q_{T'}$ in H_2^1. If the operator $\mathcal{T}'$ is compact in H_2^1, then by the Birman- Schwinger principle the number of positive eigenvalues of $\mathcal{T}'$, which are greater than $-1/\alpha$ (counting their multiplicities), is given by the formula

$$n_+(-\frac{1}{\alpha}, \mathcal{T}') := \sup\{\dim F \subset H_2^1 : Q'_T[f] > -\frac{1}{\alpha}\|f\|_{H_2^1}^2\}, \quad \alpha < 0. \tag{6.1}$$

Theorem 6.1. *Let a positive operator $T : W_2^1 \to W_2^{-1}$ satisfy the same starting conditions as in Theorem 2.1. Assume additionally that T admits a decomposition into the sum $T = T^{\text{fin}} + T'$, where $T' : H_2^1 \to H_2^{-1}$ is a bounded operator, $\|T'\| = M < \infty$. Then the negative spectrum of the operator $-\Delta_{\alpha,T}$ is finite,*

$$N_-(-\Delta_{\alpha,T}) < \infty, \text{ if } -\alpha \le M. \tag{6.2}$$

Moreover if the operator T' is compact in H_2^1, then the negative spectrum of $-\Delta_{\alpha,T}$ is finite for all non-positive α,

$$n_+(-\frac{1}{\alpha}, T') \le N_-(-\Delta_{\alpha,T}) < \infty, \ \alpha < 0. \tag{6.3}$$

A simple general finiteness criterion for the number of negative eigenvalues may be obtained on the following way.

Let $K \subset L_2$ be a finite-dimensional subspace. Define the space $H_2^{1,K}$, as the completion of C_0^∞ in the norm

$$\|f\|_{H_2^{1,K}}^2 = Q_{-\Delta}[f] + \|P_K f\|^2,$$

where $Q_{-\Delta}[f] = \int |\nabla f|^2 \, dx$ and P_K stands for the orthogonal projector onto K in L_2.

Theorem 6.2. *Let $T : W_2^1 \to W_2^{-1}$ be a positive bounded operator which belongs to the $\mathcal{H}_{-1}-$class. Assume the generalized sum $-\Delta_T = -(\Delta \widetilde{+} T)$ defines a below bounded self-adjoint operator in L_2. Then the negative spectrum of $-\Delta_T$ is finite, if and only if, there exists a finite dimension subspace $K \subset L_2$ such that the quadratic form $Q_T[f]$ generated by T is bounded on $H_2^{1,K}$. In other words the following conditions are equivalent,*

$$Q_T[f] \le Q_{-\Delta}[f] + M \|P_K f\|_{L_2}^2, \ M < \infty, \tag{6.4}$$

and

$$N_-(-\Delta_T) \le N < \infty, \tag{6.5}$$

where

$$N = \dim K, \ M = |\inf(-\Delta_T f, f)|, \ \|f\|^2 = 1. \tag{6.6}$$

Indeed (6.4) implies that the operator $-\Delta_T + M P_K$ is positive since

$$(-\Delta_T f, f) + M \|P_K f\|_{L_2}^2 = \int |\nabla f|^2 \, dx - Q_T[f] + M \|P_K f\|_{L_2}^2 \ge 0.$$

Therefore the negative spectrum of $-\Delta_T$ contains at most $N = \dim K$ eigenvalues with minimal meanings not less than $-M$. So (6.5) and (6.6) follows from (6.4).

Conversely let (6.5) hold. Put $K = \mathrm{span}\{\psi_j\}_{j=1}^N$, where ψ_j denote the eigenvectors of $-\Delta_T$, $(-\Delta_T\psi_j = E_j\psi_j, E_j < 0)$. Then for any $M \geq \max\{-E_j\}$, the inequality (6.4) is fulfilled since the quadratic form $(-\Delta_T f, f) + M \|P_K f\|_{L_2}^2$ obviously is positive.

Remark that $N_-(-\widetilde{\Delta}) < \infty$ is also true for the operator $-\widetilde{\Delta} = -\Delta_{\alpha,T}$ with an arbitrary $\alpha < 0$, if the inequality (6.4) is replaced by

$$Q_T[f] \leq -\frac{1}{\alpha}(Q_{-\Delta}[f] + M_\alpha \|P_K f\|_{L_2}^2), \tag{6.7}$$

where, of course, the bound M_α goes to infinity as $\alpha \to 0$.

We observe that (6.4) (or (6.7)), in fact means that in the case $T = T_1 + T_2$, $T_1\colon H_2^1 \to H_2^{-1}$, $T_2 : W_2^1 \to W_2^{-1}$ the condition Rank $T_2 < \infty$ is necessary for $N_-(-\widetilde{\Delta}) < \infty$. Apparently the inverse is true also and therefore in such a case if the number of negative eigenvalues of $-\Delta_{\alpha,T}$ is finite for all $\alpha < 0$, then T admits a decomposition, $T = T' + T^{\mathrm{fin}}$, where $T^{\mathrm{fin}} : W_2^1 \to W_2^{-1}$ has a finite rank and the quadratic form $Q_{T'}[f]$, $f \in W_2^1$ generates in H_2^1 a compact operator.

Acknowledgements

This work was partly supported by DFG 436 UKR 113/43 project. The author is also indebed to the Technische University at Clausthal and the Institute of Applied Mathematics at the Bonn University for hospitality.

References

[AH96]   Adams, D.R., Hedberg, L.I., Functional spaces and Potential Theory, *Springer, Berlin - Tokyo*, 1996.

[AKK95] Albeverio, S., Karwowski, W., Koshmanenko, V., Square Power of Singularly Perturbed Operators, *Math. Nachr.*, 173:5 - 24, 1995.

[AKK01] Albeverio, S., Karwowski, W., Koshmanenko, V., On Negative Eigenvalues of Generalized Laplace Operator, (sent to publication).

[AK00]   S.Albeverio, V.Koshmanenko, On Schrödinger operators perturbed by fractal potentials, *Reports on Math. Phys.*, 45:307 - 325, 2000.

[BS91]   Birman, M.S., Solomyak, M.Z., Schrödinger Operator. Estimates for number of bound states as function-theoretical problem, *AMS Transl.*, 150:1 - 54, 1991.

[Coj00]   Cojuhari, P.A., Estimates of the number of perturbed eigenvalues, *17th OT Conference Proceedings*, 97-111, 2000.

[BEKS94] Brasche, J.F.,Exner, P., Kuperin, Yu.A., Šeba, P., Schrödinger Operator with Singular Interactions, *J.Math.Anal.Appl.*, 184:112 - 139, 1994.

[EK96]   Egorov, Yu., Kondratiev, V., On Spectral Theory of Elliptic Operators, *Operator Theory. Advances and Applications*, 89, 1996.

[EK91]   Egorov, Yu. V., Kondratiev, V.A., Estimates of the negative spectrum of the elliptic operator, *AMS Transl.*, 150:111 - 140, 1991.

[Fa85]   Falconer, K.J., The geometry of fractal sets, *Cambridge Univ. Press*, 1985.

[Fa90]   Falconer, K.J. Fractal geometry, *Chichester, Wiley*, 1990.

[Hu81]   Hutchinson, J.E. Fractals and selfsimilarity, *Indiana Univ. Math. J.*, 30:713 - 747, 1981.

[KaKo99]   Karataeva, T.V., Koshmanenko, V.D., Generalized sum of operators, *Math. Notes*, 66:671 - 681, 1999.

[KK00]   Karwowski, W., Koshmanenko, V., Schrödinger operator perturbed by dynamics of lower dimension, *AMS/IP Studies in Adv. Math.*, 16:249 - 257, 2000.

[KK99]   Karwowski, W., Koshmanenko, V., Generalized Laplace Operator in $L_2(\mathbf{R}^n)$, *Proceedings of Leipzig Conference dedicated to S. Albeverio*, 1999.

[KKO98]   Karwowski, W., Koshmanenko, V., Ôta, S., Schrödinger operator perturbed by operators related to null-sets, *Positivity*, 2:77 - 98, 1998.

[Ko99]   Koshmanenko, V., Singular Quadratic Forms in Perturbation Theory, *Kluwer Academic Publishers*, 1999.

[Ko00]   Koshmanenko, V., Singular operator as a parameter of self-adjoint extensions, Proceedings of the M. Krein Conference, Odessa, (1997), *Operator Theory. Advances and Applications*, 118:205 - 223, 2000.

[Kosh00]   Koshmanenko, V., Regular approximations of singular perturbations of $\mathcal{H}_{-2}$-class, *Ukrainian Math. J.*, 52:626 - 637, 2000.

[Ma85]   Maz'ya, V.G., Sobolev Spaces, *Springer, Berlin New York*, 1985.

[PM94]   Pavlov, B., Makarov, K., Quantum scattering on a Cantor bar, *J. Math. Phys.*, 35:1522-1531, 1994.

[To79]   Tomas, L., Birman-Schwinger bounds for the Laplacian with point interactions, *J. Math. Phys.*, 20:1848-1850, 1979.

[Tr97]   Triebel, H., Fractals and Spectra, *Birkhäuser, Basel Boston Berlin*, 1997.

Address

VOLODYMYR KOSHMANENKO, Institute of Mathematics, vul. Tereshchenkivs'ka, 3, Kyiv 01601, Ukraine

E-MAIL: kosh@imath.kiev.ua

2000 Mathematics Subject Classification. Primary 47A10, 47A55; Secondary 28A80, 47B

Operator Theory:
Advances and Applications, Vol. 126
© 2001 Birkhäuser Verlag Basel/Switzerland

On $\mathcal{H}_{-4}$-Perturbations of Self-Adjoint Operators

PAVEL KURASOV AND KAZUO WATANABE

Abstract. Supersingular $\mathcal{H}_{-4}$-perturbations of positive self-adjoint operators are studied. It is proven that such singular perturbations can be described by non self-adjoint operators with real spectrum. The resolvent of the perturbed operator is calculated using generalization a of Krein's formula for self-adjoint extensions.

1 Introduction

Finite rank perturbations of self-adjoint operator are used to obtain operators with complicated structure of the spectrum which are exactly solvable. Such perturbations are called singular if the domains of the perturbed and unperturbed operators are different. Such perturbations appear for example during the investigation of point interactions [AGH-KH88,DO88]. The first rigorous mathematical treatment of such Hamiltonian was suggested by F.Berezin and L.Faddeev in 1961 [BF61]. It was realized that finite rank perturbations of self-adjoint operators defined by vectors not from the original Hilbert space can lead to interesting exactly solvable models. The well-known Krein's formula relating the resolvents of two different self-adjoint extensions of one symmetric operator plays a very important rôle in the studies of the spectral properties of these operators [Kr44,Nai43,GMT98,K99,AK00]. The perturbed operator can be defined by first restricting the original operator to a certain symmetric one and then extending it to another self-adjoint operator. Such models having numerous advantages cannot be used to model all interesting phenomena. For example the Laplace operator with point interaction in $\mathbf{R}^3$ has a nontrivial scattering matrix in the s-channel only. In order to obtain models enabling to describe more complicated scattering phenomena, one has to consider finite rank perturbations determined by extremely singular vectors. The restriction-extension procedure described above can be used in the case where the perturbation is determined by the vectors being bounded linear functionals on the domain of the original operator. It has been realized that more singular perturbations can be defined only using certain extension of the original Hilbert space. In [Sh92,Sh88,TvD91,DLShZ00] rank one supersingular perturbations were defined using self-adjoint operators acting in Pontryagin spaces. It was shown that the spectral properties of these models are described by generalized Nevanlinna functions with a finite number of negative squares [HdS97]. Similar ideas were used in [KP95,PP91,PP84,Pop92] where concrete problems of mathematical physics were attacked. Different physicists and mathematicians tried to define supersingular perturbations [DD81,DD86,Karp92,And99].

In our previous paper [KW00] supersingular rank one perturbation of positive self-adjoint operator has been defined without any use of spaces with indefinite metrics. But our approach was limited to the case of so-called $\mathcal{H}_{-3}$ perturbations, i.e. perturbations determined by vectors from the Hilbert space $\mathcal{H}_{-3}$ from the scale of Hilbert spaces associated with the original positive self-adjoint operator. [1] In the current paper we are able to make one step further and describe all supersingular perturbations from the class $\mathcal{H}_{-4}$. These perturbations can be constructed in a certain extended Hilbert space. But no self-adjoint operator can be associated with such perturbations. It is shown that such singular perturbations are described by non self-adjoint operators with real spectrum. We obtain formula for the resolvent of such operator, which is similar to celebrated Krein's formula. The spectral properties of the operators are described by generalized Nevanlinna functions.

The paper is organized as follows. In Section 2 all necessary preliminary facts concerning finite rank singular and supersingular perturbations are reviewed. Maximal operator corresponding to rank one $\mathcal{H}_{-4}$-perturbation is calculated and investigated in Section 3. The family of operators corresponding to such singular perturbation is obtained in Section 4. Some conclusions and possibilities for further developments are discussed in the last section.

2 Preliminaries

Current paper is devoted to the construction of the operator describing rank one supersingular perturbation of a given positive self-adjoint operator A acting in a certain Hilbert space $\mathcal{H}$. The perturbed operator is given by the following formal expression

$$A_\alpha = A + \alpha \langle \varphi, \cdot \rangle \varphi, \tag{2.1}$$

where φ is a vector from the scale of Hilbert spaces associated with the operator A and α is a real coupling constant describing the strength of the perturbation.[2] Consider first the case $\varphi \in \mathcal{H}$. The perturbation $\alpha \langle \varphi, \cdot \rangle \varphi$ is a bounded symmetric operator and the perturbed operator A_α is self-adjoint on the domain of the original operator A. The resolvent of the perturbed operator is given by

$$\frac{1}{A_\alpha - \lambda} = \frac{1}{A - \lambda} - \frac{1}{\frac{1}{\alpha} + \langle \varphi, \frac{1}{A-\lambda}\varphi \rangle} \left\langle \frac{1}{A - \bar\lambda}\varphi, \cdot \right\rangle \frac{1}{A - \lambda}\varphi. \tag{2.2}$$

All spectral properties of the perturbed operator A_α are described by the Nevanlinna function $Q(\lambda) = \langle \varphi, \frac{1}{A-\lambda}\varphi \rangle$ (see for example [AK00]).

---

[1] The scale of Hilbert spaces is defined rigorously in Section 2.
[2] Recent developments in this area are described in details in
   [AK97,AK97-2,GS95,KS95,KK00,KN98,Si95,AK00].

Consider now the scale of Hilbert spaces $\mathcal{H}_s$ associated with the positive operator A. The norm in each space $\mathcal{H}_s$ is defined by

$$\| U \|^2_{\mathcal{H}_s} = \langle U, (A+1)^s U \rangle,$$

where $\langle \cdot, \cdot \rangle$ is the scalar product in the original Hilbert space $\mathcal{H}$. In order to avoid misunderstanding only the scale of Hilbert spaces associated with the original operator A and the original Hilbert space $\mathcal{H}$ will be considered throughout the paper. All perturbations defined by vectors φ not from the original Hilbert space $\mathcal{H}$ are called singular. These perturbations are characterized by the fact that the domain of the perturbed operator does not coincide with the domain of the original one. In the case $\varphi \in \mathcal{H}_{-1} \setminus \mathcal{H}$ the perturbation is relatively form bounded with respect to the sesquilinear form of the operator A and the perturbed operator can be determined using the form perturbation technique. The resolvent of the perturbed operator is again given by (2.2). The main difference is that the domain of the perturbed operator does not coincide with the domain of the original operator in general, but the perturbed operator is uniquely defined as a self-adjoint operator in the original Hilbert space $\mathcal{H}$ [Si95,AK97]. Another way to define the perturbed operator is using the extension theory for symmetric operators. It is obvious that the perturbed and original operators coincide on the linear set of functions U satisfying the condition

$$\langle \varphi, U \rangle = 0. \tag{2.3}$$

Then the perturbed operator is an extension of the original operator restricted to this linear set. If $\varphi \in \mathcal{H}_{-1} \setminus \mathcal{H}$ the restricted operator is a symmetric operator with the deficiency indices $(1,1)$. Its self-adjoint extension corresponding to the formal expression (2.1) is uniquely defined. The resolvent of the perturbed operator can be described using Krein's formula [Kr44,Nai43], which coincides with (2.2) in this case.

The case $\varphi \in \mathcal{H}_{-2} \setminus \mathcal{H}_{-1}$ has to be treated using the extension theory for symmetric operators, since the perturbation is not form bounded with respect to the original operator. The restricted symmetric operator can be defined in a way similar to $\mathcal{H}_{-1}$. But the perturbed operator is not uniquely defined anymore. One can only conclude that the perturbed operator is equal to one of the self-adjoint extensions of the restricted operator. All such operators can be parametrized by one real parameter $\gamma \in \mathbf{R} \cup \{\infty\}$ as follows

$$\frac{1}{A^\gamma - \lambda} = \frac{1}{A - \lambda} - \frac{1}{\gamma + \langle \varphi, \frac{1+\lambda}{A-\lambda} \frac{1}{A+1} \varphi \rangle} \left\langle \frac{1}{A - \bar\lambda} \varphi, \cdot \right\rangle \frac{1}{A - \lambda} \varphi. \tag{2.4}$$

The relation between the real parameter γ describing the self-adjoint extensions of the restricted operator and the additive real parameter α appearing in formula (2.1) cannot be established without additional assumptions like homogeneity of the original operator and the perturbation vector. [3] The Nevanlinna function $Q(\lambda) =$

---

[3] This approach has been developed in [AK97,AK97-2].

$\langle \varphi, \frac{1+\lambda}{A-\lambda} \frac{1}{A+1} \varphi \rangle$ can be considered as a regularization of the resolvent $\langle \varphi, \frac{1}{A-\lambda} \varphi \rangle$ which is not defined in the case $\varphi \in \mathcal{H}_{-2} \setminus \mathcal{H}_{-1}$

$$Q(\lambda) = \langle \varphi, \frac{1+\lambda}{A-\lambda} \frac{1}{A+1} \varphi \rangle \overset{\text{formally}}{=} \langle \varphi, \frac{1}{A-\lambda} \varphi \rangle - \langle \varphi, \frac{1}{A+1} \varphi \rangle. \qquad (2.5)$$

Observe that the two scalar products appearing in the right hand side of the last formula are not defined for $\varphi \in \mathcal{H}_{-2} \setminus \mathcal{H}_{-1}$, but their difference is in contrast well defined. The Nevanlinna function $\langle \varphi, \frac{1+A\lambda}{A-\lambda} \frac{1}{A^2+1} \varphi \rangle$ just coincides with Krein's Q-function appearing in the formula for the difference between the resolvents of two different self-adjoint extensions of one symmetric operator with the deficiency indices $(1,1)$ [Kr44,Nai43].

The next step is to consider $\varphi \in \mathcal{H}_{-3}$. The restriction defined by (2.3) is defined only if one considers the original operator A as an operator acting in the Hilbert space $\mathcal{H}_1$. Then the domain of the unperturbed operator A coincides with the space $\mathcal{H}_3$ and the restriction (2.3) determines a symmetric operator. From another hand formula (2.2) is valid only if one considers the extended Hilbert space containing vectors $\frac{1}{A-\lambda} \varphi \in \mathcal{H}_{-1}$. It appears that such extension is in fact one-dimensional, since

$$\frac{1}{A-\lambda} \varphi - \frac{1}{A-\mu} \varphi = (\varphi - \mu) \frac{1}{(A-\lambda)(A-\mu)} \varphi \in \mathcal{H}_1.$$

Hence it is enough to include the one dimensional subspace generated by the vector $\frac{1}{A+1} \varphi$ only. Hence the perturbed operator can be defined in the Hilbert space $\mathbf{H}_{-3} = \mathcal{H}_1 \oplus \mathbf{C}$ equipped with the natural embedding ρ_{-3}

$$\begin{aligned} \rho_{-3} : \mathbf{H}_{-3} &\to \mathcal{H}_{-1} \\ \mathbf{U} = (U, u_1) &\mapsto U + u_1 \frac{1}{A+1} \varphi. \end{aligned} \qquad (2.6)$$

The perturbed operator corresponding to the formal expression (2.1) has been constructed in [KW00] by first defining certain maximal operator acting in $\mathbf{H}$ and then restricting it to a self-adjoint operator. The maximal operator determined uniquely by the equality

$$\rho \mathbf{A} \mathbf{U} = A \rho \mathbf{U}|_{\text{mod}\varphi} \qquad (2.7)$$

is similar to the adjoint operator appearing in the restriction-extension procedure used to construct $\mathcal{H}_{-2}$-perturbations. The set of self-adjoint restrictions of the maximal operator are described by one real parameter. Therefore formula (2.1) does not determine the perturbed operator uniquely, but a one parameter family of operators like in the case of $\mathcal{H}_{-2}$-perturbations. The resolvent of the perturbed operator restricted to the original Hilbert space is given by the formula

$$\rho \frac{1}{\mathbf{A}_\theta - \lambda} |_{\mathcal{H}_1} = \frac{1}{A-\lambda} - \frac{1}{(\lambda+1)\cot\theta + \langle \varphi, \frac{1}{A-\lambda} \frac{(\lambda+1)^2}{(A+1)^2} \varphi \rangle - 1} \left\langle \frac{1}{A-\bar{\lambda}} \varphi, \cdot \right\rangle \frac{1}{A-\lambda} \varphi$$

$$(2.8)$$

where $\theta \in [0, \pi)$ is the real number parametrizing the restrictions. The similarity between formulas (2.2) and (2.8) is obvious. The function

$$
Q(\lambda) \;=\; \langle \varphi, \frac{1}{A - \lambda} \frac{(\lambda + 1)^2}{(A + 1)^2} \varphi \rangle
$$

$$
\overset{\text{formally}}{=} \langle \varphi, \frac{1}{A - \lambda} \varphi \rangle - \langle \varphi, \frac{1}{A + 1} \varphi \rangle - (\lambda + 1) \langle \varphi, \frac{1}{(A + 1)^2} \varphi \rangle
$$

(2.9)

is a double regularization of the resolvent function. This function describes the spectral properties of the self-adjoint perturbed operator.

The aim of the current paper is to describe rank one singular perturbations of positive self-adjoint operators in the case $\varphi \in \mathcal{H}_{-4} \setminus \mathcal{H}_{-3}$. Our original aim was simply to generalize the ideas developed in [KW00] for the case of more singular perturbations. The main difference with the case $\varphi \in \mathcal{H}_{-3}$ is that the original operator A should be considered as an operator acting in the Hilbert space $\mathcal{H}_2$ from the scale of Hilbert spaces. Moreover this Hilbert space should be extended to include not only the vector

$$
g_1 = \frac{1}{A + 1} \varphi \in \mathcal{H}_{-2}
$$

but the vector

$$
g_2 = \frac{1}{(A + 1)^2} \varphi \in \mathcal{H}
$$

as well. Hence one has to consider the Hilbert space

$$
\mathbf{H}_{-4} = \mathcal{H}_2 \oplus \mathbf{C}^2 \tag{2.10}
$$

equipped with the standard imbedding

$$
\begin{aligned}
\rho_{-4} &: \mathbf{H}_{-4} \to \mathcal{H}_{-2} \\
\mathbf{U} &= (U, u_2, u_1) \mapsto U + u_2 \tfrac{1}{(A+1)^2} + u_1 \tfrac{1}{A+1} \varphi.
\end{aligned} \tag{2.11}
$$

The maximal operator can be defined using formula (2.7) in the way similar to $\mathcal{H}_{-3}$-perturbations. The main difference is that any symmetric restriction of the maximal operator is not self-adjoint. Hence no self-adjoint operator corresponds to formal expression (2.1). Instead one can consider the restrictions of the maximal operator that are regular operators. Regular operators are characterized by the fact that their domain coincides with the domain of their adjoint. The class of regular operators includes the self-adjoint operators. All regular restrictions of the maximal operator are parametrized by one real parameter in the way similar to $\mathcal{H}_{-3}$ perturbations. The real and imaginary parts of these operators are calculated explicitly in the current paper. The resolvent of the perturbed operator is also calculated and it is shown that the spectrum of the perturbed operator is pure real. The resolvent restricted to the original Hilbert space is given by the formula similar to Krein's formula (2.4) and all spectral properties of the perturbed operator are described by Nevanlinna function Q given by (2.9).

3 The Extended Hilbert Space and the Maximal Operator

Following the ideas expressed in Section 2, consider the Hilbert space [4] $\mathbf{H} \equiv \mathbf{H}_{-4} = \mathcal{H}_2 \oplus \mathbf{C}^2$ equipped with the scalar product

$$\ll \mathbf{U}, \mathbf{V} \gg = \langle U, V \rangle_{\mathcal{H}_2} + \bar{u}_2 v_2 + \bar{u}_1 v_1$$
$$= \langle U, (1+A)^2 V \rangle + \bar{u}_2 v_2 + \bar{u}_1 v_1. \tag{3.1}$$

One can consider different scalar product in $\mathbf{H}$ but we decided to study the simplest case in order to avoid not necessary complications. The maximal operator $\mathbf{A}$ being analog of the adjoint operator in the extension theory and acting in $\mathbf{H}$ is defined using formula (2.7). The operator A appearing in this formula is the extension of the original operator A to the space $\mathcal{H}_{-2}$. This operator maps the space $\mathcal{H}_{-2}$ onto the space $\mathcal{H}_{-4}$. To describe the domain of the operator $\mathbf{A}$ one needs to consider the vector

$$g_3 = \frac{1}{A+1} g_2 = \frac{1}{(A+1)^3} \varphi. \tag{3.2}$$

Lemma 3.1. *Let $\varphi \in \mathcal{H}_{-4}$, then the maximal operator $\mathbf{A}$ determined by (2.7) in $\mathbf{H}$ is defined on the domain*

$$\mathrm{Dom}(\mathbf{A}) = \{\mathbf{U} = (U_r + u_3 g_3, u_2, u_1), U_r \in \mathcal{H}_4, u_3, u_2, u_1 \in \mathbf{C}\} \tag{3.3}$$

by the formula

$$\mathbf{A} \begin{pmatrix} U_r + u_3 g_3 \\ u_2 \\ u_1 \end{pmatrix} = \begin{pmatrix} AU_r - u_3 g_3 \\ u_3 - u_2 \\ u_2 - u_1 \end{pmatrix}. \tag{3.4}$$

Proof. Consider any vector $\mathbf{U}$ from the domain of the operator $\mathbf{A}$ and let us denote its image by $\mathbf{V} = (V, v_2, v_1)$. Then equality (2.7) can be written as follows

$$V + v_2 g_2 + v_1 g_1 \overset{\mathrm{mod}\ \varphi}{=} AU - u_2 g_2 + (u_2 - u_1) g_1 + u_1 \varphi. \tag{3.5}$$

Taking into account that this equality holds modulus φ and that $AU \in \mathcal{H}_0$ we conclude that

$$v_1 = u_2 - u_1. \tag{3.6}$$

This implies

$$V + v_2 g_2 = AU - u_2 g_2,$$

---

[4] We are going to drop the subindex -4 in the notations for the extended Hilbert space and standard imbedding introduced in (2.10) and (2.11).

where the usual equality sign is used, since the functions appearing on both sides of the equation belong to the original Hilbert space $\mathcal{H}$. This equality can be written as

$$V + U + v_2 g_2 + u_2 g_2 = (A+1)U$$

and therefore

$$U = \frac{1}{A+1}(V - U) + (v_2 - u_2)\frac{1}{A+1}g_2.$$

It follows that the element U possesses the following representation

$$U = U_r + u_3 g_3,$$

where $U_r \in \mathcal{H}_4$, $u_3 \in \mathbf{C}$ and g_3 is given by (3.2). Then equality (3.5) can be written as

$$V + v_2 g_2 + v_1 g_1 = AU_r - u_3 g_3 + (u_3 - u_2)g_2 + (u_2 - u_1)g_1,$$

and one can deduce that

$$v_2 = u_3 - u_2, \quad V = AU_r - u_3 g_3.$$

The Lemma is proven. $\square$

The spectrum of the operator $\mathbf{A}$ covers the whole complex plane. Really consider any complex number λ. Then the element

$$\mathbf{U} = \begin{pmatrix} \dfrac{(1+\lambda)^3}{A-\lambda}g_3 + (1+\lambda)^2 g_3 \\ (1+\lambda) \\ 1 \end{pmatrix} = \begin{pmatrix} (1+\lambda)^2\dfrac{A+1}{A-\lambda}g_3 \\ (1+\lambda) \\ 1 \end{pmatrix}$$

solves the equation $\mathbf{AU} = \lambda\mathbf{U}$. Note that the last formula reads as follows in the special case $\lambda = -1$

$$\mathbf{A}\begin{pmatrix} 0 \\ 0 \\ 1 \end{pmatrix} = -\begin{pmatrix} 0 \\ 0 \\ 1 \end{pmatrix}.$$

Let us calculate the adjoint operator $\mathbf{A}^*$.

Lemma 3.2. *The operator $\mathbf{A}^*$, adjoint to $\mathbf{A}$, is defined on the domain*

$$\mathrm{Dom}\,(\mathbf{A}^*) = \{\mathbf{U} = (U_r, u_2, u_1); U_r \in \mathcal{H}_4, u_2, u_1 \in \mathbf{C}, \tag{3.7}$$
$$u_2 = \langle \varphi, U_r \rangle\}$$

by the formula

$$\mathbf{A}^*\begin{pmatrix} U_r \\ u_2 \\ u_1 \end{pmatrix} = \begin{pmatrix} AU_r \\ u_1 - u_2 \\ -u_1 \end{pmatrix}. \tag{3.8}$$

Proof. Consider arbitrary elements $\mathbf{U} \in \mathrm{Dom}\,(\mathbf{A})$ and $\mathbf{V} = (V, v_2, v_1) \in \mathbf{H}$. The sesquilinear form of the operator $\mathbf{A}$ can be written as follows

$$\ll (\mathbf{A} + 1)\mathbf{U}, \mathbf{V} \gg \,= \langle (A + 1)U_r, (1 + A)^2 V \rangle + \bar{u}_3 v_2 + \bar{u}_2 v_1$$

$$= \langle U_r + u_3 g_3, (1 + A)^3 V \rangle + \bar{u}_3 \left\{ -\langle g_3, (1 + A)^3 V \rangle + v_2 \right\}$$

$$+ \bar{u}_2 v_1.$$

$$(3.9)$$

Consider the subset of elements $\mathbf{U} \in \mathrm{Dom}\,(\mathbf{A})$ with $u_1 = u_2 = u_3 = 0$. Then the first term in the last formula is a bounded functional with respect to $\mathbf{U} \in \mathbf{H}$ if and only if $V = V_r \in \mathcal{H}_4$. Consider next arbitrary $\mathbf{U} \in \mathrm{Dom}\,(\mathbf{A})$. The first and the last two terms are bounded linear functionals with respect to $\mathbf{U} \in \mathbf{H}$, but $\mathbf{U} \mapsto u_3$ is not a bounded linear functional. Therefore we conclude that $\ll \mathbf{AU}, \mathbf{V} \gg$ is a bounded linear functional with respect to $\mathbf{U}$ only if the following (boundary) condition is satisfied

$$v_2 = \langle g_3, (1 + A)^3 V_r \rangle \equiv \langle \varphi, V_r \rangle. \tag{3.10}$$

We conclude that the domain of the adjoint operator $\mathbf{A}^*$ consists of elements $\mathbf{V} \in \mathbf{H}$ possessing the representation

$$\mathbf{V} = (V_r, v_2, v_1), \quad V_r \in \mathcal{H}_4, v_{1,2} \in \mathbf{C}$$

and satisfying (3.10). Under these conditions formula (3.9) reads as follows

$$\ll \mathbf{AU}, \mathbf{V} \gg \,= \langle U_r + u_3 g_3, (1 + A)^2 A V_r \rangle + \bar{u}_2 (v_1 - v_2) + \bar{u}_1 (-v_1).$$

This formula implies that the action of the adjoint operator is given by (3.8) and the formulated conditions on the elements from its domain are not only necessary but also sufficient. The Lemma is proven. $\square$

The domain of $\mathbf{A}^*$ is included in the domain of $\mathbf{A}$. But the operator $\mathbf{A}^*$ is not a restriction of $\mathbf{A}$. Therefore no restriction of the operator $\mathbf{A}$ is self-adjoint like in the case of $\mathcal{H}_{-3}$-perturbations. One can prove this fact directly using the boundary form of the operator $\mathbf{A}$.

Lemma 3.3. *The boundary form of the maximal operator* $\mathbf{A}$ *is given by*

$$\ll \mathbf{AU}, \mathbf{V} \gg - \ll \mathbf{U}, \mathbf{AV} \gg \,= \left\langle \begin{pmatrix} 0 & -1 & 0 & 0 \\ 1 & 0 & -1 & 0 \\ 0 & 1 & 0 & -1 \\ 0 & 0 & 1 & 0 \end{pmatrix} \begin{pmatrix} \langle \varphi, U_r \rangle \\ u_3 \\ u_2 \\ u_1 \end{pmatrix}, \begin{pmatrix} \langle \varphi, V_r \rangle \\ v_3 \\ v_2 \\ v_1 \end{pmatrix} \right\rangle$$

$$(3.11)$$

Proof. The following straightforward calculations prove the Lemma

$$\ll \mathbf{AU}, \mathbf{V} \gg - \ll \mathbf{U}, \mathbf{AV} \gg$$
$$= \ll \begin{pmatrix} AU_r - u_3 g_3 \\ u_3 - u_2 \\ u_2 - u_1 \end{pmatrix}, \begin{pmatrix} V_r + v_3 g_3 \\ v_2 \\ v_1 \end{pmatrix} \gg - \ll \begin{pmatrix} U_r + u_3 g_3 \\ u_2 \\ u_1 \end{pmatrix}, \begin{pmatrix} AV_r - v_3 g_3 \\ v_3 - v_2 \\ v_2 - v_1 \end{pmatrix} \gg$$

$$= \langle AU_r - u_3 g_3, (1+A)^2 (V_r + v_3 g_3) \rangle + (\bar{u}_3 - \bar{u}_2)v_2 + (\bar{u}_2 - \bar{u}_1)v_1$$
$$- \langle U_r + u_3 g_3, (1+A)^2 (AV_r - v_3 g_3) \rangle - \bar{u}(v_3 - v_2) - \bar{u}_1(v_2 - v_1)$$

$$= -\bar{u}_3 \langle \varphi, V_r \rangle + \langle U_r, \varphi \rangle v_3 + \bar{u}_3 v_2 + \bar{u}_2 v_1 - \bar{u}_2 v_3 - \bar{u}_1 v_2.$$

We have used that $\varphi = (1+A)^3 g_3$ in these calculations. The Lemma is proven. $\square$

The sesquilinear boundary form of the maximal operator can also be presented in the form

$$\ll \mathbf{AU}, \mathbf{V} \gg - \ll \mathbf{U}, \mathbf{AV} \gg$$
$$= \bar{u}_3 \left(v_2 - \langle \varphi, V_r \rangle \right) - \overline{(u_2 - \langle \varphi, U_r \rangle)} v_3 + \bar{u}_2 v_1 - \bar{u}_1 v_2. \tag{3.12}$$

The matrix describing the boundary form

$$\begin{pmatrix} 0 & -1 & 0 & 0 \\ 1 & 0 & -1 & 0 \\ 0 & 1 & 0 & -1 \\ 0 & 0 & 1 & 0 \end{pmatrix}$$

is symplectic and has rank four. [5] Any symmetric restriction of the operator $\mathbf{A}$ is described by at least two boundary conditions. Such restriction cannot be self-adjoint, since the kernel of the operator $\mathbf{A} - \lambda$, $\Im\lambda \neq 0$ has dimension 1. Therefore no restriction of the operator $\mathbf{A}$ is a self-adjoint operator in the Hilbert space $\mathbf{H}$. It follows that no self-adjoint operator corresponds to formal expression (2.1) in the case $\varphi \in \mathcal{H}_{-4} \setminus \mathcal{H}_{-3}$. In what follows we are going to show that the operator corresponding to (2.1) can be determined in the class of regular operators.

4 Supersingular Perturbation as a Regular Operator

In this section we are going to define an operator corresponding to (2.1) in the class of regular operators. We call an operator acting in the Hilbert space **regular** if it is densely defined and the domain of the operator coincides with the domain of the adjoint one. The operators corresponding to (2.1) will be defined by restricting the maximal operator $\mathbf{A}$ to a certain regular operator.

---

[5] Its characteristic determinant is equal to $\lambda^4 + 3\lambda^2 + 1$.

Theorem 4.1. *All regular restrictions of the operator* $\mathbf{A}$ *are described by the following boundary condition*

$$a\langle \varphi, U_r\rangle + bu_3 + cu_2 = 0, \tag{4.1}$$

where (a, b, c) *is a three dimensional nonzero vector* $(a, b, c) \in \mathbf{R}^3$ *orthogonal to the vector* $(1, 0, 1)$.

Proof. The domain $\mathrm{Dom}\,(\mathbf{A})$ consists of all elements $\mathbf{U} \in \mathbf{H}$ possessing the following representation

$$\mathbf{U} = (U_r + u_3 g_3, u_2, u_1),$$

where $U_r \in \mathcal{H}_4$, $u_1, u_2, u_3 \in \mathbf{C}$. The domain $\mathrm{Dom}\,(\mathbf{A}^*)$ of the adjoint operator is a subdomain of $\mathrm{Dom}\,(\mathbf{A})$ described by the boundary conditions

$$\begin{cases} u_3 = 0, \\ u_2 = \langle \varphi, U_r\rangle. \end{cases}$$

Therefore the quotient space $\mathrm{Dom}\,(\mathbf{A})/\mathrm{Dom}\,(\mathbf{A}^*)$ has dimension 2 and any nontrivial restriction [6] of the domain $\mathrm{Dom}\,(\mathbf{A})$ to a linear subset containing $\mathrm{Dom}\,(\mathbf{A}^*)$ is described by the boundary condition

$$a\langle \varphi, U_r\rangle + bu_3 + cu_2 = 0, \tag{4.2}$$

where a, b, and c are arbitrary complex numbers, not all equal to zero simultaneously $|a|^2 + |b|^2 + |c|^2 \neq 0$. Let us denote the corresponding subspace by $\mathbf{D}$ and the restriction of the operator $\mathbf{A}$ to this subspace by $\mathbf{A}|_{\mathbf{D}}$. The operator $\mathbf{A}|_{\mathbf{D}}$ is a restriction of the operator $\mathbf{A}$. Therefore the adjoint operator $\mathbf{A}|_{\mathbf{D}}^*$ is an extension of the operator $\mathbf{A}^*$. The sesquilinear form of the operator $\mathbf{A}|_{\mathbf{D}}$ is given by formula (3.9), where now $\mathbf{U} \in \mathrm{D}$. Consider vectors $\mathbf{U}$ with $u_2 = u_3 = \langle \varphi, U_r\rangle = 0$. Then the scalar product

$$\langle U_r, (A+1)^3 V\rangle$$

generates a bounded linear functional with respect to $(U_r, 0, 0) \in \mathbf{H}$ and the standard norm in $\mathbf{H}$ if and only if the following representation holds [7]

$$(A+1)^3 V = c\varphi + \tilde{f},$$

where $c \in \mathbf{C}$, $\tilde{f} \in \mathcal{H}_{-2}$. This implies that

$$V = cg_3 + \frac{1}{(A+1)^3}\tilde{f},$$

---

[6] Nontrivial restriction in this context is the one having domain different from both $\mathrm{Dom}\,(\mathbf{A}^*)$ and $\mathrm{Dom}\,(\mathbf{A})$.

[7] Remember that U_r is orthogonal to φ.

and it follows that the vector V possesses the representation

$$V = V_r + v_3 g_3,$$

where $V_r \in \mathcal{H}_4, \quad v_3 \in \mathbf{C}$. Then the sesquilinear form is given by

$$\ll (\mathbf{A}+1)\mathbf{U}, \mathbf{V} \gg$$
$$= \langle U_r + u_3 g_3, (A+1)^3 V_r \rangle - \bar{u}_3 \langle \varphi, V_r \rangle + v_3 \langle U_r, \varphi \rangle + \bar{u}_3 v_2 + \bar{u}_2 v_1.$$

Let us consider separately three cases covering all possible values of the parameters a, b, and c.

1. Consider first the general case $a \neq 0$. Taking into account

$$\langle \varphi, U_r \rangle = -\frac{b}{a} u_3 - \frac{c}{a} u_2$$

one can get

$$\ll (\mathbf{A}+1)\mathbf{U}, \mathbf{V} \gg$$
$$= \langle U_r + u_3 g_3, (A+1)^3 V_r \rangle + \bar{u}_3 \left(-\langle \varphi, V_r \rangle - \frac{\bar{b}}{\bar{a}} v_3 + v_2 \right) + \bar{u}_2 \left(v_1 - \frac{\bar{c}}{\bar{a}} v_3 \right).$$

The last expression determines a bounded linear functional if and only if the following relation holds

$$\bar{a}\langle \varphi, V_r \rangle + \bar{b} v_3 - \bar{a} v_2 = 0.$$

This condition coincides with (4.2) if and only if

$$a = -c$$

and the complex numbers a and b have the same phase. Hence without loss of generality the constants a, b, c can be chosen real and such that $c = -a$, i.e. $\langle (a, b, c), (1, 0, 1) \rangle = 0$.

2. Consider now the case $a = 0, b \neq 0$. The boundary condition has the following form

$$u_3 = -\frac{c}{b} u_2. \tag{4.3}$$

Hence the sesquilinear form of the operator is given by

$$\ll (\mathbf{A}+1)\mathbf{U}, \mathbf{V} \gg$$
$$= \langle U_r + u_3 g_3, (A+1)^3 V_r \rangle + \bar{u}_2 \left[\frac{\bar{c}}{\bar{b}} (\langle \varphi, V_r \rangle - v_2) + v_1 \right] + \langle U_r, \varphi \rangle v_3.$$

This form defines bounded linear functional with respect to $\mathbf{U} \in \mathbf{H}$ if and only if

$$v_3 = 0,$$

since $\langle U_r, \varphi \rangle$ is not a bounded linear functional. The last condition coincides with (4.3) only if $c = 0$. Then condition (4.3) reads as follows

$$bu_3 = 0.$$

Without loss of generality the constant b can be chosen real.

3. Consider the last possible case $a = 0 = b$, $c \neq 0$. The sesquilinear form given by

$$\ll (\mathbf{A} + 1)\mathbf{U}, \mathbf{V} \gg = \langle U_r + u_3 g_3, (A + 1)^3 V_r \rangle + \bar{u}_3 (v_2 - \langle \varphi, V_r \rangle) + v_3 \langle U_r, \varphi \rangle,$$

determines bounded linear functional if and only if the following conditions are satisfied

$$v_3 = 0 \quad \text{and} \quad v_2 = \langle \varphi, V_r \rangle.$$

These conditions never coincide with the condition $u_2 = 0$. Hence this boundary condition does not define any regular restriction of the operator $\mathbf{A}$.

The boundary conditions described in 1 and 2 cover all boundary conditions of the form (4.1) with nonzero real vectors (a, b, c) orthogonal to $(1, 0, 1)$. The Theorem is proven. $\square$

The last theorem states that all regular restrictions of the operator $\mathbf{A}$ are described by real three dimensional vectors (a, b, c) subject to the orthogonality condition $(a, b, c) \perp (1, 0, 1)$. The length of the vector (a, b, c) plays no rôle and therefore all boundary conditions can be parametrized by one real parameter - "angle" $\theta \in [0, \pi)$ as follows:

$$\sin \theta \langle \varphi, U_r \rangle + \cos \theta u_3 - \sin \theta u_2 = 0. \tag{4.4}$$

The following definition will be used.

Definition 4.1. The operator $\mathbf{A}_\theta$ is the restriction of the maximal operator $\mathbf{A}$ to the set of functions satisfying boundary conditions (4.4).

Let us calculate the operator adjoint to $\mathbf{A}_\theta$. The domain of this operator coincides with the domain $\mathrm{Dom}\,(\mathbf{A}_\theta)$. The sesquilinear form of the operator $\mathbf{A}_\theta$ can be presented by the following expression using the fact, that the functions from the domains of the operators $\mathbf{A}_\theta$ and $\mathbf{A}_\theta^*$ satisfy (4.4)

$$\ll (\mathbf{A}_\theta + 1)\mathbf{U}, \mathbf{V} \gg = \langle U_r + u_3 g_3, (A + 1)^3 V_r \rangle + \bar{u}_2 v_3 + \bar{u}_2 v_1.$$

Hence the action of the operator $\mathbf{A}_\theta^*$ is given by

$$\mathbf{A}_\theta^* \begin{pmatrix} V_r + v_3 g_3 \\ v_2 \\ v_1 \end{pmatrix} = \begin{pmatrix} AV_r - v_3 g_3 \\ v_3 + v_1 - v_2 \\ -v_1 \end{pmatrix}. \tag{4.5}$$

The operator $\mathbf{A}_\theta$ and its adjoint are related by

$$\mathbf{A}_\theta - \mathbf{A}_\theta^* = i \begin{pmatrix} 0 & 0 & 0 \\ 0 & 0 & i \\ 0 & -i & 0 \end{pmatrix}. \tag{4.6}$$

Thus the regular operator corresponding to the formal expression (2.1) is not defined uniquely. Like in the case of $\mathcal{H}_{-2}$ and $\mathcal{H}_{-3}$-perturbations one parameter family of operators has been constructed. The real and imaginary parts of the operator $\mathbf{A}_\theta$ are given by

$$\mathbf{A}_\theta = \Re\mathbf{A}_\theta + i\Im\mathbf{A}_\theta;$$

$$(\Re\mathbf{A}_\theta) \begin{pmatrix} U_r + u_3 g_3 \\ u_2 \\ u_1 \end{pmatrix} = \begin{pmatrix} AU_r - u_3 g_3 \\ u_3 - u_2 + \frac{1}{2}u_1 \\ \frac{1}{2}u_2 - u_1 \end{pmatrix};$$

$$\Im\mathbf{A}_\theta = \frac{1}{2} \begin{pmatrix} 0 & 0 & 0 \\ 0 & 0 & i \\ 0 & -i & 0 \end{pmatrix}. \tag{4.7}$$

The imaginary part of $\mathbf{A}_\theta$ is a bounded operator. Therefore the operator $\Re\mathbf{A}_\theta$ given by (4.7) is a self-adjoint operator on the domain $\mathrm{Dom}(\mathbf{A}_\theta)$. The following theorem proves that the spectrum of the operator $\mathbf{A}_\theta$ is pure real even if the operator itself is not self-adjoint.

Theorem 4.2. *The resolvent of the operator $\mathbf{A}_\theta$ for all nonreal λ is given by the 3×3 bounded matrix operator*

$$\frac{1}{\mathbf{A}_\theta - \lambda} = \mathbf{R}(\lambda), \tag{4.8}$$

having the following components corresponding to the orthogonal decomposition of the Hilbert space $\mathbf{H} = \mathcal{H}_2 \oplus \mathbf{C} \oplus \mathbf{C} \ni \mathbf{U} = (U, u_2, u_1)$

$$R_{00}(\lambda) = \frac{1}{A - \lambda} - (1 + \lambda)^2 \frac{\sin\theta}{\Delta} \frac{1}{A - \lambda} g_2 \langle \frac{1}{A - \bar{\lambda}} \varphi, \cdot \rangle;$$

$$R_{20}(\lambda) = -(1 + \lambda)\frac{\sin\theta}{\Delta} \langle \frac{1}{A - \bar{\lambda}} \varphi, \cdot \rangle;$$

$$R_{10}(\lambda) = -\frac{\sin\theta}{\Delta}\langle\frac{1}{A-\bar\lambda}\varphi,\cdot\rangle;$$

$$R_{02}(\lambda) = -(1+\lambda)\frac{\sin\theta}{\Delta}\frac{1}{A-\lambda}g_2;$$

$$R_{22}(\lambda) = -\frac{1}{\Delta}\left((1+\lambda)\langle\varphi,\frac{1+\lambda}{A-\lambda}g_3\rangle\sin\theta + (1+\lambda)\cos\theta\right);$$

$$R_{12}(\lambda) = -\frac{1}{\Delta}\left(\langle\varphi,\frac{1+\lambda}{A-\lambda}g_3\rangle\sin\theta + \cos\theta\right);$$

$$R_{01}(\lambda) = R_{21}(\lambda) = 0;$$

$$R_{11}(\lambda) = -\frac{1}{\Delta}\left((1+\lambda)\langle\varphi,\frac{1+\lambda}{A-\lambda}g_3\rangle\sin\theta + (1+\lambda)\cos\theta - \sin\theta\right),$$

where the subindex 0 *corresponds to the component* $U \in \mathcal{H}_2$ *and*

$$\Delta(\lambda,\theta) = (1+\lambda)^2\left\{\left(\langle\varphi,\frac{1+\lambda}{A-\lambda}g_3\rangle - \frac{1}{1+\lambda}\right)\sin\theta + \cos\theta\right\}. \qquad (4.9)$$

Proof. Consider arbitrary $\mathbf{F} = (F, f_2, f_1) \in \mathbf{H}$. Then the resolvent equation

$$(\mathbf{A} - \lambda)\begin{pmatrix} U_r + u_3 g_3 \\ u_2 \\ u_1 \end{pmatrix} = \begin{pmatrix} F \\ f_2 \\ f_1 \end{pmatrix} \qquad (4.10)$$

together with the boundary condition (4.4) imply that

$$\begin{pmatrix} 1 & -\langle\varphi,\frac{1+\lambda}{A-\lambda}g_3\rangle & 0 & 0 \\ 0 & 1 & -(1+\lambda) & 0 \\ 0 & 0 & 1 & -(1+\lambda) \\ \sin\theta & \cos\theta & -\sin\theta & 0 \end{pmatrix}\begin{pmatrix} \langle\varphi,U_r\rangle \\ u_3 \\ u_2 \\ u_1 \end{pmatrix} = \begin{pmatrix} \langle\varphi,\frac{1}{A-\lambda}F\rangle \\ f_2 \\ f_1 \\ 0 \end{pmatrix}$$

The determinant of the matrix appearing in the last equation given by (4.9). The determinant is equal to zero for nonreal λ only if the expression in brackets is equal to zero. The imaginary part of this expression can be calculated explicitly

$$I = y\left\{\langle g_2, \frac{(A+1)^2}{(A-x)^2 + y^2}g_2\rangle + \frac{1}{(1+x)^2 + y^2}\right\},$$

where we used the following notations for the imaginary and real parts of λ : $\lambda = x + iy$. Both items are positive and their sum cannot be equal to zero. Hence the determinant of the matrix is different from zero for all nonreal λ. Therefore the linear system has always unique solution for those λ:

$$\langle \varphi, U_r \rangle = \frac{-1}{\Delta} \left\{ ((1+\lambda)\sin\theta - (1+\lambda)^2\cos\theta)\langle \varphi, \frac{1}{A-\lambda}F\rangle \right.$$
$$\left. + \langle \varphi, \frac{1+\lambda}{A-\lambda}g_3\rangle \sin\theta f_2 \right\};$$

$$u_3 = \frac{-1}{\Delta} \left\{ (1+\lambda)^2\sin\theta\langle \varphi, \frac{1}{A-\lambda}F\rangle + (1+\lambda)\sin\theta f_2 \right\};$$

$$u_2 = \frac{-1}{\Delta} \left\{ (1+\lambda)\sin\theta\langle \varphi, \frac{1}{A-\lambda}F\rangle \right.$$
$$\left. + ((1+\lambda)\langle \varphi, \frac{1+\lambda}{A-\lambda}g_3\rangle + (1+\lambda)\cos\theta)f_2 \right\}; \qquad (4.11)$$

$$u_1 = \frac{-1}{\Delta} \left\{ \sin\theta\langle \varphi, \frac{1}{A-\lambda}F\rangle + (\cos\theta + \sin\theta\langle \varphi, \frac{1+\lambda}{A-\lambda}g_3\rangle))f_2 \right.$$
$$\left. \left((1+\lambda)\cos\theta - \sin\theta - (1+\lambda)\langle \varphi, \frac{1+\lambda}{A-\lambda}g_3\rangle \sin\theta \right) f_1 \right\}.$$

The component U can be calculated from the first equation (4.10)

$$U = u_r + u_3 g_3$$
$$= \frac{1}{A-\lambda}F + (1+\lambda)u_3\frac{1}{A-\lambda}g_3 + u_3 g_3$$
$$= \frac{1}{A-\lambda}F + u_3\frac{1}{A-\lambda}g_2.$$

Hence the resolvent of the operator $\mathbf{A}$ is given by formula (4.8) for all nonreal λ. The Theorem is proven. $\square$

The theorem implies that the spectrum of the operator $\mathbf{A}_\theta$ is real. Consider the restriction of the resolvent to the subspace $\mathcal{H}_2 \subset \mathbf{H}$ combined with the embedding ρ

$$\rho\frac{1}{\mathbf{A}_\theta - \lambda}\Big|_{\mathcal{H}_2} = \frac{1}{A-\lambda}$$

$$- \frac{1}{(\lambda+1)^2\left\{\cot\theta + \langle \varphi, \frac{1+\lambda}{A-\lambda}\frac{1}{(A+1)^3}\varphi\rangle - \frac{1}{1+\lambda}\right\}} \left\langle \frac{1}{A-\bar\lambda}\varphi, \cdot \right\rangle \frac{1}{A-\lambda}\varphi. \qquad (4.12)$$

The last formula is analogous to Krein's formula connecting the resolvents of two self-adjoint extensions of one symmetric operator and is very similar to formula (2.8) describing the restricted resolvent of the self-adjoint operator corresponding to the singular $\mathcal{H}_{-3}$-perturbation. The spectral properties of the operator $\mathbf{A}_\theta$ are described by the function

$$Q(\lambda) \;\;=\;\; \langle \varphi, \frac{1}{A-\lambda}\frac{(\lambda+1)^3}{(A+1)^3}\varphi \rangle$$

$$\overset{\text{formally}}{=} \langle \varphi, \frac{1}{A-\lambda}\varphi \rangle - \langle \varphi, \frac{1}{A+1}\varphi \rangle - (1+\lambda)\langle \varphi, \frac{1}{(A+1)^2}\varphi \rangle$$

$$-(1+\lambda)^2 \langle \varphi, \frac{1}{(A+1)^3}\varphi \rangle,$$

which is a triple regularized resolvent function.

5   Conclusions

Rank one supersingular $\mathcal{H}_{-4}$-perturbation of a positive self-adjoint operator has been determined in the class of regular operator. It has been shown that such operator cannot be defined in the class of self-adjoint operators. The spectrum of the perturbed operator is pure real and it follows that the properties of this operator are similar to those of self-adjoint operators. One has to study the question whether this regular operator is similar to a certain self-adjoint operator. The method suggested in this paper can be generalized to determine even more singular perturbations of positive self-adjoint operator corresponding to vectors $\varphi \in \mathcal{H}_{-n}$, $n \geq 5$. This program will be carried out in one of the forthcoming publications.

Acknowledgements

The work of P.Kurasov was supported by Sweden-Japan Sasakawa Foundation, The Royal Swedish Academy of Sciences, Stockholm University and Gakushuin University. The work of K.Watanabe was supported by Gakushuin University. The authors are grateful to S.Albeverio, H.Langer, S.Naboko, and B.Pavlov for stimulating discussions.

References

[AGH-KH88]   S.Albeverio, F.Gesztesy, R.Høegh-Krohn and H.Holden, Solvable models in quantum mechanics, Springer, 1988.

[AK97]   S.Albeverio and P.Kurasov, Rank one perturbations, approximations, and selfadjoint extensions, *J. Funct. Anal.*, **148**, 1997, 152–169.

[AK97-2] S.Albeverio and P.Kurasov, Rank one perturbations of not semibounded operators, *Int. Eq. Oper. Theory*, **27**, 1997, 379–400.

[AK00] S.Albeverio and P.Kurasov, Singular perturbations of differential operators, Cambridge Univ. Press, 2000 (London Mathematical Society Lecture Notes 271).

[And99] I.Andronov. Zero-range potential model of a protruding stiffener, *J. Phys. A*, **32** , 1999, L231–238.

[BF61] F.A.Berezin and L.D.Faddeev, Remark on the Schrödinger equation with singular potential, *Dokl. Akad. Nauk SSSR*, **137**, 1011–1014. 1961.

[DO88] Yu.N.Demkov and V.N.Ostrovsky, Zero-range potentials and their applications in atomic physics, Plenum Press, New York and London, 1988.

[DD81] Yu.N.Demkov and G.F.Drukarev, Loosely bound particle with nonzero orbital angular momentum in an electric or magnetic field, Sov. Phys. JETP, **54** (4), 1981, 650–656.

[DD86] Yu.N.Demkov and G.F.Drukarev, Potential curves for the p-state of the system AB^- in the approximation of small radius potentials, in: Problems of atomic collisions, vyp. 3, Leningrad Univ. Press, Leningrad, 1986, 184–195.

[TvD91] J.F.van Diejen and A.Tip, Scattering from generalized point interactions using self–adjoint extensions in Pontryagin spaces, *J. Math. Phys.*, **32**, 1991, 630–641.

[DLShZ00] A.Dijksma, H.Langer, Yu.Shondin and C.Zeinstra, Self–adjoint operators with inner singularities and Pontryagin spaces, to be published in Proceedings of Krein conference.

[GMT98] F.Gesztesy, K.Makarov, and Tsekanovsky, An addendum to Krein's formula, *J. Math. Anal. Appl.*, **222** (1998), 594–606.

[GS95] F.Gesztesy and B.Simon, Rank-one perturbations at infinite coupling, *J. Funct. Anal.*, **128**, 1995, 245–252.

[HdS97] S.Hassi and H. de Snoo, One-dimensional graph perturbations of self–adjoint relations, *Ann. Acad. Sci. Fenn. Math.*, **22**, 1997, 123–164.

[Karp92] Yu.E.Karpeshina, Zero-Range Model of p-scattering by a Potential Well, Preprint Forschungsinstitut für Mathematik, ETH, Zürich, 1992.

[KP95] A.Kiselev and I.Popov, An indefinite metric and scattering by regions with a small aperture, *Mat. Zametki*, **58**, 1995, 837–850, 959.

[KS95] A.Kiselev and B.Simon, Rank one perturbations with infinitesimal coupling, *J. Funct. Anal.*, **130**, 1995, 345–356.

[K99] V.Koshmanenko, Singular Quadratic Forms in Perturbation Theory, Kluwer, Dordrecht, 1999.

[Kr44] M.Krein, On Hermitian operators whose deficiency indices are 1, *Comptes Rendus (Doklady) Acad. Sci. URSS (N.S.)*, **43**, 131-134 1944.

[KK00] P.Kurasov and S.T.Kuroda, Krein's formula in perturbation theory, in preparation.

[KW00] P.Kurasov, K.Watanabe, On rank one $\mathcal{H}_{-3}$-perturbations of positive self-adjoint operators, to be published in Proceedings of S.Albeverio Conference, AMS 2000.

[KN98] S.T.Kuroda, H.Nagatani, $\mathcal{H}_{-2}$-construction and some applications, in: Mathematical Results in Quantum Mechanics (QMath 7, Prague, 1998), J.Dittrich, P.Exner, M.Tartar (eds.), *Operator Theory: Advances and Applications*, **108** Birkhäuser Verlag, 1999.

[Nai43]   M.Naimark, On spectral functions of a symmetric operator, *Bull. Acad. Sciences URSS*, **7**, 285-296, 1943.

[Pav87]   B.Pavlov, The theory of extensions and explicitly solvable models, *Uspekhi Mat. Nauk*, **42**, 99-131, 1987.

[Pav88]   B.Pavlov, Boundary conditions on thin manifolds and the semiboundedness of the three-body Schrödinger operator with point potential, *Mat. Sb. (N.S.)*, **136**, 163-177, 1988.

[PP84]   B.S.Pavlov and I.Popov, Scattering by resonators with small and point holes, *Vestnik Leningrad. Univ. Mat. Mekh. Astronom.*, vyp. 3, 1984, 116–118.

[PP91]   B.S.Pavlov and I.Popov, An acoustic model of zero-width slits and the hydrodynamic stability of a boundary layer, *Teoret. Mat. Fiz.*, **86**, 1991, 391–401.

[Pop92]   I.Popov, The Helmholtz resonator and operator extension theory in a space with indefinite metric, *Mat. Sb.*, **183**, 1992, 3–27.

[Sh92]   Yu.Shondin, Perturbation of differential operators on high-codimension manifold and the extension theory for symmetric linear relations in an indefinite metric space, *Teoret. Mat. Fiz.*, **92**, 1992, 466–472.

[Sh88]   Yu.Shondin, Quantum mechanical models in $\mathbf{R}^n$ connected with extensions of the energy operator in a Pontryagin space, *Teoret. Mat. Fiz.*, **74**, 1988, 331–344.

[Si95]   B.Simon, Spectral analysis of rank one perturbations and applications, in: Mathematical quantum theory. II. Schrödinger operators (Vancouver, BC, 1993), 109–149, *CRM Proc. Lecture Notes*, **8**, AMS, Providence, RI, 1995.

Addresses

PAVEL KURASOV, Dept. of Math., Stockholm University, 10691 Stockholm, Sweden

E-MAIL: pak@matematik.su.se

KAZUO WATANABE, Dept. of Mathematics, Gakushuin Univ., 1-5-1 Mejiro, Toshima-ku, Tokyo, 171-8588 Japan

E-MAIL: kazuo.watanabe@gakushuin.ac.jp

2000 Mathematics Subject Classification. Primary 47B25, 47A20, 81Q10; Secondary 35P05

Operator Theory:
Advances and Applications, Vol. 126
© 2001 Birkhäuser Verlag Basel/Switzerland

Global Attractor for Generalized 2D Ginzburg–Landau Equation

YONGSHENG LI AND BOLING GUO

Abstract. In this article we consider the 2D generalized Ginzburg–Landau equations and prove the existence of the global attractor in some weighted Sobolev space.

1 Introduction

There has been extensive work on the Ginzburg–Landau equations (GLE) of type

$$u_t = \gamma u + (1 + i\nu)\Delta u - (1 + i\mu)|u|^{2\sigma}u, \tag{1.1}$$

which describes various pattern formations and the onset of instabilities in non equilibrium fluid dynamical systems, as well as in the theory of phase transitions and superconductivity. Here $i = \sqrt{-1}$, γ, ν, and μ are given real numbers. $\gamma > 0$ is the modulational instability parameter and ν the dispersive parameter of both signs. Doering et al [DGHN88], Ghidaglia and Hèron [GhH87], Promislow [Pr90], etc. studied the finite dimensional global attractor and related dynamical issues for the 1D and 2D (i.e. one and two spatial dimensional) GLE with cubic nonlinearity ($\sigma = 1$). Bartuccelli et al [BCDGG90] investigated "strong" and "weak" turbulence of (1.1) in the case of cubic nonlinearity in 2D case. Doering et al [DGL94] showed the global existence and regularity of solutions of (1.1) in all cases (of spatial dimensions and parameters). Ginibre and Velo [GV96] studied the solutions in local spaces, that is, the spaces of functions satisfying only the local regularity condition but with no restrictions on their behavior at infinity. Mielke [Mi97] discussed the solution of (1.1) in weighted L^p space and derived some new bounds and investigated some properties of attractors.

A 1D generalized GLE has the form (see [EI89,Do90,DE91])

$$u_t = \alpha_0 u + \alpha_1 u_{xx} + \alpha_2|u|^2u + \alpha_3|u|^2u_x + \alpha_4 u^2\bar{u}_x - \alpha_5|u|^4u,$$

where $\alpha_0 > 0, \alpha_j = a_j + ib_j, j = 1,...,5, a_1 > 0, a_5 > 0$. The global existence of solutions and the long time behavior have been studied in [DH94,DHT93,GG95].

In this paper, we study the long time behavior of the Cauchy problem for the 2D generalized Ginzburg–Landau equation

$$u_t = \gamma u + (1 + i\nu)\Delta u - (1 + i\mu)|u|^{2\sigma}u + \lambda_1 \cdot \nabla(|u|^2u) + (\lambda_2 \cdot \nabla u)|u|^2, \tag{1.2}$$

$$u(0, x) = u_0(x), \quad x \in \mathbf{R}^2. \tag{1.3}$$

Here λ_1 and λ_2 are constant vectors with complex components. The most important and interesting case in physics is $\sigma = 2$. Guo and Wang [GW95] proved the existence of a finite dimensional global attractor of (1.2) with periodic boundary conditions by assuming that there exists a positive number $\delta > 0$ so that

$$\left(\sqrt{1 + \left(\frac{\mu - \nu\delta^2}{1 + \delta^2}\right)^2} - 1 \right)^{-1} \geq \sigma \geq 3. \tag{1.4}$$

Later the results on the existence of solution of (1.2) were improved by Gao and Duan [GD97] by replacing the right half part inequality of (1.4) with $\sigma \geq (1 + \sqrt{10})/2$. Li and Guo [LG00] considered (1.2)(1.3) and proved the global existence of solutions for the parameters σ, ν, and μ satisfying

$$(\mathbf{A}) \qquad \text{(i)} \ \ \sigma > 2 \ \ \text{and} \ \ \text{(ii)} \ \ -1 - \nu\mu < \frac{\sqrt{2\sigma + 1}}{\sigma} |\nu - \mu|.$$

Since (1.2) is translationally invariant, we cannot expect a nontrivial attractor that is compact in a Banach space with a uniform (i.e. translationally invariant) norm. Therefore we shall base our analysis on some weighted function spaces so that the compactness is available. We shall prove

Main Theorem *Assume that σ, ν, and μ satisfy* $(\mathbf{A})$. *Let* $p(\sigma) = \max\{\frac{3}{2}, \frac{4\sigma + 2}{2\sigma + 3}\}$, *and* $p_0 \in (p(\sigma), 2)$. *Then* (1.2)(1.3) *generates a semigroup* $S(t)$ *which possesses a global attractor* $\mathcal{A}$ *in* W^{1,p_0}_{lu} *with the properties*

(1) *(invariance)* $S(t)\mathcal{A} = \mathcal{A}$, $\forall t \geq 0$;

(2) *(compactness)* $\mathcal{A}$ *is bounded in* H^2_{lu} *and thus compact in the space* H^1_ρ;

(3) *(attractivity)* $\mathcal{A}$ *attracts points of* W^{1,p_0}_{lu} *and bounded subsets of* H^1_{lu}. *That is,*

$$\lim_{t \to \infty} \mathrm{dist}_\rho(S(t)u_0, \mathcal{A}) \stackrel{\triangle}{=} \lim_{t \to \infty} \inf_{v \in \mathcal{A}} \|S(t)u_0 - v\|_{H^1_\rho} = 0, \ \forall u_0 \in u_0 \in W^{1,p_0}_{lu},$$

$$\lim_{t \to \infty} \mathrm{dist}_\rho(S(t)B, \mathcal{A}) = \lim_{t \to \infty} \sup_{v \in B} \mathrm{dist}_\rho(S(t)v, \mathcal{A}) = 0, \ \forall B \subset H^1_{lu} \text{ bounded.}$$

(4) *(Regularity)* For each $q > p(\sigma), \mathcal{A} \subset W^{2,q}_{lu}$.

Remark (i) For the definitions of weighted spaces $W^{m,p}_\rho, H^m_\rho$ and $W^{m,p}_{lu}, H^m_{lu}$ see Section 2. dist_ρ is the distance in the space H^1_ρ.

(ii) The inequality of $(\mathbf{A}\text{ii})$ is the same as [BGO96, (15)], which indicates the "soft" turbulence for (1.1). The first and the third quadrants of the (ν, μ)-plane, which indicate the modulational stability of (1.1), are contained in the area between the pair of hyperbolae of $(\mathbf{A}\text{ii})$.

(iii) The assumptions $(\mathbf{A}\text{i})$ and $(\mathbf{A}\text{ii})$, the class of initial data for the global existence are more general than that used in [GD97,GW95]. Note that, for any $\sigma > 0$, the left half inequality of (1.4) are commensurate in fact with the union of two strips $|\nu| < \frac{\sqrt{2\sigma+1}}{\sigma}$ and $|\mu| < \frac{\sqrt{2\sigma+1}}{\sigma}$ in the (ν, μ)-plane, which are contained in the area between the pair of hyperbolae of (**A**ii).

(iv) When $\sigma = 2$, if we assume further that $|\lambda_1|$ and $|\lambda_2|$ are suitably small, the main theorem is also true.

2 Existence of Solutions and Attractor

Let $\rho > 0$ be a suitable weight function of $x \in \mathbf{R}^2$ with

$$|\nabla \rho(x)|, |\Delta \rho(x)| \le \rho_0 \rho(x), \quad \text{and} \quad \int \rho(x)\, dx = \rho_0 < +\infty. \tag{2.1}$$

For example, $\rho = \dfrac{1}{\cosh|x|}$, or $\rho = e^{-|x|}$ (maybe up to suitable coefficients). Denote by $T_y\rho(x) = \rho(x - y)$ the translated weight. We shall use the weighted L^p-norm $\|u\|_{p,\rho} = \left(\int \rho |u|^p\, dx\right)^{1/p}$, $1 < p < \infty$, and the uniformly local norm $|u\|_{p,lu} = \sup_{y \in \mathbf{R}^2} \|u\|_{p,T_y\rho}$. For $1 < p < q < \infty$, $\|u\|_{p}, \rho \le \rho_0^{\frac{1}{p} - \frac{1}{q}} \|u\|_{q,\rho}$. We denote by L_ρ^p the weighted L^p-space consisting of all u with $\|u\|_{p,\rho} < +\infty$, and by L_{lu}^p the uniformly local space consisting of all u with $\|u\|_{p,lu} < +\infty$, and $\|T_y u - u\|_{p,lu} \to 0$ as $y \to 0$. Then both $(L_\rho^p, \|\cdot\|_{p,\rho})$ and $(L_{lu}^p, \|\cdot\|_{p,lu})$ are Banach spaces. We shall also use the weighted Sobolev space $W_\rho^{m,p}$ with the norm $\|u\|_{W_\rho^{m,p}} = \left(\sum_{k \le m} \|D^k u\|_{p,\rho}^p\right)^{1/p}$ and uniformly local Sobolev space $W_{lu}^{m,p}$ = the completion of C_b^∞ under the norm $\|u\|_{W_{lu}^{m,p}} = \sup_{y \in \mathbf{R}^2} \|u\|_{W_{T_y\rho}^{m,p}}$. Especially, $H_\rho^m = W_\rho^{m,2}$, $H_{lu}^m = W_{lu}^{m,2}$. We note that the weighted inequality of Gagliardo–Nirenberg type holds, and that $W_{lu}^{1,p}$ is a Banach algebra for $p > 2$. We have also the compact imbedding $W_{lu}^{m,p} \hookrightarrow W_\rho^{j,r}$ if $\frac{1}{r} > \frac{1}{p} - \frac{m-j}{2}, 1 < p, r < \infty$. We remark that, due to the translation invariance, the imbedding of $W_{lu}^{m,p} \hookrightarrow W_{lu}^{j,r}$ could not be compact.

For any $1 < p < \infty$, we let $X_p = L_{lu}^p$ and define a linear operator $A_p \colon D(A_p) \subset X_p \to X_p$ by $A_p u = (1 + i\nu)\Delta u$, $D(A_p) = W_{lu}^{2,p}$. It has been proved in [Mi97] that for $p \ge 2$, A_p generates an analytical semigroup on L_{lu}^p. We can show that such analytic generation result holds true for $1 < p < 2$.

Let $B_p = A_p - (R_1 + 1)$ with $R_1 > 0$ being sufficiently large, then 0 is contained in the resolvent of B_p, and B_p generates an analytic semigroup $e^{B_p t}$ $(t \ge 0)$ in X_p. Moreover we can define the fractional powers $(-B_p)^\alpha$ of $-B_p$ with the domains of definition $D((-B_p)^\alpha)$ for every $0 < \alpha < 1$, $D((-B_p)^{1/2}) = W_{lu}^{1,p}$.

We write (1.2)(1.3) in a functional setting

$$u_t = B_p u + F(u), \qquad u(0) = u_0, \tag{2.2}$$

$$F(u) = (\gamma + R_1 + 1)u + (1 + i\mu)|u|^{2\sigma}u + (\lambda_1 \cdot \nabla)(|u|^2 u) + (\lambda_2 \cdot u)|u|^2.$$

It is not difficult to see that the nonlinear map $F(u)$ is locally Lipschitz continuous from $D((-B_p)^{1/2}) = W_{lu}^{1,p}$ into $X_p = L_{lu}^p$ for $p > 2$. Therefore, for every $u_0 \in D((-B_p)^{1/2}) = W_{lu}^{1,p}$, there exists a unique local solution of the abstract Cauchy problem (2.2). However we would rather prove the local existence in $W_{lu}^{1,p}$ $(p < 2)$, a wider class of initial data. In fact we can prove by contraction mapping

Theorem 2.1. *If $u_0 \in W_{lu}^{1,p_0}$ (see the main theorem for p_0), then there exists a unique solution of the Cauchy problem (2.2) $u(t) \in C([0, T_*), W_{lu}^{1,p_0}) \cap C((0, T_*), W_{lu}^{2,p_0})$. If $T^* < +\infty$, then $\lim_{t \to T^*} \|u(t)\|_{W_{lu}^{1,p_0}} = +\infty$.*

Since $1 < p_0 < 2, H_{lu}^1 \hookrightarrow W_{lu}^{1,p_0}$, it suffices to prove the time uniform boundedness of solutions in weighted space H_{lu}^1 to obtain the global existence of the solution.

Lemma 2.1. *Assume that $(\mathbf{A})$ and $2 \le p < \frac{2\sqrt{1+\nu^2}}{\sqrt{1+\nu^2}-1}$ hold. Then there exist a C independent of R and a $t_0(R) > 0$ such that $\|u\|_{p,\rho}^p \le C, t \ge t_0(R)$, whenever $\|u_0\|_{p,\rho} \le R$.*

Proof. We multiply (1.2) by $\rho|u|^{p-2}\overline{u}$, integrate over $\mathbf{R}^2$ and then take real parts. Note that $\mathrm{Re} \int (1 + i\nu)\rho|u|^{p-2}\overline{u}\Delta u\, dx$ can be written as

$$-\mathrm{Re} \int (1 + i\nu)\nabla\rho|u|^{p-2}\overline{u}\nabla u\, dx - \frac{1}{4}\int \rho|u|^{p-4} \sum_{j=1}^2 (\overline{u}\partial_j u, u\partial_j\overline{u}) M(\nu, p) \begin{pmatrix} u\partial_j\overline{u} \\ \overline{u}\partial_j u \end{pmatrix} dx,$$

where $M(\nu, p) = \overline{M(\nu, p)}^{tr} = \begin{pmatrix} p & (1 + i\nu)(p - 2) \\ * & p \end{pmatrix}$. Under the assumption on p, the smaller eigenvalue of $M(\nu, p)$

$$\lambda_M(\nu, p) = p - |p - 2|\sqrt{1 + \nu^2} > 0, \tag{2.3}$$

so $M(\nu, p)$ is positively definite, therefore we can obtain

$$\frac{1}{p}\frac{d}{dt}\|u\|_{p,\rho}^p \le C\|u\|_{p,\rho}^p - \frac{1}{2}\|u\|_{p+2\sigma,\rho}^{p+2\sigma} \le -\frac{C}{p}\|u\|_{p,\rho}^p + \frac{C}{p}.$$

By Gronwall inequality we have the lemma. $\qquad\square$

Remark For $\sigma = 2$, we assume further that $6|\lambda_1| + 2|\lambda_2| < \lambda_M(\nu, p)$, then the lemma is also valid.

Lemma 2.2. *Under the assumption $(\mathbf{A})$, there exist a C independent of R and a $t_1(R) > 0$ such that $\|\nabla u(t)\|_{2,\rho}^2 \le C, t \ge t_1(R)$, whenever $\|u_0\|_{H_\rho^1} \le R$.*

Proof. Similar to the proof of Lemma 2.1, for any α with $|\alpha| < \frac{\sqrt{2\sigma+1}}{\sigma}$, the smaller eigenvalue $\lambda_M(\alpha, 2\sigma + 2)$ of $M(\alpha, 2\sigma + 2)$ is positive and thus $M(\alpha, 2\sigma + 2)$ is definitely positive. Thus we have

$$\mathrm{Re} \int (1 + i\alpha)\rho|u|^{2\sigma}\overline{u}\Delta u\, dx + \lambda_M(\alpha, 2\sigma) \int \rho|u|^{2\sigma}|\nabla u|^2\, dx$$

$$- \rho_\alpha \int \rho|u|^{2\sigma+1}|\nabla u|\, dx \leq 0. \tag{2.4}$$

Define $V_\delta(u(t)) = \int \rho\left(\frac{1}{2}|\nabla u|^2 + \frac{\delta}{2\sigma+2}|u|^{2\sigma+2}\right) dx$, multiply (2.4) by $-\eta$ (both $\delta > 0$ and $\eta > 0$ will be suitably chosen), then add it to $\frac{d}{dt}V_\delta(u(t))$ and get

$$\frac{d}{dt}V_\delta(u(t)) \leq \gamma(\|\nabla u\|_2^2 + \delta\|u\|_{2\sigma+2}^{2\sigma+2}) - (1 - \kappa)(\|\Delta u\|_2^2 + \delta\|u\|_{4\sigma+2}^{4\sigma+2})$$

$$- \eta\lambda_M(\alpha, 2\sigma + 2) \int \rho|u|^{2\sigma}|\nabla u|^2\, dx$$

$$+ \frac{1}{2}\mathrm{Re} \int \rho\big(|u|^{2\sigma}u, \Delta u\big) \cdot N \cdot \begin{pmatrix} |u|^{2\sigma}\overline{u} \\ \Delta\overline{u} \end{pmatrix} dx$$

$$+ (\rho_\mu + \eta\rho_\alpha) \int \rho|u|^{2\sigma+1}|\nabla u|\, dx + \rho_\nu \int \rho|\nabla u||\Delta u|\, dx$$

$$+ (3|\lambda_1| + |\lambda_2|)\left\{\rho_0 \int \rho|u|^2|\nabla u|^2\, dx + \delta \int \rho|u|^{2\sigma+3}|\nabla u|\, dx\right.$$

$$\left. + \int \rho|u|^2|\nabla u||\Delta u|\, dx\right\}, \tag{2.5}$$

where $\rho_s = \sqrt{1 + s^2}\rho_0$ (see (2.1) for ρ_0), $0 \leq \kappa < 1$ is to be determined, and the matrix $N = \overline{N}^{tr} = \begin{pmatrix} -2\delta\kappa & 1 + \delta - \eta - i(\delta\nu - \mu - \alpha\eta) \\ * & -2\kappa \end{pmatrix}$. When σ, ν and μ satisfy the assumption (**A**), we can choose suitable δ, η positive, $\kappa \in [0, 1)$ and $|\alpha| < \frac{\sqrt{2\sigma+1}}{\sigma}$ such that N is nonpositive (see [Mi97,LG00]). Hence

$$\mathrm{Re} \int \rho\big(|u|^{2\sigma}u, \Delta u\big) \cdot N \cdot \begin{pmatrix} |u|^{2\sigma}\overline{u} \\ \Delta\overline{u} \end{pmatrix} dx \leq 0.$$

Noting (2.1) and $\|u\|_{2,\rho} \leq C$, carefully estimating the last five integrals of (2.5), we have

$$\frac{d}{dt}V_\delta(u(t)) \leq -2\gamma V_\delta(u(t)) + C.$$

by Gronwall inequality we complete the proof of the lemma. $\square$

Corollary 2.1. *Under the same assumption of* (**A**), $\|u(t)\|_{H^1_{lu}} \leq L$, $t \geq t_1(R)$, *whenever* $\|u_0\|_{H^1_{lu}} \leq R$, *where L is independent of R.*

Remark When $\sigma = 2$ and $3|\lambda_1| + |\lambda_2|$ is small enough, Lemma 2.2 and Corollary 2.1 are also valid.

Theorem 2.2. *Assume that* $(\mathbf{A})$ *holds,* $p(\sigma) = \max\{\frac{3}{2}, \frac{4\sigma+2}{2\sigma+3}\} < p_0 < 2$. *Then*

(i) *For each* $u_0 \in W_{lu}^{1,p_0}$, *(1.2)(1.3) have a unique solution*

$$u(t) \in C([0, +\infty); W_{lu}^{1,p_0}) \bigcap C((0, +\infty); W_{lu}^{2,p_0}). \tag{2.6}$$

The semigroup $S(t)$ *generated by (1.2),(1.3) is continuous in* W_{lu}^{1,p_0} *for* $t \geq 0$.

(ii) $S(t)$ *is point dissipative in* W_{lu}^{1,p_0} *and bounded dissipative in* H_{lu}^1.

Proof. The global existence of solutions and bounded dissipativity in H_{lu}^1 are direct consequences of Theorem 2.1 and Corollary 2.1. The point dissipativity of $S(t)$ follows from the regularity of the solutions. $\qquad\qquad\qquad\qquad\qquad\qquad\square$

The main theorem follows immediately from the compact imbedding of $W_{lu}^{2,p}$ into $W_{\rho}^{1,p}$ and the standard existence theory of global attractors. The global attractor $\mathcal{A}$ is the ω-limit set of the bounded absorbing set $B(0, L) \subset H_{lu}^1$ (see Corollary 2.1) under the action of the semigroup $S(t)$ in the topology of H_{ρ}^1. $\mathcal{A}$ attracts points in W_{lu}^{1,p_0} and bounded subsets of H_{lu}^1 with respect to the metric induced by the norm of H_{ρ}^1. By the regularity of $S(t)$, $\mathcal{A}$ is contained in $W_{lu}^{2,q}$ for any $p(\sigma) < q < \infty$.

Acknowledgements

This work is supported by the National Science Foundation of China under Grant Number 10001013.

References

[BCDGG90]   M. V. Bartuccelli, P. Constantin, C. Doering, J. Gibbon and M. Gisselfält, On the possibility of soft and hard turbulence in the complex Ginzburg–Landau equation, *Physica D* **44** (1990), 421–444.

[BGO96]   M. V. Bartuccelli, J. Gibbon and M. Oliver, Length scales in solutions of the complex Ginzburg–Landau equation, *Physica D* **89** (1996), 167–286.

[Do90]   A. Doelman, On the nonlinear evolution of patterns (modulation equations and their solutions), Ph.D. Thesis, University of Utrecht, the Netherland, 1990.

[DE91]   A. Doelman and W. Eckhaus, Periodic and quasi-periodic solutions of degenerate modulation equations, *Physica D* **53** (1991), 249–266.

[DGHN88]   C. Doering, J. D. Gibbon, D. Holm and B. Nicolaenko, Low-dimensional behavior in the complex Ginzburg–Landau equation, *Nonlinearity* **1** (1988), 279-309.

[DGL94]   C. R. Doering, J. D. Gibbon and C. D. Levermore, Weak and strong solutions of the complex Ginzburg–Landau equation, *Physica D* **71** (1994), 285-318.

[DH94]   J. Duan and P. Holmes, On the Cauchy problem of a generalized Ginzburg–Landau equation, *Nonl. Anal. TMA* **22** (1994), 1033-1040.

[DHT93]   J. Duan, P. Holmes and E. S. Titi, Regularity approximation and asymptotic dynamics for a generalized Ginzburg–Landau equation, *Nonlinearity* **6** (1993), 915-933.

[EI89]   W. Eckhaus and G. Iooss, Strong selection or rejection of spatially periodic patterns in degenerate bifurcations, *Physica D* 39 (1989), 124–146.

[GG95]   H. Gao and B. Guo, Finite dimensional inertial forms for 1D generalized Ginzburg–Landau equation, *Sci. in China, Ser. A* **25**(12) (1995), 1233-1247.

[GD97]   Hongjun Gao and J. Duan, On the initial value problem for the generalized 2D Ginzburg–Landau equation, *J. Math. Anal. Appl.* **216** (1997), 536–548.

[GhH87]   J-M. Ghidaglia and B. Hèron, Dimension of the attractor associated to the Ginzburg–Landau equation, *Physica D* **28** (1987), 282-304.

[GV96]   G. Ginibre and J. Velo, The Cauchy problem in local spaces for the complex Ginzburg–Landau equation, I. Compactness Methods, *Physica D* **95** (1996), 191–228.

[GW95]   Boling Guo and Bixiang Wang, Finite dimensional behavior for the derivative Ginzburg–Landau equation in two spatial dimensions, *Physica D* **89** (1995), 83–99.

[LG00]   Yongsheng Li and Boling Guo Global Existence of Solutions to the Derivative 2D Ginzburg–Landau Equation, *J. Math. Anal. Appl.* **249** (2000), 412–432.

[Mi97]   A. Mielke, The complex Ginzburg–Landau equation on large and unbounded domains: sharper bounds and attractors, *Nonlinearity* **10** (1997), 199–222.

[Pr90]   K. Promislow, Induced trajectories and approximate inertial manifolds for the Ginzburg–Landau partial differential equation, *Physica D* **41** (1990), 232–252.

Addresses

Yongsheng Li, Department of Mathematics, Huazhong University of Science and Technology, Wuhan, Hubei 430074, China

E-mail: liys@public.wh.hb.cn

BOLING GUO, Institute of Applied Physics and Computational Mathematics, P. O. Box 8009, Beijing 100088, China

E-MAIL: js8s@mail.iapcm.ac.cn

2000 Mathematics Subject Classification. Primary 35R35, 35K55; Secondary 58F39

Operator Theory:
Advances and Applications, Vol. 126
© 2001 Birkhäuser Verlag Basel/Switzerland

Local Asymptotic Properties of Multifractional Brownian Motion

S.C. LIM AND S.V. MUNIANDY

Abstract. One direct way to generalize fractional Brownian motion to multifractional Brownian motion is to replace the constant Hölder exponent by a certain deterministic function of time. In this paper local asymptotic properties of two types of multifractional Brownian motion will be considered.

1   Introduction

Fractional Brownian motion and some of its generalization can be regarded as Gaussian processes that belong to a larger class of processes known as the elliptic Gaussian processes. The covariance of an elliptic Gaussian process is given by the Green function of an elliptic pseudodifferential operator A . The local scaling properties of such a process can be characterized by the principal symbol of A [BJR97]. Let $\sigma(t,\omega)$ denotes the symbol of A, an elliptic symmetric positive pseudodifferential operator :

$$(Af)(t) = \int e^{i\omega t}\sigma(t,\omega)\widehat{f}(\omega)d\omega, \tag{1.1}$$

where $\widehat{f}$ is the Fourier transform of f. First, note that a Gaussian Markov process is associated to a differential operator with polynomial symbol [Pit71]. Elliptic Gaussian processes are natural extensions of Markov processes when A becomes non-local. When the symbol does not depend on time, the associated Gaussian process has stationary increments. In particular, when $\sigma(t,\omega) = |\omega|^{2H-1}, 0 < H < 1$, the resulting process is a fractional Brownian motion (fBm) indexed by H.

A class of elliptic Gaussian random processes can be obtained when the symbol is allowed to be time-dependent such that for large $|\omega|$,

$$\sigma(t,\omega) = a(t)|\omega|^{2H+1} + b(t)|\omega|^{2H+1}$$

and with some additional conditions imposed on $\sigma(t,\omega)$ (see [BCIJ98] for details). These processes, known as filtered white noises, can be regarded as generalization of fBm as they behaves locally like fBm of order H. Another way of generalizing fBm is to allow non-constant index $H(t)$. The resulting process is called the multifractional Brownian motion (mBm) which is not an elliptic Gaussian random process. However, Benassi *et al.* [BJR97,BCIJ97] have shown that the techniques used in studying the local scaling properties of filtered white noise can still be applied to mBm.

In this paper we study the local properties of a particular type of mBm using a different approach. The properties of locally asymptotical stationarity for the increment process, and the locally asymptotical self-similarity for the process itself can be verified by direct computation of the local variance for the increment process and the local covariance of mBm. These properties lead to the conclusion that mBm behaves locally like a fBm.

2 Two Equivalent Classes of Multifractional Brownian Motions

The standard definition of fractional Brownian motion which is widely accepted is the moving average version introduced by Mandelbrot and van Ness [MVN68]:

$$
B_H(t) = \frac{1}{\Gamma(H + 1/2)} \left\{ \int_{-\infty}^{0} \left[(t - s)^{H-1/2} - (-s)^{H-1/2} \right] dB(s) \right.
$$

$$
\left. + \int_{0}^{t} (t - s)^{H-1/2} dB(s) \right\}, \tag{2.1}
$$

where the Hölder exponent is restricted in the interval $0 < H < 1$. The standard fBm given in (2.1) is self-similar and its increments are stationary $-$ the two properties that allow one to associate a generalized power-law spectrum of the form $\frac{1}{\omega^\alpha}$ where $\alpha = 2H + 1$. The covariance of fBm has the following simple form

$$
\langle B_H(t_1) B_H(t_2) \rangle = \frac{V_H}{2} \left(|t_1|^{2H} + |t_2|^{2H} - |t_1 - t_2|^{2H} \right)
$$

where

$$
V_H = var(B_H(1)) = \frac{\Gamma(2 - 2H) \cos(\pi H)}{\pi H (2H - 1)}.
$$

There also exist another equivalent definition (up to a multiplicative constant) of fBm in the form of harmonizable representation [Yag58]:

$$
B_H(t) = \frac{1}{2\pi} \int_{-\infty}^{\infty} \frac{e^{it\omega} - 1}{|\omega|^{H+1/2}} dB(\omega). \tag{2.2}
$$

A direct generalization of fBm to mBm can be carried out by replacing the Hölder exponent H by a time-varying function, $H(t)$, satisfying $H : [0, \infty) \to (0, 1)$. Accordingly, two versions of mBm have been introduced independently by Peltier and Levy-Vehel [PLV95] and Benassi *et al.* [BJR97] based on the moving average and harmonizable representations of fBm respectively. The moving average representation is defined as [PLV95]

$$B_{H(t)}(t) = \frac{1}{\Gamma(H(t)+1/2)} \left\{ \int_{-\infty}^{0} \left[(t-s)^{H(t)-1/2} - (-s)^{H(t)-1/2} \right] dB(s) \right.$$

$$\left. + \int_{0}^{t} (t-s)^{H(t)-1/2} dB(s) \right\}, \tag{2.3}$$

where the time-dependent Hölder exponent $H(t)$ is restricted in the interval $0 < H(t) < 1$ and assumed to be smooth. On the other hand, mBm in the harmonizable representation takes the form [BJR97]

$$\bar{B}_{H(t)} = \frac{1}{2\pi} \int_{-\infty}^{\infty} \frac{e^{it\omega}-1}{|\omega|^{H(t)+1/2}} dB(\omega). \tag{2.4}$$

These two versions of mBms have been shown to be equivalent up to a multiplicative deterministic function of time by Cohen [Coh99] using the method of Fourier transform and by Lim and Muniandy [LM00] through a direct computation of their covariances. It can be shown that the both definitions of mBm when appropriately normalized have the following covariance function :

$$\langle B_{H(t_1)}(t_1) B_{H(t_2)}(t_2) \rangle = \frac{C\left(\frac{H(t_1)+H(t_2)}{2}\right)}{2V_{H(t_1)}V_{H(t_2)}} \times$$

$$+ |t_2|^{H(t_1)+H(t_2)} - |t_1 - t_2|^{H(t_1)+H(t_2)} \Big) \tag{2.5}$$

where

$$V_{H(t)} = \frac{\Gamma(2-2H(t))\cos(\pi H(t))}{\pi H(t)(2H(t)-1)}$$

and

$$C\left(\frac{H(t_1)+H(t_2)}{2}\right) = \frac{\Gamma(2-H(t_1)-H(t_2))\cos\left(\pi \frac{(H(t_1)-H(t_2))}{2}\right)}{\pi\left(\frac{H(t_1)+H(t_2)}{2}\right)(H(t_1)+H(t_2)-1)}.$$

Due to the fact that the Hölder exponent is time-dependent, mBm fails to satisfy the global self-similarity property and the increment process of mBm does not satisfy the stationary property. Instead, standard mBm now satisfies the locally asymptotically self-similarity [BJR97] and its increment process is locally stationary.

Definition 2.1. A mBm, $X(t)$ indexed by the Hölder exponent $H(t) \in C^\beta$ such that $H : [0, \infty) \to (0, 1)$ for $t \in R$ and some $\beta > \sup(H(t))$ is said to be locally asymptotically self-similar (*lass*) at point t_o if

$$\lim_{\rho \to 0_+} \left(\frac{X(t_o + \rho u) - X(t_o)}{\rho^{H(t_o)}} \right)_{u \in R} \equiv \left(B_{H(t_o)}(u) \right)_{u \in R} \qquad (2.6)$$

where the equality in law is up to a multiplicative deterministic function of time and $B_{H(t_o)}$ is the fBm indexed by $H(t_o)$.

With the assumption that $H(t)$ is β-Hölder function such that $0 < \inf(H(t)) \le \sup(H(t)) < \min(1, \beta)$, one may approximate $H(t+\rho u) \approx H(t)$ as $\rho \to 0$. Therefore the local covariance function of the normalized mBm has the following limiting form

$$\langle B_{H(t)}(t + \tau) B_{H(t)}(t) \rangle \sim \frac{1}{2} \left(|t + \tau|^{2H(t)} + |t|^{2H(t)} - |\tau|^{2H(t)} \right), \qquad \tau \to 0. \quad (2.7)$$

The variance of the increment process becomes

$$\langle [B_{H(t)}(t + \tau) - B_{H(t)}(t)]^2 \rangle \sim |\tau|^{2H(t)}, \qquad \tau \to 0$$

implying that the increment processes of mBm is locally stationary. It follows that the local Hausdorff dimension of the graphs of mBm is given by [PLV95]

$$\dim\{B_{H(t)}(t), \quad t \in [a, b]\} = 2 - \min\{H(t), \quad t \in [a, b]\}, \qquad (2.8)$$

for each interval $[a, b] \subset R^+$. In the next section, we introduce an alternative definition of mBm based on the fractional Brownian motion of Riemann-Liouville type (RL-fBm) and study its local properties.

3 RL-mBm and its Local Properties

The fractional Brownian motion based on the Riemann-Liouville fractional integral is defined as [BA66]:

$$X_H(t) = \frac{1}{\Gamma(H + 1/2)} \int_0^t (t - s)^{H-1/2} dB(s) \qquad t \ge 0, \qquad (3.1)$$

which holds for $H > 0$ but can extended to the range $-1/2 < H < 0$ as a generalized stochastic process. Due to constraint that the process to begin at $t = 0$, its increment process is not stationary, thus failed to have a generalized spectral density of power-law type. This is the main reason for the lack of interest for using such a process among physicists and engineers in modelling of physical phenomena with power-law spectral behaviors.

On the other hand, we shall show that the RL-mbm when generalized to RL-mBm may offer some advantages over the standard mBm. We consider the RL-mBm as the generalization of (3.1) defined as

$$X_{H(t)}(t) = \frac{1}{\Gamma(H(t) + 1/2)} \int_0^t (t - s)^{H(t)-1/2} dB(s) \tag{3.2}$$

with the following covariance [LM00]

$$\langle X_{H(t_1)}(t_1) X_{H(t_2)}(t_2) \rangle = \frac{2t_1^{H(t_1)+1/2} t_2^{H(t_2)-1/2}}{2H(t_1) + 1/2)\Gamma(H(t_1) + 1/2)\Gamma(H(t_2) + 1/2)} \times$$
$$_2F_1(1/2 - H(t_2), 1, H(t_1) + 3/2, t_1/t_2). \tag{3.3}$$

for $0 < t_1 < t_2$ and $_2F_1$ is Gauss hypergeometric function. It follows from (3.3) that the variance of RL-mBm

$$\langle [X_{H(t)}(t)]^2 \rangle \sim |t|^{2H(t)},$$

exhibits a similar time-dependence as the variance of standard mBm. Despite the complex form of the covariance, one can verify the local properties of RL-mBm with simple arguments based on the variance of its increment process.

In order to carry out explicit calculation on the process $X_{H(t)}$, it is necessary to assume $H(t)$ to be continuous and smooth such that $H(t+\tau) \approx H(t)$ for sufficiently small τ. The variance of the increment process of RL-mBm is given by

$$\langle (X_{H(t+\tau)}(t + \tau) - X_{H(t)}(t))^2 \rangle$$

$$= \frac{1}{(\Gamma(H(t) + 1/2))^2} \left\{ \int_0^t \left[(t + \tau - u)^{H(t)-1/2} - (t - u)^{H(t)-1/2} \right]^2 du \right.$$

$$\left. + \int_t^{t+\tau} (t + \tau - u)^{2H(t)-1} du \right\}$$

$$= \frac{\tau^{2H(t)}}{(\Gamma(H(t) + 1/2))^2} \left\{ \int_0^{t/\tau} \left[(1 + u)^{H(t)-1/2} - u^{H(t)-1/2} \right]^2 du \right.$$

$$\left. + \int_0^1 u^{2H(t)-1} du \right\}$$

$$= \frac{\tau^{2H(t)}}{(\Gamma(H(t) + 1/2))^2} \left\{ I + \frac{1}{2H(t)} \right\}. \tag{3.4}$$

It is easy to show that I will be a function of $H(t)$ only and does not depend on t explicitly iff $t/\tau \to 0$ or $t/\tau \to \infty$. The condition $t/\tau \to 0$ can be satisfied only for very large time lag τ (which is inconsistent with our assumption τ small) or for t very small. In real situations, such requirements are too stringent and therefore of little physical interest. For $t/\tau \to \infty$, it can be fulfilled with either $t \to \infty$ or $\tau \to 0$. To be consistent with our assumption τ is very small to get the approximation $H(t+\tau) \approx H(t)$, we shall consider this as the necessary condition for I to be independent of t explicitly. A direct calculation gives the exact value of I as

$$I = \frac{\Gamma(1 - 2H(t))\cos(\pi H(t))}{(\Gamma(H(t) + 1/2))^2 \pi H(t)}$$

such that for $\tau \to 0$

$$\left\langle (X_{H(t)}(t + \tau) - X_{H(t)}(t))^2 \right\rangle = D_{H(t)} \tau^{2H(t)} \tag{3.5}$$

where

$$D_{H(t)} = \frac{\Gamma(1 - 2H(t))\cos(\pi H(t))}{\pi H(t)} + \frac{1}{2H(t)(\Gamma(H(t) + 1/2)^2}.$$

It is in this sense one says that the increment process of RL-mBm is locally asymptotically stationary.

The local covariance of RL-mBm can now be obtained by using (3.5). For $\tau \to 0$

$$\left\langle X_{H(t)}(t + \tau) X_{H(t)}(t) \right\rangle = \frac{1}{2H(t)(\Gamma(H(t) + 1/2))^2} \left[|t + \tau|^{2H(t)} + |t|^{2H(t)} \right.$$
$$\left. - 2H(t)(\Gamma(H(t) + 1/2))^2 D_{H(t)} \tau^{2H(t)} \right] \tag{3.6}$$

which differs from the local covariance of the standard mBm (2.7) in the coefficient of $\tau^{2H(t)}$. This difference in the local covariance does not affect the similarity of their local properties since the local variance of increment processes for the two types of mBm has the same τ-dependence, i.e. $\tau^{2H(t)}$. Now we proceed to prove the locally asymptotically self-similar property for RL-mBm. One gets by using (3.5)

$$\left\langle \frac{X_{H(t_o)}(t_o + \epsilon u) X_{H(t_o)}(t_o + \epsilon v)}{\epsilon^{2H(t_o)}} \right\rangle$$
$$= D_{H(t_o)} \left[|u|^{2H(t_o)} + |v|^{2H(t_o)} - |u - v|^{2H(t_o)} \right] \tag{3.7}$$

which verifies the locally asymptotically self-similar condition:

$$\lim_{\epsilon \to 0_+} \left(\frac{X_{H(t_o)}(t + \epsilon u) - X_{H(t_o)}(t_o)}{\epsilon^{2H(t_o)}} \right)_{u \in R^+} = \left(B_{H(t_o)}(u) \right)_{u \in R^+}.$$

It is interesting to remark that the locally asymptotic stationarity property (3.5) and the form of local covariance (3.6) allow one to study the local power-law behavior of the spectrum of RL-fBm and RL-mBm based on the Wigner-Ville distribution and the time-scale distribution of the wavelet scalogram [LM01]. For the local time-scale analysis, consider the continous wavelet transform defined as

$$T_X(t,a) = \frac{1}{\sqrt{a}} \int_{-\infty}^{\infty} X(t) \Psi^* \left(\frac{t-s}{a} \right) ds,$$

where $\Psi(t)$ is the mother wavelet and $a > 0$ is the scaling parameter [Mal98,Fla99]. Recalling that the local scaling of the increment processes of RL-mBm behave as

$$\langle |X_{H(t)}(t'+\tau) - X_{H(t)}(t')|^2 \rangle \sim |\tau|^{2H(t)}, \quad 0 < \tau < \epsilon << 1, \tag{3.8}$$

for $t' \in [t - \epsilon/2, t + \epsilon/2]$ and by using the vanishing moment condition of the wavelet, one can express the wavelet coefficiets as

$$T_X(t,a) = \frac{1}{\sqrt{a}} \int_{-\infty}^{\infty} [X_{H(t)}(t'+\tau) - X_{H(t)}(t')] \Psi^* \left(\frac{\tau - (t-t')}{a} \right) d\tau.$$

Thereby, the wavelet scalogram can be shown to exhibit power-law scaling behavior of the form

$$\langle |T_X(t,a)|^2 \rangle \approx a^{2H(t)+1} C_\Psi(t), \qquad a = O(\tau) \to 0_+, \tag{3.9}$$

where $C_\Psi(t)$ is a function that depends on the correlation of two wavelets with overlapping supports. Based on the scaling behavior shown in (3.9), one may infer that the graphs of R-L mBms are locally asymptotically self-similar with the local Hausdorff dimensions given in (2.8).

Finally, we give some comments regarding various differential equations proposed to describe fBm. There exist a number of suggestions of modified Fokker-Planck (FP) equations which are assumed to describe fBm. For example, the modified FP equation proposed by Wang and Lung [WL90]

$$\frac{\partial P(x,t)}{\partial t} = 2HDt^{2H-1} \frac{\partial^2 P(x,t)}{\partial x^2}, \tag{3.10}$$

where $P(x,t)$ is the probability density function (pdf) and D is the diffusion constant. If the initial condition is takes as $P(x,0) = \delta(x)$, (3.10) has the following solution

$$P(x,t) = \frac{1}{\sqrt{4\pi Dt^{2H}}} \exp \left[-\frac{x^2}{4Dt^{2H}} \right], \tag{3.11}$$

which is the pdf for fBm.

On the other hand, there exists a fractional-time FP equation [MK00]

$$\frac{\partial^{2H} P(x,t)}{\partial t^{2H}} = D_H \frac{\partial^2 P(x,t)}{\partial x^2} \tag{3.12}$$

with initial condition $P(x,t) = \delta(x)$ and $\frac{\partial P(x,t)}{dt}|_{t=0} = 0$. The solution of the (3.12) can be obtained by Fourier transform techniques and it is expressed in the form of Mittag-Leffler function, E_{2H} namely,

$$\widetilde{P}(k,t) = E_{2H}\left[-D_H t^{2H} k^2\right]$$
$$= \sum_{j=0}^{\infty} \frac{(-D_H t^{2H} k^2)^j}{\Gamma(1+2Hj)}, \tag{3.13}$$

where $\widetilde{P}(k,t)$ is the Fourier transform of $P(x,t)$ with respect to x. By inverse Fourier transform of $\widetilde{P}(k,t)$, the solution of (3.12) can be expressed in terms of Fox H-function

$$P(x,t) = \frac{1}{\sqrt{4\pi D_H t^{2H}}} H_{1,2}^{2,0}\left[\frac{x}{4D_H t^{2H}} \,\middle|\, \begin{matrix}(1-H,2H)\\(0,1),(\frac{1}{2},1)\end{matrix}\right] \tag{3.14}$$

which gives the following mean square displacement

$$\langle x^2(t)\rangle \propto t^{2H}$$

which has the same time dependence as that for fBm. However, the pdf (3.14) is non-Gaussian in contrast to the Gaussian pdf for fBm. We remark that the probablity density function alone does not determine the random process completely. In particular, two different random Gaussian processes which have the same variance will lead to same probability density function. The claim that (3.10) and (3.12) describe fBm is valid up to leading term in $t \to \infty$ limit.

In fact, the exact equation that describe fBm of RL type is defined in term of fractional differential equation

$$\frac{d^{2H} X_H(t)}{dt^{2H}} = W(t), \tag{3.15}$$

subjected to the initial condition $X_H(0) = 0$ and $W(t)$ is the Gaussian white noise. Similarly, one may expect that the fractional differential equation for RL-mBm to be

$$\frac{d^{2H(t)} X_{H(t)}(t)}{dt^{2H(t)}} = W(t). \tag{3.16}$$

However, the fractional differential operator now becomes time dependent, therefore the standard fractional calculus techniques cannot be used. One possibility is to consider the generalized fractional derivative and integral of RL type in the form

$$\frac{d^{\alpha(t)} f(t)}{dt^{\alpha(t)}} = \int_a^t \frac{(t-\tau)^{-(\alpha(t)-n)}}{\Gamma(1-\alpha(t)+n)} \frac{d^{n+1}}{d\tau^{n+1}} f(\tau)d\tau \tag{3.17}$$

with its inverse operator which can be used to define RL-mBm. Another way to generalizing derivative to that of variable order is

$$\frac{d^{\alpha(t)}f(t)}{dt^{\alpha(t)}} = \int_a^t \frac{(t-\tau)^{\alpha(\tau)}}{\Gamma(1-\alpha(\tau)+n)}\frac{d^{n+1}f(\tau)}{d\tau^{n+1}}d\tau. \qquad (3.18)$$

For some initial investigations of the generalized fractional calculus involving variable order, one may refer to [Sam95,Hoh00].

4 Conclusion

In this paper we have studied the local properties of mBm. In particular, we showed that the R-L mBm exhibits many similar behaviors as the standard mBm when examined at the locally asymptotic conditions. These observations suggest that the RL-mBm may be useful as an alternative model in the studies of locally self-similar processes that begin at time origin $t = 0$, with the mean square displacement in the form $\sim t^{2H(t)}$.

Acknowledgements

This research is supported by the Malaysian Ministry of Science, Technology and Environment under IRPA Grant No. 09-02-02-0092. SCL would like to thank Deutscher Akademischer Austauschdienst (DAAD) for an award which supported his research visit to Institut für Mathematik, Technische Universität Clausthal and his host Michael Demuth for the hospitality.

References

[BA66]   Barnes, J.A., Allen, D.W., A statistical model of Flicker noise. *Proc. IEEE*, 54: 176-178, 1966.

[BCI98]   Benassi, A., Cohen, S., Istas, J., Identifying the multifractional function of a Gaussian process, *Statistic and Probabilty Lett.*, 39:337-345, 1998.

[BCIJ97]   Benassi, A., Cohen, S., Istas, J., Jaffard, S., Identification of EGR processes, In *Fractals in Engineering*, Level-Vehel, J., Lutton E., Tricot, C. (Eds.), pp. 115-123, Springer, Berlin, 1997.

[BCIJ98]   Benassi, A., Cohen, S., Istas, J., Jaffard, S., Identification of filtered white noises, *Stochastic Processes and Their Applications*, 75:31-49, 1998.

[BJR97]   Benassi, A., Jaffard, S., Roux, D., Elliptic gaussian random processes, *Revista Matematica Iberoamericana*, 13:19-90, 1997.

[Coh99]   Cohen, S., From self-similarity to local self-similarity: the estimation problem. In, *in Fractals: Theory and Applications in Engineering*, Dekking, M., Levy Vehel, J., Lutton,E., Tricot, C. (Eds.), pp. 3-17, Springer, Berlin, 1999.

[Fla99]   Flandrin, P., *Time-Frequency/Time-Scale Analysis*, Academic Press, San Diego, 1999.

[Hoh00]   Hoh, W., Pseudo differential operator with negative definite symbols of variable order, *Revista Matematica Iberoamerica*, 16: 219-241, 2000.

[LM00]   Lim, S.C., Muniandy, S.V., On some possible generalizations of fractional Brownian motion, *Phys. Lett. A*, 266: 140-145, 2000.

[LM01]   Lim, S.C. Fractional Brownian motion and Riemann-Liouville Multifractional Brownian motion, *J. Phys. A*, (in press) 2001.

[Mal98]   Mallat, S. *A Wavelet Tour of Signal Processing*, Academic Press, San Diego, 1998.

[MVN68]   Mandelbrot, B.B., van Ness, J. Fractional Brownian motion, fractional noises and applications, *SIAM Rev.*, 10:422-437, 1968.

[MK00]   Metzler, R., Klafter, J. The random walk's guide to anomalous diffusion: a fractional dynamic approach, *Phys. Rep.*, 339: 1-77, 2000

[PLV95]   Peltier, R.F., Levy-Vehel, J., Multifractional Brownian motion: definition and preliminary results, *INRIA Preprint* , 2645: 1-40, 1995.

[Pit71]   Pitt, L.D., A Markov property for Gaussian processes with a multidimensional parameter, *Arch. Rational Mech Anal.*, 43:367-391, 1971.

[Sam95]   Samko, S., Fractional integration and differentiation of variable order, *Analy. Mathematica*, 21: 213-236, 1995.

[WL90]   Wang, K.G., Lung, C.W., Long-time correlation effects and fractal Brownian motion, *Phys. Lett.*, 151:119-121, 1990.

[Yag58]   Yaglom, A.M., Correlation theory of processes with random stationary n-th increments, *Am. Math. Soc. Trans.*, 8:87-141, 1958.

Addresses

S.C. LIM, School of Applied Physics, Universiti Kebangsaan Malaysia, 43600 UKM Bangi, Selangor, Malaysia

E-MAIL: sclim@pkrisc.cc.ukm.my

S.V. MUNIANDY, School of Applied Physics, Universiti Kebangsaan Malaysia, 43600 UKM Bangi, Selangor, Malaysia

E-MAIL: msithi@pkrisc.cc.ukm.my

2000 Mathematics Subject Classification. Primary 60G15, 26A33; Secondary 60G17, 60G18

Operator Theory:
Advances and Applications, Vol. 126
© 2001 Birkhäuser Verlag Basel/Switzerland

Strong Uniqueness for Dirichlet Operators with Singular Potentials

VITALI LISKEVICH AND OLEKSIY US

Abstract. We study the problem of strong uniqueness in L^2 for the Dirichlet operator perturbed by a singular complex-valued potential. We reveal sufficient conditions on the logarithmic derivative β of the measure ρdx and the potential q, which ensure that the operator $(\Delta + \beta \cdot \nabla - q) \restriction C_0^\infty(\mathbb{R}^d)$ has a unique extension generating a C_0-semigroup on L^2. The method of a-priori estimates of solutions of the corresponding elliptic equations is employed.

1 Introduction and Main Result

In this paper we study the operator $H = -\Delta - \beta \cdot \nabla + q$, $D(H) = C_0^\infty(\mathbb{R}^d) =: C_0^\infty$, in the space $L^2 := L^2(\mathbb{R}^d, \rho dx)$, where β is the logarithmic derivative of the measure ρdx and $q = V + iW$, with V, W measurable and real-valued. We assume that $\rho > 0$ a.e., $\rho \in L^1_{loc}(\mathbb{R}^d, dx)$, $\beta \in L^2_{loc}$ and $V, W \in L^2_{loc}$, $V = V^+ - V^-$, $V^\pm \geq 0$. Assuming that H is accretive we seek for conditions on β and q ensuring that its closure is m-accretive, which is equivalent to the fact that $-H$ has a unique extension generating a C_0-semigroup on L^2. This problem is referred to as the strong uniqueness problem in L^2, and for $W = 0$ it is equivalent to the essential self-adjointness of H.

The strong uniqueness problem for Dirichlet operators in weighted spaces was studied in a number of papers (see e.g. [LS92], [BKR97], [Li99], [Eb00] and references therein). For the Schrödinger operators with singular potentials having form-bounded negative part, the most optimal results on strong uniqueness in L^p were obtained in [BS90]. In the present paper we provide a criterion of strong uniqueness in L^2 for the Dirichlet operator perturbed by a complex-valued potential, extending the corresponding results of [Li99] and [BS90].

Our strategy to prove the main result is as follows. First the problem is reduced ("localized") to a degenerate operator with coefficients vanishing outside a ball. Afterwards we study the operator on the ball using the method of a-priori estimates, which was developed in [LS92], and a perturbation technique inherited from [BS90].

We use the following notations: $\| \cdot \|_p$ is the norm in L^p, $\langle f \rangle$ stands for the integral of f w.r.t. the measure ρdx. By $\mathcal{F}_R$ we denote the class of spherically symmetric functions $\eta \in C_0^\infty$ such that $0 \leq \eta \leq 1$ and $\eta = 1$ on a ball B_R of radius R centered at the origin. We set $V_\eta := V\eta$ and $W_\eta := W\eta$. Let $\mathcal{L}_\eta$ and H_η be the operators in L^2 associated with the closure of the forms $\langle \eta \nabla u, \eta \nabla v \rangle$ and

$$\tau_\eta(u, v) := \langle \eta \nabla u, \eta \nabla v \rangle + \langle V_\eta u, v \rangle + i \langle W_\eta u, v \rangle, \ \ u, v \in C_b^1(\mathbb{R}^d),$$

respectively.

Next we list the additional conditions on the coefficients of the operator H.

(A1) there is an $R < \infty$ such that for every $R_1 > R$ there is a constant $C = C(R_1)$ such that for all $\varphi \in C_0^\infty$ we have

$$\| \, |\beta| \varphi \mathbb{1}_{B_{R_1} \setminus B_R} \|_2^2 \leq C \big(\| \, |\nabla \varphi| \, \|_2^2 + \|\varphi\|_2^2 \big),$$

where $\mathbb{1}_{B_{R_1} \setminus B_R}$ is the characteristic function of the set $B_{R_1} \setminus B_R$;
(A2) for every $\eta \in \mathcal{F}_R$ there are constants $a(\eta) > 0$ and $c(\eta) \in \mathbb{R}$ such that for all $\varphi \in C_0^\infty$

$$|\langle W_\eta \varphi, \varphi \rangle| \leq a(\eta) \| (\mathcal{L}_\eta \dotplus V_\eta^+)^{\frac{1}{2}} \varphi \|_2^2 + c(\eta) \|\varphi\|_2^2;$$

(A3) there exists a constant $\bar{c}$ such that for every $\eta \in \mathcal{F}_R$ the operator

$$V_\eta^- \leq \alpha(\eta)(\mathcal{L}_\eta \dotplus V_\eta^+) + \bar{c}$$

with $\alpha(\eta) \in (0, 1)$ (inequality in the form sense);
(A4) $V_\eta^- \in L^{k(\eta)}$, where $k(\eta) := 1 + \dfrac{1}{\sqrt{1 - \alpha(\eta)}}$.

Condition (A1) is identical to that of [Li99] and conditions (A3), (A4) are modelled in accordance with those of [BS90]. Condition (A2) combined with (A3) guarantees that the form τ_η is sectorial. Observe that $k(\eta) > 2$ and $k(\eta) \to 2$ if $\alpha(\eta) \to 0$.

The main result of the paper reads as follows.

Theorem 1.1. *Let $\beta \in L_{loc}^4$ and $V, W \in L_{loc}^2$. Assume that conditions (A1)-(A4) hold. Then the closure of the operator H is m-accretive.*

2 Proof of Main Result

We start with the localization theorem which is an extension of the corresponding result of [SI78] to the case of weighted spaces and complex-valued potentials.

Theorem 2.1. *Let $\beta, V, W \in L_{loc}^2$ and conditions (A1)-(A3) hold. Assume that the closure in L^2 of the operator $H_\eta \restriction C_0^\infty$ is m-accretive. Then the closure of H is m-accretive.*

Proof. It follows from (A3) that the operator $\lambda + H$ is accretive for all $\lambda \geq \bar{c}$. Therefore by Lumer-Phillips' theorem it suffices to verify that the range of $(\lambda + H)$ is dense in L^2, i.e. that if $u \in L^2$ and

$$\langle (\lambda + H)\varphi, u \rangle = 0 \quad for \ all \ \varphi \in C_0^\infty, \tag{2.1}$$

then $u = 0$.

Claim. $u\xi \in \mathcal{D} := \mathcal{D}(\mathcal{L}_\eta^{\frac{1}{2}}) \cap \mathcal{D}((V_\eta^+)^{\frac{1}{2}})$ for all $\xi, \eta \in \mathcal{F}_R$, such that $\eta = 1$ on $supp\,\xi$. Indeed, it is easy to check that

$$\langle (\lambda + H)\varphi, u\xi \rangle = \langle (\lambda + H_\eta)\varphi, u\xi \rangle \quad (\varphi \in C_0^\infty(\mathbb{R}^d)).$$

Since $\xi\varphi \in C_0^\infty(\mathbb{R}^d)$, (2.1) implies that

$$\langle (\lambda + H)\varphi, u\xi \rangle = 2\langle \nabla\xi \cdot \nabla\varphi, u \rangle + \langle (\Delta\xi + \beta \cdot \nabla\xi)\varphi, u \rangle. \tag{2.2}$$

Applying the Schwarz inequality and (A1) we have

$$\|\nabla\xi \cdot \nabla\varphi\|_2, \ \|(\Delta\xi)\varphi\|_2, \ \|(\beta \cdot \nabla\xi)\varphi\|_2 \leq C_\xi \|(1 + \mathcal{L}_\eta)^{\frac{1}{2}}\varphi\|_2.$$

Observing that by (A3) one can find a constant C_η depending on η such that

$$\|(1 + \mathcal{L}_\eta)^{\frac{1}{2}}\varphi\|_2 \leq C_\eta \|(\lambda + \mathcal{L}_\eta + V_\eta)^{\frac{1}{2}}\varphi\|_2,$$

we conclude that

$$|\langle (\lambda + H_\eta)\varphi, u\xi \rangle| \leq C_{u,\xi,\eta} \|(\lambda + \mathcal{L}_\eta + V_\eta)^{\frac{1}{2}}\varphi\|_2. \tag{2.3}$$

Estimate (2.3) implies that the LHS of (2.2) defines a linear bounded functional on $\mathcal{D}$, i.e. there is a $v \in \mathcal{D}$ such that

$$\langle (\lambda + H_\eta)\varphi, u\xi \rangle = \langle (\lambda + \mathcal{L}_\eta + V_\eta)^{\frac{1}{2}}\varphi, (\lambda + \mathcal{L}_\eta + V_\eta)^{\frac{1}{2}}v \rangle. \tag{2.4}$$

Since the form τ_η is sectorial and the operator $\lambda + \mathcal{L}_\eta + V_\eta \geq 0$ for all $\lambda \geq \bar{c}$ it follows (see e.g. [Ka84], Th. VI.3.2) that

$$\lambda + H_\eta = (\lambda + \mathcal{L}_\eta + V_\eta)^{\frac{1}{2}}(Id + i\,B)(\lambda + \mathcal{L}_\eta + V_\eta)^{\frac{1}{2}}$$

where B is a bounded self-adjoint operator in L^2. The operator $Id - iB : L^2 \to L^2$ is clearly bijective, and the mapping $(\lambda + \mathcal{L}_\eta + V_\eta)^{\frac{1}{2}} : \mathcal{D} \to L^2$ is known to be isomorphic. Therefore for every $v \in \mathcal{D}$ one can find a (unique) $w \in \mathcal{D}$ such that

$$(Id - iB)(\lambda + \mathcal{L}_\eta + V_\eta)^{\frac{1}{2}}w = (\lambda + \mathcal{L}_\eta + V_\eta)^{\frac{1}{2}}v. \tag{2.5}$$

Combining (2.4) and (2.5) we get

$$\langle(\lambda + H_\eta)\varphi,\, u\xi\rangle = \langle(Id + iB)(\lambda + \mathcal{L}_\eta \dotplus V_\eta)^{\frac{1}{2}}\varphi,\, (\lambda + \mathcal{L}_\eta \dotplus V_\eta)^{\frac{1}{2}}w\rangle.$$

We employ the strong uniqueness of $H_\eta \!\restriction_{C_0^\infty}$ and obtain the equality $\langle\psi,\, u\xi\rangle = \langle\psi,\, w\rangle$ for all $\psi \in L^2$. Hence, $w = u\xi$ μ-a.e. and the Claim follows.

Making use of equality (2.2) and the Claim one derives the following estimate for all $\xi \in \mathcal{F}_R$:

$$\|u\xi\|_2^2 \leq \frac{1}{\lambda - \bar{c}}\|u|\nabla\xi|\|_2^2$$

(for details see the proof of Theorem 2 in [Li99]). Therefore $u = 0$. $\qquad\square$

Before proceeding further we discuss some properties of degenerate operators with smooth coefficients. Let Ω be the interior of $supp\,\eta$ and $b \in C^\infty(\overline{\Omega}, \mathbb{R}^d)$. Let us consider the operator $\mathcal{A}_\eta = -(\nabla + b)\eta^2\nabla$ in $C(\overline{\Omega})$ with the domain $\mathcal{D}(\mathcal{A}_\eta) = C^2(\overline{\Omega})$. By a result of Taira (cf. [Ta93], Th.1) the closure $\overline{\mathcal{A}_\eta}$ generates a Feller semigroup $\exp(-t\overline{\mathcal{A}_\eta})$ on $C(\overline{\Omega})$. Furthermore, by ([Ta93], Th.2) there exists a λ_0 such that for all $\lambda > \lambda_0$ and $f \in C(\overline{\Omega})$ one has $(\lambda + \overline{\mathcal{A}_\eta})^{-1}f \in C^2(\overline{\Omega})$.

In order to complete the proof of Theorem 1.1 it remains to show that C_0^∞ is a core for the operator H_η. In the next theorem ([Li99], Th.3) we formulate an a-priori estimate which is crucial in the proof of the uniqueness for the operator H_η.

Theorem 2.2. *Let $\beta \in L_{loc}^4$. Let u be the solution of the equation $\lambda u - (\nabla + b)\eta^2\nabla u = f$ with $b \in C^\infty(\overline{\Omega}, \mathbb{R}^d)$, $f \in C^\infty(\overline{\Omega})$ and $\lambda > 0$ sufficiently large. Then there are constants C_1 and C_2 such that*

$$\||\eta|\nabla u|\,\|_4^4 \leq C_1\lambda^{-2}\|f\|_\infty^4\big(\||\beta\eta|\,\|_4^4 + \||b\eta|\,\|_4^4 + \||\nabla\eta|\,\|_4^4\big)$$
$$+ C_2\lambda^{-2}\|f\|_2^2\big(\|\eta|\beta|\,\|_2^2 + \|f\|_\infty^2\big) + C_2\|f\|_\infty\|f\|_1.$$

Finally we present the uniqueness result for the degenerate operator H_η.

Theorem 2.3. *Let $\beta \in L_{loc}^4$ and V^+, $W \in L_{loc}^2$. We assume that conditions (A2)-(A4) hold. Then $C_0^\infty(\mathbb{R}^d)$ is a core for the operator H_η.*

Proof. Let $\widehat{\mathcal{L}_\eta}$ and $\widehat{H_\eta}$ be the operators in $L_\eta^2 := L^2(\Omega, \mu)$ which are defined in the same way as $\mathcal{L}_\eta$ and H_η respectively.

Step 1. We claim that the set $\mathcal{D}_1 := (\lambda + \widehat{\mathcal{L}_\eta} \dotplus U_\eta)^{-1}C_0^\infty(\Omega)$ is a core for $\widehat{H_\eta}$, where $U_\eta := V_\eta^+ + iW_\eta$ (one can verify that $\mathcal{D}_1 \subset \mathcal{D}(\widehat{\mathcal{L}_\eta}) \cap \mathcal{D}(V_\eta + iW_\eta)$). Let $u \in L_\eta^2$ and $\langle(\lambda + \widehat{H_\eta})\varphi,\, u\rangle = 0$ for all $\varphi \in \mathcal{D}_1$. Observe that here and below $\langle g\rangle := \int_\Omega g\,d\mu$. Taking $\varphi = (\lambda + \widehat{\mathcal{L}_\eta} \dotplus U_\eta)^{-1}\psi$, $\psi \in C_0^\infty(\Omega)$ we obtain the equality

$$\langle\psi,\, u\rangle = \langle(\lambda + \widehat{\mathcal{L}_\eta} \dotplus \overline{U}_\eta)^{-1}\psi,\, V_\eta^- u\rangle \ \textit{for all} \ \psi \in C_0^\infty(\Omega). \tag{2.6}$$

We note that due to (A4) $V_\eta^- u \in L^q$ with $q = \frac{2k(\eta)}{2+k(\eta)}$. Thus (2.6) implies that $u = (\lambda + \widehat{\mathcal{L}}_\eta + \overline{U}_\eta)_q^{-1} V_\eta^- u$ in L_η^q (the subscript q by the operator means that the latter is considered as an operator in L_η^q). Set $z = 1 + \frac{2}{q'}$, where $q' = \frac{q}{q-1}$. Then $u|u|^{z-2} \in L^{q'}$. Therefore

$$\langle u|u|^{z-2}, (\lambda + \widehat{H}_{\eta,q}^+)u\rangle = \langle u|u|^{z-2}, V_\eta^- u\rangle, \qquad (2.7)$$

where $\widehat{H}_{\eta,q}^+$ is the generator of a the C_0-semigroup in L_η^q associated with (the closure of) the sectorial form $\langle \eta\nabla u, \eta\nabla v\rangle + \langle (V_\eta^+ - iW_\eta)u, v\rangle$.

Let $(u_k)_{k\in\mathbb{N}} \subset C_0^\infty(\Omega)$ and $u_k \to u$ strongly in L_η^2. Let $T_q(t), t \geq 0$ stand for the C_0-semigroup generated by $\widehat{H}_{\eta,q}^+$. For $n \in \mathbb{N}$ we set $u_{n,k} := T_q(1/n)u_k$. Then $u_{n,k} \in \mathcal{D}(\widehat{H}_\eta^+) \cap L_\eta^\infty$ and $\widehat{H}_\eta^+ u_{n,k} = \widehat{H}_{\eta,q}^+ u_{n,k}$ (here and below $H_\eta^+ := H_{\eta,2}^+$. Set $\varphi_{n,k} := u_{n,k}|u_{n,k}|^{z-2}$. One can verify that

$$\lim_n \lim_k \langle \varphi_{n,k}, (\lambda + \widehat{H}_\eta^+)u_{n,k}\rangle = \langle \varphi, (\lambda + \widehat{H}_{\eta,q}^+)u\rangle,$$
$$\lim_n \lim_k \langle \varphi_{n,k}, V_\eta^- u_{n,k}\rangle = \langle \varphi, V_\eta^- u\rangle. \qquad (2.8)$$

Set $\widehat{\mathcal{D}} := \mathcal{D}((\widehat{\mathcal{L}}_r)^{\frac{1}{2}}) \cap \mathcal{D}((V_\eta^+)^{\frac{1}{2}})$. We introduce functions $\varphi_{n,k,\varepsilon} := u_{n,k}(|u_{n,k}| \vee \varepsilon)^{z-2}$ and $v_{n,k,\varepsilon} := u_{n,k}(|u_{n,k}| \vee \varepsilon)^{\frac{z-2}{2}}$, $\varepsilon > 0$. It follows from the dominated convergence theorem that $\varphi_{n,k,\varepsilon} \to \varphi_{n,k}$ in $L_\eta^{q'}$ and $v_{n,k,\varepsilon} \to v_{n,k}$ in L_η^2 as $\varepsilon \to 0$. Then it follows from (2.8) that

$$\lim_n \lim_k \lim_\varepsilon \langle \varphi_{r,k,\varepsilon}, (\lambda + \widehat{H}_\eta^+)u_{n,k}\rangle = \lim_n \lim_k \lim_\varepsilon \langle V_\eta^- |v_{n,k,\varepsilon}|^2\rangle. \qquad (2.9)$$

The set $\widehat{\mathcal{D}}$ is a Dirichlet space and functions $y(|y| \vee \varepsilon)^\gamma$, $y \in \mathbb{C}$ are normal contractions of $\mathbb{C}$ when $\gamma < 0$, therefore $v_{n,k,\varepsilon}, \varphi_{n,k,\varepsilon} \in \widehat{\mathcal{D}}$. A straightforward computation shows that

$$Re \, \langle \varphi_{n,k,\varepsilon}, \widehat{H}_\eta^+ u_{n,k}\rangle \geq \frac{4(z - 1)}{z^2}\||\eta|\nabla v_{n,k,\varepsilon}| \|_2^2 + \langle V_\eta^+ |v_{n,k,\varepsilon}|^2\rangle. \qquad (2.10)$$

Combining (2.9) and (2.10) and making use of (A3) we arrive at the inequality $\lim_n \lim_k \lim_\varepsilon (\lambda - \bar{c})\|v_{n,k,\varepsilon}\|_2^2 = (\lambda - \bar{c})\|u\|_z^z \leq 0$. Hence $u = 0$.

Step 2. We claim that $\cup_{m\geq 1}(m + \widehat{\mathcal{L}}_\eta)^{-1}C^\infty(\overline{\Omega})$ is a core. Indeed, it is easy to see that the set $\cup_{m\geq 1}(m + \widehat{\mathcal{L}}_\eta)^{-1}(\lambda + \widehat{\mathcal{L}}_\eta + U_\eta)^{-1}L_\eta^\infty$ is a domain of strong uniqueness for $\widehat{H}_\eta$, and therefore so is $\cup_{m\geq 1}(m + \widehat{\mathcal{L}}_\eta)^{-1}L_\eta^\infty$. For every element $f \in L_\eta^\infty$ one can find a sequence $(\varphi_k)_{k\in\mathbb{N}} \subset C^\infty(\overline{\Omega})$ such that $\|f - \varphi_k\|_2 \to 0$ and $\sup_k \|\varphi_k\|_\infty < \infty$. The operator $\widehat{\mathcal{L}}_\eta(m + \widehat{\mathcal{L}}_\eta)^{-1}$ is clearly bounded in L^2. Hence the claim follows.

Step 3. Let $b^n \in C^\infty(\overline{\Omega}, \mathbb{R}^d)$ and $b^n \to \beta \mathbb{1}_\Omega$ in L_η^4. For $f \in C^\infty(\overline{\Omega})$ we set $\phi_n := (\lambda + \overline{A}_{\eta;n})^{-1}f$, where $\overline{A}_{\eta;n}$ is the operator $\overline{A}_\eta$ with $b = b^n$. It follows from

Theorem 2.2 that

$$\|\phi_n - (\lambda + \widehat{\mathcal{L}}_\eta)^{-1}f\|_2 \leq C(\eta, \beta, f)\| \eta|b^n - \beta| \|_4,$$
$$\|(\lambda + \widehat{\mathcal{L}}_\eta))(\phi_n - (\lambda + \widehat{\mathcal{L}}_\eta)^{-1})f\|_2 \leq C(\eta, \beta, f)\| \eta|b^n - \beta| \|_4. \tag{2.11}$$

It follows from (2.11) that $C^2(\overline{\Omega})$ is a core for $\widehat{H}_\eta$. The last statement is equivalent to the fact that $C_b^2(\mathbb{R}^d) \cap L^2$ is a core for the operator H_η. To show that C_0^∞ is also a core we use standard approximation for functions from $C_b^2(\mathbb{R}^d)$ by elements of C_0^∞ (we omit the details, see e.g. [Li99]). $\square$

Acknowledgements

Financial support of EPSRC through grant GR/L76631 and British Council through The British-German Academic Research Collaboration Program (Project 1040) is gratefully acknowledged. The authors are also thankful to Z. Sobol for valuable discussions and comments.

References

[BKR97] V.Bogachev, N.Krylov and M.Röckner. *Elliptic Regularity and Essential Self-adjointness of Dirichlet Operators on $\mathbb{R}^n$*. Ann.Scuola Norm. Sup. Pisa **24**, 451-462 (1997).

[BS90] A.Beliy and Yu.Semenov. *On the L^p-theory of the Schrödinger semigroups.II*, Siberian Math. J.**31**,16-26 (1990).

[Eb00] A.Eberle. *L^p-uniqueness of Non-symmetric Diffusion Operators with Singular Drift Coefficients. I. Finite-Dimensional Case.* J.Funct.Anal.**173**, 328-342 (2000).

[Ka84] T.Kato. *Perturbation Theory for Linear Operators.* Second Corrected Printing of the Second Edition. Springer-Verlag, Berlin-Heidelberg-New York-Tokio, 1984.

[Li99] V.Liskevich. *On the Uniqueness Problem for Dirichlet Operators.* J. Funct. Anal.**162**, 1-13 (1999).

[LS92] V.Liskevich and Yu.Semenov. *Dirichlet Operators: a Priori Estimates and Uniqueness Problem.* J.Funct.Anal.**109**, 199-213 (1992).

[SI78] C.Simader. *Essential Self-adjointness of Schrödinger Operators Bounded from Below.* Math.Z. **159**, N 1, 47-50 (1978).

[Ta93] K.Taira. *On Existence of Feller Semigroups with Dirichlet Conditions.* Tsukuba J. Math. **17**, 377-427 (1993).

Address

VITALI LISKEVICH, OLEKSIY US, School of Mathematics, University cf Bristol, Bristol, BS8 1TW, UK

E-MAIL: V.Liskevich@bristol.ac.uk, A.Us@bristol.ac.uk

2000 Mathematics Subject Classification. Primary 47B44 ; Secondary 31C25, 35J70

Operator Theory:
Advances and Applications, Vol. 126
© 2001 Birkhäuser Verlag Basel/Switzerland

Hardy Type Inequalities, Mourre Estimate and A-priori Decay for Eigenfunctions

MARIUS MĂNTOIU AND RADU PURICE

Abstract. We address the problem of finding upper bounds on the decay at infinity for the eigenfunctions of a large class of self-adjoint operators, covering many interesting quantum Hamiltonians and valid even for eigenfunctions associated to embedded eigenvalues. Inspired by [AHS89], [AMG87], [FH82] and [FHHO82], we put into evidence a strategy [MP00a] that starts from a local Mourre inequality with respect to a conjugate operator (that is not supposed to be self-adjoint) and produces a local weighted estimation; from this, a cut-off procedure (both on the test function and on the weights [Ag82], [AMG87]) leads to a Hardy type estimation and a simple argument [MP00a] gives then an a-priori bound (see Proposition 1.1 below) for the decay of eigenfunctions associated to eigenvalues belonging to the interval on which the initial Mourre inequality is valid. We apply this method to a large class of perturbations of convolution operators in $\mathbb{R}^n$ and among them a class of perturbed periodic Schrödinger Hamiltonians.

1   Introduction

Let

$$\mathcal{H} = L^2(\mathbb{R}^n), \qquad H = H^*, \qquad E \in \mathbb{R},$$

$$Q_j f(x) := x_j f(x), \; \mathcal{D}(Q_j) := \left\{ f \in L^2 \; \Big| \; \int |x_j|^2 |f(x)|^2 dx < \infty \right\}$$

and let D_j be the unique commuting self-adjoint extensions of the operators:

$$D_j f := -i \frac{\partial f}{\partial x_j}, \forall f \in C_0^\infty(\mathbb{R}^n).$$

Suppose given two *weight functions* w, $\tilde{w}$, i.e. two growing, strictly positive and unbounded functions (we shall suppose that $w \leq \tilde{w}$). We denote:

$$\tilde{W} = \tilde{w}(Q), \qquad W = w(Q), \qquad Q := (Q_1, ... Q_n),$$

The following result can easily be proved [MP00a]:

Proposition 1.1. *Let H be a self-adjoint operator in $\mathcal{H}$, let E be an eigenvalue of H with eigenvector $g \in \mathcal{H}$. Suppose:*

1. *$\exists R > 0, \exists C > 0$ such that we have the inequality:*

$$\|Wu\| \leq C \left\|\tilde{W}(H - E)u\right\|$$

for any $u \in \mathcal{D}(H)$ with $\mathrm{supp}\, u \cap \{|x| \leq R\} = \emptyset$;

2. *$\forall u \in \mathcal{D}(H)\ \forall \varphi \in C_0^\infty(\mathbb{R}^n),\ \varphi u \in \mathcal{D}(H)$ and $H\varphi u \in \mathcal{D}(\tilde{W})$;*

Then the eigenvector g belongs to $\mathcal{D}(W)$.

2 Main Results

2.1 Analytic Case

Definition 2.1. 1. For a finite complex measure ν on $\mathbb{R}^n$ we denote by $|\nu|$ its total variation and by $\mathcal{M}(\mathbb{R}^n)$ the space of finite complex measures on $\mathbb{R}^n$. Let $\mathcal{F}\mathcal{M}(\mathbb{R}^n)$ be the space of Borel functions on $\mathbb{R}^n$ that are Fourier transforms of measures in $\mathcal{M}(\mathbb{R}^n)$. For a function $\mu \in \mathcal{F}\mathcal{M}(\mathbb{R}^n)$ we denote by $\hat{\mu}(dk)$ its Fourier transform.

2. For $\delta > 0$, $\mathbb{C}_\delta^n := \{z \in \mathbb{C}^n \mid |\mathcal{I}mz_j| < \delta, \quad \forall\, j \in \{1, ..., n\}\}$,

3. For any real s we define the function spaces $S_0^s(\mathbb{R}^n)$ as the sets:

$$\left\{\rho \in C_{pol}^\infty(\mathbb{R}^n) \mid |(\partial^\alpha \rho)(x)| \leq C_\alpha \min\{< x >^{s-|\alpha|}, < \rho(x) >\}, \forall \alpha \in \mathbb{N}^n\right\}$$

and then:

$$\mathcal{O}_0^s(\mathbb{C}_\delta^n) := \{\rho \in \mathcal{O}(\mathbb{C}_\delta^n) \mid \rho(\cdot + iy) \in S_0^s(\mathbb{R}^n)\}$$

(with uniform estimations for $\max_j |y_j| < \delta$).

4. Let $\mathcal{O}^s(\mathbb{C}_\delta^n)$, with $s \geq 0$ be the set of analytic functions on the strip $\mathbb{C}_\delta^n$, of the form $\lambda = \mu\rho$, where:

    (a) μ is analytic on $\mathbb{C}_\delta^n$ and $\forall y \in \mathbb{R}^n$ with $|y_j| < \delta$, for any $j \in \{1, ..., n\}$, the function $\mathbb{R}^n \ni x \longmapsto \mu(x + iy) \in \mathbb{C}$ is of class $\mathcal{F}\mathcal{M}(\mathbb{R}^n)$;

    (b) $\rho \in \mathcal{O}_0^s(\mathbb{C}_\delta^n)$;

    (c) there exists a positive constant κ such that $\kappa |\rho| \leq (1 + |\lambda|)$.

5. we denote $\mathcal{G}$ the domain of $\lambda(D)$ with the norm:

$$\|f\|_{\mathcal{G}}^2 := \|f\|^2 + \|\lambda(D)f\|^2.$$

Definition 2.2. For a function $\lambda \in \mathcal{O}^s(\mathbb{C}_\delta^n)$ we define its set of *regular values*:

$$\mathcal{E}(\lambda) := \left\{t \in \mathbb{R} \mid \exists \epsilon > 0, \exists \kappa > 0 \text{ s.t. } |\nabla\lambda(k)| \geq \kappa\ \forall k \in \lambda^{-1}((t - \epsilon, t + \epsilon))\right\}.$$

We call *generalized critical value* a point in the complement of $\mathcal{E}(\lambda)$ in $\mathbb{R}$.

Theorem 2.1. *Let $\delta > 0$, $\lambda = \rho\mu \in \mathcal{O}^s(\mathbb{C}^n_\delta)$ and $E \in \mathcal{E}(\lambda)$. Suppose V is the operator of multiplication with a real function that is relatively bounded with respect to $\lambda(D)$ and satisfies the decay condition:*

$$\lim_{R\to\infty} \|\chi(|Q| > R) < Q > V(Q)(\lambda(D) + i)^{-1}\| = 0.$$

Let $H := \lambda(D) + V$ be the self-adjoint operator sum. Then there is a strictly positive constant $\gamma_0 < \delta$

such that for any $\gamma \in (0, \gamma_0)$ there are positive constants R and C, such that for any $f \in \mathcal{G}$ with support in $\{|x| \geq R\}$:

$$\left\|e^{\gamma<Q>}f\right\|_{\mathcal{G}} \leq C\left\|\sqrt{<Q>}e^{\gamma<Q>}(H - E)f\right\|.$$

A generalization of this result to convolution with analytic operator valued functions allows to obtain a similar result for a class of perturbed periodic Schrödinger operators (in the Floquet representation) [MP00b].

2.2 C^k - Case

Definition 2.3. We shall denote by $\mathcal{M}^{(p)}(\mathbb{R}^n)$ the space of measures ν on $\mathbb{R}^n$ such that $< x >^p |\nu|$ defines a finite measure on $\mathbb{R}^n$ (with $|\nu|$ the total variation measure associated to ν) and with $\mathcal{F}\mathcal{M}^{(p)}(\mathbb{R}^n)$ the space of functions having Fourier transform of class $\mathcal{M}^{(p)}(\mathbb{R}^n)$.

Theorem 2.2. *For $p > 5$ and $s \geq 0$ let $\lambda = \rho\mu$ with $\mu \in \mathcal{F}\mathcal{M}^{(p)}(\mathbb{R}^n)$, $\rho \in S^s_0(\mathbb{R}^r)$, such that there exists a positive constant κ with $\kappa\,|\rho| \leq (1+|\lambda|)$, and $E \in \mathcal{E}(\lambda)$ be a regular value for λ. Suppose V is the operator of multiplication with a real function that is relatively bounded with respect to $\lambda(D)$ and satisfies the decay condition:*

$$\lim_{R\to\infty} \|\chi(|Q| > R) < Q > V(Q)(\lambda(D) + i)^{-1}\| = 0.$$

Let $H := \lambda(D) + V$ be the self-adjoint operator sum. For $r \in [0, (1/2)(p-5))$; there are two constants R and C_r (depending on λ and on E), such that for any $f \in \mathcal{G}$ with support disjoint of $\{|x| \geq R\}$ we have the estimation:

$$\|< Q >^r f\|^2_{\mathcal{G}} \leq C_r \|< Q >^{r+1} (H - E)f\|^2.$$

3   The Role of a Mourre Estimation

The main ingredient in the proofs of the above results consists in proving a weighted local estimate for the unperturbed "convolution type operator" and in order to do this, the starting point is a "Mourre estimate" with respect to a well-suited conjugate operator [MP00a].

3.1 The Weights

Due to our cut-off procedure, we have to work with a class of weights containing the exponential weight we are interested in and a family of bounded weights that approach it.

Definition 3.1. For any $\gamma \in (0, \delta)$ and any $m \geq 1$ we define the class of functions $\Phi_{\gamma,m}$ given by:

$$\left\{ \varphi \in C^\infty \left([1, \infty) ; \mathbb{R}\right) \mid 0 \leq \varphi' \leq \gamma; \; |\varphi''(t)| \leq \frac{\gamma}{t}; \; \left|\varphi^{(l)}(t)\right| \leq \gamma, \; \forall l \leq m + 1 \right\},$$

and we consider weight functions of the form $w(x) := e^{\varphi(<x>)}$ with $\varphi \in \Phi_{\gamma,m}$.

For any weight function $w(x) := e^{\varphi(<x>)}$ with $\varphi \in \Phi_{\gamma,m}$ we denote:

$$X(x) := \nabla(\varphi(x)) = \frac{x}{<x>}\varphi'(< x >).$$

3.2 The Conjugate Operator

The most natural choice as a conjugate operator for the convolution operator $\lambda(D)$ would be [GN98a], [GN98b]:

$$A_0 := \frac{1}{2} \sum_{j=1}^n \{(\partial_j\lambda)(D)Q_j + Q_j(\partial_j\lambda)(D)\}$$

This choice is useful in order to obtain polynomialy weighted Hardy type inequalities, but does not allow for obtaining exponentially weighted inequalities. Formally, in this case we need the following conjugate operator:

$$A := \frac{1}{2} \sum_{j=1}^n \int_0^1 \{(\partial_j\lambda)(D - isX(Q))Q_j + Q_j(\partial_j\lambda)(D + isX(Q))\}\, ds.$$

3.3 A Pseudodifferential Calculus

We use a functional calculus procedure based on the unitary group $U(x)$ generated by the family $D = (D_1, ...D_n)$ in $\mathcal{H}$. We shall deal with operators that formally can be written as:

$$(\lambda \diamond G)(Q, D) = \int_{\mathbb{R}^n} \hat{\lambda}(y)G(y, Q)U(y)dy$$

for regular functions λ and functions G of class C^∞ having some specific growth properties at infinity.

Remark: Formally these operators are of pseudodifferential type, having a symbol $\lambda(\xi)G(y-x,x)$:

Hypothesis 3.1. $G \in C^\infty(\mathbb{R}^n \times \mathbb{R}^n)$ and there exist W and Ω two even, non-decreasing, strictly positive functions of class $C^\infty(\mathbb{R}^n)$ that satisfy $W(x+y) \leq CW(x)W(y)$, $\Omega(x+y) \leq C\Omega(x)\Omega(y)$, (for some $C < \infty$) and such that G verifies:

$$\forall m \in \mathbb{N},\ \exists C_m < \infty \text{ such that: } \max_{|\alpha| \leq m} \left| \left(\partial_y^\alpha G \right)(y,z) \right| \leq C_m W(y)\Omega^{-1}(z).$$

In our developments $W(x) = e^{\gamma|x|}$, (for $\gamma > 0$) or $W(x) =< x >^p$, (for $p > 0$) and $\Omega(x) =< x >^r$, (for $r \geq 0$).

Proposition 3.1. *Let* $W \in C_0^\infty(\mathbb{R}^n)$, $\Omega \in C_0^\infty(\mathbb{R}^n)$ *and* $G \in C^\infty(\mathbb{R}^n \times \mathbb{R}^n)$ *satisfy the above Hypothesis; let* $\lambda = \mu\rho \in C^\infty(\mathbb{R}^n \times \mathbb{R}^n)$ *with* $\mu \in \mathcal{FM}(\mathbb{R}^n)$, $\rho \in S_0^s(\mathbb{R}^n)$ *and verifying the following bounds (for some $m > s$):*

$$\||\mu\||_{(1)} := \|\Omega W \hat{\mu}\|_{\mathcal{M}} < \infty; \quad \||\rho\||_{(2)} := \max_{|\alpha|=m} \sup_{0 \leq t \leq 1} \left\| W_t \Omega \widehat{\rho^{(c)}} \right\|_{L^-} < \infty$$

where $W_t(x) := W(tx)$. *Then for any function g in $C_0^\infty(\mathbb{R}^n)$ one has:*

$$\|(\lambda \diamond G)(Q,D)g\|_{L^2} \leq$$
$$\leq C_m(G)\||\mu\||_{(1)} \left\{ \max_{|\alpha|<m} \left\|\Omega^{-1}(Q)\rho^{(\alpha)}(D)g\right\|_{L^2} + \||\rho\||_{(2)} \left\|\Omega^{-1}(Q)g\right\|_{L^2} \right\}$$

with $C_m(G)$ a constant depending only on G and its derivatives up to order m with respect to the first variable.

Acknowledgements

Research partially supported by the Swiss National Science Foundation and the grant ANSTI-555/2000

References

[Ag82]   S. Agmon: "Lectures on Exponential Decay of Solutions of Second Order Elliptic Equations", Princeton Univ. Press, (1982).

[AHS89]   S. Agmon, I. Herbst, E. Skibsted: "Perturbation of Embedded Eigenvalues in the Generalized N-Body Problem", *Comm. Math. Phys.* **122**, 411 - 438, (1989).

[AMG87]   W. Amrein, Anne Boutet de Monvel, V. Georgescu: "Hardy Type Inequalities for Abstract Differential Operators", *Memoirs of the American Mathematical Society*, **375**, 1-119, (1987).

[FH82]   R. Froese, I. Herbst: "Exponential Bounds and Absence of Positive Eigenvalues for N-Body Schrödinger Operators, *Comm. Math. Phys.*, **87**, 429-447, (1982).

[FHHO82] R. Froese, I. Herbst, Maria Hoffmann - Ostenhof, T. Hoffmann - Ostenhof: "L^2-Exponential Lower Bounds to Solutions of the Schrödinger Equation", *Comm. Math. Phys.*, **87**, 265-286, (1982).

[GN98a]   Ch. Gérard, F. Nier: "The Mourre Theory for Analytically Fibred Operators", *J. Func. Anal.* **152** (1), 202 - 219, (1998).

[GN98b]   Ch. Gérard, F. Nier: "Scattering Theory for the Perturbations of Periodic Schrödinger Operators", *J. Math. Kyoto Univ.* **38** (4), 595 - 634, (1998).

[MP00a]   M. Măntoiu, R. Purice: "Weighted Estimations from a Conjugate Operator", *Lett. Math. Phys.* **51**, 17 - 35, 2000.

[MP00b]   M. Măntoiu, R. Purice: "A-Priori Decay for Eigenfunctions of Perturbed Periodic Schrödinger Operators", preprint Univ. de Geneve UGVA-DPT 2000/02-1071

Addresses

MARIUS MĂNTOIU, Institute of Mathematics, Romanian Academy, P.O. Box 1-764, RO-70700 Bucharest, Romania
Present Address: Departement de Physique Théorique; Université de Genève; 32, bd. d'Yvoy; CH-1211 Genève 4; Switzerland

E-MAIL: Marius.Mantoiu@physics.unige.ch

RADU PURICE, Institute of Mathematics, Romanian Academy, P.O. Box 1-764, RO-70700 Bucharest, Romania

E-MAIL: Radu.Purice@imar.ro

2000 Mathematics Subject Classification. Primary 35B40 ; Secondary 81Q10

Operator Theory:
Advances and Applications, Vol. 126
© 2001 Birkhäuser Verlag Basel/Switzerland

Surgery and the Relative Index in Elliptic Theory

VLADIMIR NAZAIKINSKII AND BORIS STERNIN

Abstract. We prove a general theorem on the behavior of the relative index under surgery for a wide class of Fredholm operators, including relative index theorems for elliptic operators due to Gromov–Lawson, Anghel, Teleman, Booß-Bavnbek–Wojciechowski, et al. as special cases. In conjunction with additional conditions (like symmetry conditions), this theorem permits one to compute the analytical index of a given operator. In particular, we obtain new index formulas for elliptic pseudodifferential operators and quantized canonical transformations on manifolds with conical singularities

1 Introduction

Recently there have been a number of results concerning elliptic operators with symmetry conditions on manifolds with singularities. These results in particular include index theorems for general elliptic pseudodifferential operators and quantized contact transformations (e.g., see [SSS98], [FST99], [FST98], [NSS98], [NSS99a]). The main tools for obtaining these results in the cited papers include gluing together two copies of the manifold and extending the operators to the double. Furthermore, the double is a smooth closed manifold, on which the Atiyah–Singer index theory [AS63] (for the case of pseudodifferential operators) or the Epstein–Melrose–Leichtnam–Nest–Tsygan theory [EM98], [LNT00] (for the case of quantized canonical transformations) can be applied, and the main problem is to express the difference between twice the index of the original operator and the index of the extended operator on the double via invariants of the conormal symbol. Gluing a copy of the manifold followed by the extension of the operator or, at least, the symbol to the double (or, more generally, to a manifold containing the original manifold as a part) is in fact quite an old idea. Indeed, this technique was successfully applied to the index problem for boundary value problems in early Soviet papers on elliptic theory (e.g., see Agranovich [Agr65] and Dezin [Dez64] as well as Agranovich's review [Agr97], where more detailed references can be found) and was later widely used by many authors. We mention the papers [Hsi72], [Hsi76], [Sto74] by Hsiung and Stong, who used gluing by an orientation-reversing automorphism of the boundary to obtain the index of the signature operator, as well as the papers [GS83], [GS83a] by Gilkey and Smith. Naturally, the list can be continued.

The fact that the extension to a wider manifold (in particular, to the double) proves to be fruitful in index problems is a consequence of the so-called *index locality principle* or a *relative index theorem*. For smooth closed compact manifolds, this principle trivially follows from the so-called *local index formula* (e.g., see [Gil84] or [BW93, Chap. 25]): the index of an elliptic operator D on a closed manifold M is given by the integral of a "local density" that depends only on the principal symbol $\sigma(D)$ and its derivatives. Known proofs of the locality principle in some other special cases are based on techniques related to the fundamental solution of the heat equation or the formula

$$\operatorname{ind} D = \operatorname{Trace}(1 - RD) - \operatorname{Trace}(1 - DR), \tag{1.1}$$

where R is an almost-inverse of D modulo trace class operators. A method based on Eq. (1.1) was used, for example, by Gromov and Lawson [GL83] in the proof of a relative index theorem for Dirac operators on complete noncompact Riemannian manifolds and later by Anghel [Ang93] in a generalization of that theorem to arbitrary essentially self-adjoint supersymmetric Fredholm elliptic first-order operators. The heat kernel method was used by Donnelly [Don87] in the proof of a relative index theorem for the signature operator and also by Bunke [Bun92]. We also mention the book [BW93] by Booß-Bavnbek and Wojciechowski and the paper [Tel84], where Teleman used a subtle homotopy technique to prove a relative index theorem for signature operators on Lipschitz manifolds. We do not try to give an exhaustive list of related publications. Let us only mention that the well-known Agranovich and Agranovich–Dynin theorems [Agr65], [Dyn61], [AD62] that express the relative index of two boundary value problems with the same boundary conditions and different operators (coinciding on the boundary) or with the same operator but different boundary conditions can essentially be interpreted as a statement of the locality principle for boundary value problems.

The technique of attaching a copy of the manifold and passing to the double has also been successfully used in the index theory of pseudodifferential operators on manifolds with singularities. The index theorem for operators with a symmetric conormal symbol was obtained by this method in [SSS98]. Later, the ideas of [SSS98] (combined with the use of an orientation-reversing automorphism of the boundary) were applied in [FST98] to obtain an index theorem for operators that satisfy a symmetry condition involving this automorphism.[1] This theorem applies to the Cauchy–Riemann operator on a two-dimensional surface as well as (just as in [Hsi72], [Hsi76]) to the signature operator provided that the base of the cone possesses the above-mentioned automorphism. In [FST99], an index theorem for two-dimensional surfaces is obtained under a symmetry condition involving an arbitrary diffeomorphism of the base of the cone. Note that in all above-mentioned papers the symmetry condition is imposed on the *entire* conormal symbol.

The aim of the present paper is to prove the locality principle in a sufficiently general case so as to ensure that it can be applied to elliptic pseudodifferential operators as well as elliptic Fourier integral operators on manifolds with singularities.

---

[1] In the earlier paper [Hsi72], the same gluing as in [FST98] was applied in the case of manifolds with boundary.

Thus, we must introduce a new class of Fredholm operators for which the locality principle holds and which includes both pseudodifferential operators and Fourier integral operators. Note that the class of abstract elliptic operators introduced by Atiyah [Ati69] (which served as a starting point for the development of KK-theory) cannot be used here, since the commutators of Fourier integral operators with operators of multiplication by functions are not compact in general.

The preliminary version of this paper was published in [NS00], [NS99]. The complete version containing the proofs is to appear in [NS01].

2 Bottleneck Spaces and the General Relative Index Theorem

Definition 2.1. (a) A *bottleneck space* is a separable Hilbert space H equipped with the structure of a module over the commutative algebra $C^\infty([-1,1])$ (the action is continuous in the topology of $C^\infty([-1,1])$, and the unit function $1 \in C^\infty([-1,1])$ acts as the identity operator in H).

(b) The *support* of an element $h \in H$ is the closed set

$$\operatorname{supp} h = \bigcap \varphi^{-1}(0) \subset [-1,1], \tag{2.1}$$

where $\varphi^{-1}(0)$ is the preimage of the point 0 and the intersection is taken over all elements $\varphi \in C^\infty([-1,1])$ such that $\varphi h = 0$. For an arbitrary subset $F \subset [-1,1]$, we denote by $H(F)$ the closure in H of the set of elements $h \in H$ supported in F.

Example 2.1. Let M be a manifold (possibly noncompact and with singularities and/or boundary), and let $\chi : M \to [-1,1]$ be a continuous function that is smooth in the smooth part of M and satisfies the following condition: $\chi^{-1}((-1,1))$ is a compact set that does not contain any singularities and has an empty intersection with the boundary. The Sobolev spaces $H^s(M)$ (with arbitrary weight functions near singularities and at infinity) can be equipped with a natural structure of bottleneck spaces as follows: one sets

$$(\varphi h)(x) = \varphi(\chi(x))\, h(x), \quad x \in M \tag{2.2}$$

for $h \in H^s(M)$ and $\varphi \in C^\infty([-1,1])$. Since χ is locally constant outside a compact set and near singularities, it follows that the operator of multiplication by $\varphi(\chi(x))$ is continuous in $H^s(M)$ regardless of the choice of the weights. If M has a nonempty boundary, then the Sobolev spaces $H^s(\partial M)$ also bear a natural structure of bottleneck spaces. This structure is given by the same formula (2.2), where now $h \in H^s(\partial M)$ and $x \in \partial M$.

The construction of this example is most important in applications. To describe it unambiguously, it suffices to specify the manifold M, the Sobolev spaces, and the function χ.

Definition 2.2. Let H_1 and H_2 be bottleneck spaces, and let $F \subset [-1,1]$ be a given subset. We say that H_1 *coincides* with H_2 on F and write $H_1 \overset{F}{=} H_2$ if an isomorphism $H_1(F) \approx H_2(F)$ (not necessarily isometric) is given. In this case, we also say that H_1 and H_2 are *modifications* of each other on $[-1,1]\backslash F$ and write $H_1 \overset{[-1,1]\backslash F}{\longleftrightarrow} H_2$. A square

$$
\begin{array}{ccc}
H_1 & \overset{B}{\longleftrightarrow} & H_2 \\
{\scriptstyle A}\updownarrow & & \updownarrow{\scriptstyle C} \\
H_3 & \overset{D}{\longleftrightarrow} & H_4
\end{array}
$$

of modifications is said to *commute* if the corresponding diagram

$$
\begin{array}{ccc}
H_1(F) & \approx & H_2(F) \\
\wr\wr & & \wr\wr \\
H_3(F) & \approx & H_4(F),
\end{array}
\qquad\qquad F = [-1,1]\backslash(A \cup B \cup C \cup D),
$$

of isomorphisms commutes.

Definition 2.3. (a) A *proper operator* in bottleneck spaces H and G is a continuous (with respect to the uniform operator topology) family of bounded linear operators

$$
D_\delta : H \to G, \quad \delta > 0, \tag{2.3}
$$

such that for each $\varepsilon > 0$ there exists a $\delta_0 > 0$ with the following property:

$$
\operatorname{supp} D_\delta h \subset U_\varepsilon(\operatorname{supp} h) \qquad \forall h \in H \tag{2.4}
$$

for every $\delta < \delta_0$. Here $U_\varepsilon(F)$ is the ε-neighborhood of F.

(b) An *elliptic operator* in bottleneck spaces H and G is a proper operator (2.3) such that D_δ is Fredholm for each δ and has an almost-inverse $D_\delta^{[-1]}$ such that the family $D_\delta^{[-1]}$ is also a proper operator.

As a rule, we omit the parameter δ in the notation of a proper operator.

Now let

$$
A_i : H_i \to G_i, \quad i = 1, 2,
$$

be proper operators in bottleneck spaces, and let $F \subset [-1,1]$ be an *open* subset (in the topology of $[-1,1]$).

Definition 2.4. We say that A_1 *coincides* with A_2 on F and write $A_1 \overset{F}{=} A_2$ if $H_1 \overset{F}{=} H_2$, $G_1 \overset{F}{=} G_2$, and the following condition is satisfied for each compact subset $K \subset F$: there exists a $\delta_0 > 0$ such that

$$A_{1\delta}h = A_{2\delta}h \quad \text{for} \quad \delta < \delta_0 \text{ and } \operatorname{supp} h \subset K. \tag{2.5}$$

(This is well defined, since $h \in H_1(F) \approx H_2(F)$ and, F being open, $A_{1\delta}h, A_{2\delta}h \in G_1(F) \approx G_2(F)$ for small δ.) In this case, we also say that A_1 is obtained from A_2 by a *modification* on $[-1,1]\backslash F$ and write $A_1 \overset{[-1,1]\backslash F}{\longleftrightarrow} A_2$. The notion of a *commutative square* of modifications is defined in an obvious way.

Now we are in a position to state our main result.

Theorem 2.1. *Suppose that the following commutative diagram of modifications of elliptic operators in bottleneck spaces holds:*

$$
\begin{array}{ccc}
D & \overset{\{-1\}}{\longleftrightarrow} & D_- \\
{\scriptstyle\{1\}}\updownarrow & & \updownarrow{\scriptstyle\{1\}} \\
D_+ & \overset{\{-1\}}{\longleftrightarrow} & D_{+-}
\end{array}.
$$

Then

$$\operatorname{ind}(D) - \operatorname{ind}(D_-) = \operatorname{ind}(D_+) - \operatorname{ind}(D_{+-}).$$

3 An index theorem for pseudodifferential operators on manifolds with conical singularities

Let N be a compact closed manifold with conical singularities. Without loss of generality, we assume that there is only one conical point α (the base Ω of the corresponding cone is not assumed to be connected). Next, let

$$\widehat{D} \colon H^s(N, E) \longrightarrow H^{s-m}(N, F)$$

be an elliptic pseudodifferential operator in Sobolev spaces on N with conormal symbol $D_0(p)$.

Suppose that $g\colon \Omega \to \Omega$ is a diffeomorphism and

$$\mu_E : E|_\Omega \longrightarrow g^*(E|_\Omega), \qquad \mu_F : F|_\Omega \longrightarrow g^*(F|_\Omega)$$

are bundle isomorphisms. Let the following condition be satisfied.

Condition A. *The conormal symbols $D_0(p)$ and $\mu_F^{-1} g^* D_0(-p)(g^*)^{-1}\mu_E$ are homotopic in the class of conormal symbols.*[2]

We take some homotopy and denote it by D_{0t},

$$D_{00}(p) = D_0(p), \qquad D_{01}(p) = \mu_F^{-1} g^* D_0(-p)(g^*)^{-1}\mu_E.$$

Let us construct a closed manifold $\mathcal{N}$ and bundles $\mathcal{E}$ and $\mathcal{F}$ over it as follows. By cutting away a small neighborhood of the conical point in N, we obtain a manifold $\widetilde{N}$ with boundary $\partial \widetilde{N} = \Omega$. Let us attach two copies of $\widetilde{N}$ to the opposite ends of the cylinder $\Omega \times [0,1]$, using the identity mapping for gluing at the left end and the mapping g at the right end. We lift the bundles $E|_\Omega$ and $F|_\Omega$ to $\Omega \times [0,1]$ in the natural way and glue them to the corresponding bundles over the two copies of $\widetilde{N}$ using the identity isomorphism at the left end and the isomorphisms μ_E and μ_F at the right end. The resulting bundles over $\mathcal{N}$ will be denoted by $\mathcal{E}$ and $\mathcal{F}$, respectively.

Theorem 3.1. *The following index formula holds under the above-mentioned conditions:*

$$\operatorname{ind} \widehat{D} = \frac{1}{2}\left\{\operatorname{ind} \mathcal{D} + \operatorname{sf} D_{0t}\right\},$$

where $\mathcal{D}\colon H^s(\mathcal{N}, \mathcal{E}) \longrightarrow H^{s-m}(\mathcal{N}, \mathcal{F})$ is an elliptic operator on the closed manifold $\mathcal{N}$ whose principal symbol coincides with the principal symbol of $\widehat{D}$ on the copies of $\widetilde{N}$ and with the principal symbol of the homotopy D_{0t} on the cylinder $\Omega \times [0,1]$.

The proof is based on Theorem 2.1.

4   An index theorem for Fourier integral operators on manifolds with conical singularities

The index problem for Fourier integral operators (or quantized contact transformations) was posed by Weinstein [Wei76], [Wei77] and recently solved (for operators on smooth manifolds) in [EM98], [LNT00].

In this section we state an index theorem for Fourier integral operators on manifolds with conical singularities.

Let N_1 and N_2 be manifolds with conical singularities, and let $g : T^* N_1 \backslash \{0\} \to T^* N_2 \backslash \{0\}$ be an $\mathbf{R}_+$-homogeneous canonical transformation of their compressed cotangent bundles. By $L(g)$ we denote the graph of g. Let Λ be the Maslov line

---

[2] The case of symmetry with respect to a point $p_0 \neq 0$ can be treated in a completely similar way.

bundle over $L(g)$ and $a : L(g) \to \Lambda$ a nonvanishing section homogeneous of order m. Consider the Fourier integral operator

$$T(g, a) = H^{s,\gamma}(N_1) \to H^{s-m,\gamma}(N_2)$$

(the construction can be found in [NSS98], [NSS99a]). We say that $T(g, a)$ is *elliptic* (on a given weight line $\{\mathrm{Im}\, p = \gamma\}$) if its conormal symbol $T_0(p)$ is invertible everywhere on the weight line. In what follows, we assume that $\gamma = 0$ to simplify the notation.

Theorem 4.1. *Let $T(g, a)$ be an elliptic Fourier integral operator such that the conormal symbols $T_0(p)$ and $T_0(-p)$ are homotopic in the class of conormal symbols of formally elliptic Fourier integral operators. Then*

$$\mathrm{ind}\, T(g, a) = \frac{1}{2}\{\mathrm{ind}\, \mathcal{T} + \mathrm{sf}\, T_{0t}\},$$

where T_{0t} is a homotopy joining $T_0(p)$ with $T_0(-p)$ and the Fourier integral operator $\mathcal{T} : H^s(\mathcal{N}_1) \to H^s(\mathcal{N}_2)$ and the closed manifolds $\mathcal{N}_1$ and $\mathcal{N}_2$ are obtained from $T(g, a)$, N_1 and N_2 by the construction similar to that in Theorem 3.1.

The proof is based on Theorem 2.1.

Acknowledgements

The authors are grateful to Dr. Savin and Prof. Shatalov for useful discussions and Prof. Schulze for his attention and support.

References

[AD62]   Agranovich, M. S., Dynin, A. S., General elliptic boundary value problems for elliptic systems in higher-dimensional domains, *Dokl. Akad. Nauk SSSR*, 146:511–514, 1962.

[Agr65]   Agranovich, M. S., Elliptic singular integro-differential operators, *Uspekhi Matem. Nauk*, 20(5):3–20, 1965.

[Agr97]   Agranovich, M. S.,. Elliptic boundary problems, in Partial Differential Equations IX. Elliptic Boundary Value Problems, Vol. 79 in Encyclopaedia of Mathematical Sciences, *Springer-Verlag*, Berlin–Heidelberg, 1997, pp. 1–144.

[Ang93]   Anghel, N., An abstract index theorem on non-compact Riemannian manifolds, *Houston J. of Math.*, 19:223–237, 1993.

[AS63]   Atiyah, M., Singer, I., The index of elliptic operators on compact manifolds, *Bull. Amer. Math. Soc.*, 69:422-433, 1963.

[Ati69]   Atiyah, M., Global theory of elliptic operators, *Proc. of the Int. Symposium on Functional Analysis*, 21–30, 1969, *University of Tokyo Press*, Tokyo

[Bun92]   Bunke, U., Relative index theory, *J. Funct. Anal.*, 105:63–76, 1992.

[BW93]   Booß-Bavnbek, B., Wojciechowski, K., Elliptic Boundary Problems for Dirac Operators, *Birkhäuser*, Boston–Basel–Berlin, 1993.

[Dez64]   Dezin, A. A., Invariant Differential Operators and Boundary Value Problems, American Mathematical Society, 1964.

[Don87]   Donnelly, H., Essential spectrum and the heat kernel, *J. Funct. Anal.*, 75:362–381, 1987.

[Dyn61]   Dynin, A. S., Multidimensional elliptic boundary problems with one unknown function, *Dokl. Akad. Nauk SSSR*, 141:285–287, 1961.

[EM98]   Epstein, C., Melrose, R., Contact degree and the index of Fourier integral operators, *Math. Res. Lett.*, 5(3):363–381, 1998.

[FST98]   Fedosov, B. V., Schulze, B. W., Tarkhanov, N. N., A Remark on the Index of Symmetric Operators, Preprint 98/4, Univ. Potsdam, Institut für Mathematik, Potsdam, 1998.

[FST99]   Fedosov, B. V., Schulze, B. W., Tarkhanov, N. N., The index of higher order operators on singular surfaces, *Pacific J. of Math.*, 191(1):25–48, 1999.

[Gil84]   Gilkey, P. B., Invariance Theory, the Heat Equation and the Atiyah–Singer Index Theorem, *Publish or Perish*, Wilmington, 1984.

[GL83]   Gromov, M., Lawson Jr., H. B., Positive scalar curvature and the Dirac operator on complete Riemannian manifolds, *Publ. Math. IHES*, 58:295–408, 1983.

[GS83]   Gilkey, P. B., Smith, L., The eta invariant for a class of elliptic boundary value problems, *Comm. Pure Appl. Math.*, 36:85–132, 1983.

[GS83a]   P. B. Gilkey and L. Smith. The twisted index problem for manifolds with boundary, *J. Diff. Geometry*, 18(3):393–44, 1983.

[Hsi72]   Hsiung, C.-C., The signature and *G*-signature of manifolds with boundary, *J. Diff. Geometry*, 6:595–598, 1972.

[Hsi76]   Hsiung, C.-C. A remark on cobordism of manifolds with boundary, *Arch. Math.*, 27:551–555, 1976.

[LNT00]   Leichtnam, E., Nest, R., Tsygan, B., Local formula for the index of a Fourier integral operator, Preprint Math. DG/0004022, 2000.

[NS99]   Nazaikinskii, V., Sternin, B., Surgery and the Relative Index of Elliptic Operators, Universität Potsdam, Institut für Mathematik, Potsdam, 1999.

[NS00]   Nazaikinskii, V., Sternin, B., Localization and surgery in index theory of elliptic operators, *Dokl. Ross. Akad. Nauk*, 370(1):19–23, 2000.

[NS01]   Nazaikinskii, V., Sternin, B., The index locality principle in elliptic theory, *Funktsional'nyi Analiz i Prilozhen.*, 2001 (to appear).

[NSS98]   Nazaikinskii, V., Schulze, B.-W., Sternin, B., The Index of Quantized Contact Transformations on Manifolds with Conical Singularities, Preprint 98/16, Univ. Potsdam, Institut für Mathematik, Potsdam, 1998.

[NSS99a]   Nazaikinskii, V., Schulze, B.-W., Sternin, B., The index of quantized contact transformations on manifolds with conical singularities, *Dokl. Ross. Akad. Nauk*, 368(5):598–600, 1999.

[SSS98]   Schulze, B.-W., Sternin, B., Shatalov, V., On the index of differential operators on manifolds with conical singularities, *Annals of Global Analysis and Geometry*, 16(2)141–172, 1998.

[Sto74]   Stong, R. E., Manifolds with reflecting boundary, *J. Diff. Geometry*, 9:465–474, 1974.

[Tel84]   Teleman, N., The index of signature operators on Lipschitz manifolds, *Publ. Math. IHES*, 58:39–78, 1984.

[Wei76]   Weinstein, A., Fourier integral operators, quantization, and the spectrum of a Riemannian manifold, *Géométrie Symplectique et Physique Mathématique*, 237:289–298, 1976.

[Wei77]   Weinstein, A. Some questions about the index of quantized contact transformations, *RIMS Kôkûryuku*, 104:1–14, 1977.

Addresses

VLADIMIR NAZAIKINSKII, Institute for Problems in Mechanics, Russian Academy of Sciences, Vernadskogo 101-1, 117526 Moscow, Russia

E-MAIL: nazaikinskii@mtu-net.ru

BORIS STERNIN, Department of Computational Mathematics and Cybernetics, Moscow State University, Vorob'evy Gory, Moscow 119899, Russia

E-MAIL: sternine@mtu-net.ru

2000 Mathematics Subject Classification. Primary 58J20; Secondary 58J30, 58J40, 19K56

Operator Theory:
Advances and Applications, Vol. 126
© 2001 Birkhäuser Verlag Basel/Switzerland

Propagation of Wave Packets and its Applications

Takashi Ōkaji

Abstract. The purpose of this talk is to study propagation of microlocal singularities of solutions to the Schrödinger equations with magnetic vector or electric scalar potential. The Hamiltonian may grow quadratically at infinity. We will present a new powerful approach based on a microlocal conservation law in terms of the Wigner transformation of the solutions. Our method enables us to give precise information on the evolution of the wave packets of solutions in the sense of A. Cordoba and C. Fefferman. As a result, we can show reconstruction of microlocal singularities and creation of singularities from oscillatory initial data as well as smoothing effects of solutions.

1   Introduction

Let us consider the Schrödinger operator

$$H_0 = \frac{1}{2}\left\{-(\nabla - iAx)^2 + \langle Ex, x\rangle\right\},$$

where A is a real skew symmetric matrix and E is a real symmetric one.

Suppose that H is a self adjoint operator on $L^2(\mathbf{R}^n)$ as a second order perturbation of H_0 such that the symbol of $\tilde{H} = H - H_0$, $\sigma(\tilde{H}) = \sum_{|\alpha|\leq 2} a_\alpha(x)\xi^\alpha$ satisfies that there exists a real number $\kappa < 2$ such that $a_\alpha(x) \in S(\langle x\rangle^{\kappa-|\alpha|}; \langle x\rangle^{-2}dx^2)$. We consider the Cauchy problem

$$\frac{1}{i}\partial_t u + Hu = 0, \ (x,t) \in \mathbf{R}^n \times \mathbf{R}, \quad u(0) = u_0.$$

The regularity of the fundamental solution of the Schrödinger equation associated with the Hamiltonian $h(x,\xi) = |\xi|^2/2 + V(x)$ was studied by D. Fujiwara ([F79]) and K. Yajima ([Y96]) for the various class of potentials $V \in C^\infty(\mathbf{R}^n; \mathbf{R})$. S. Zelditch and A. Weinstein ([Z83], [W85], [WZ82]) studied the same problem when $h(x,\xi) = (|\xi|^2 + |x|^2)/2 + V(x)$, where V is a smooth bounded perturbation (i.e. $V \in S^0(\mathbf{R}^n)$). For more general perturbation V, there is a related work due to L. Kapitanski, I. Rodnianski and K. Yajima [KRY97]. As for the Hamiltonian related to a Riemannian manifold, when it may grow at most of sublinear order at infinity, the microlocal smoothing effect in the C^∞ or analytic category has been studied in [CKS95], [D96], [K96], [RZ99], [W99-1] and [W99-2].

2 Main Results

We shall treat the case where A and E are commutative: $AE = EA$. Thus, essentially, we may assume that $n = 2$ and

$$A = \begin{pmatrix} 0 & \mu \\ -\mu & 0 \end{pmatrix} \quad \text{and} \quad E = \varepsilon Id$$

with a nonnegative constant μ and a real number ε. Note that the magnetic field is equal to $-2A$. To state our result, we define the life span of smoothing effects as follows:

$$t_c = \inf\{t > 0;\ e^{-itH_0}u_0 \notin C^\infty(\mathbf{R}^2) \text{ with } \exists u_0(x) \in \mathcal{E}' \cap L^2(\mathbf{R}^2)\}.$$

Especially, we define $t_c = +\infty$ if $e^{-itH_0}u_0$ is smooth at every time $t > 0$ for all initial data $u_0(x) \in \mathcal{E}' \cap L^2(\mathbf{R}^2)$. In what follows, let us denote $T^*(\mathbf{R}^2)\backslash\{0\}$ by $\dot{T}^*(\mathbf{R}^2)$. Our results on smoothing effect and reconstruction of singularities are the followings.

Theorem 2.1. $t_c = \pi/\sqrt{\mu^2 + \varepsilon}$ if $\mu^2 + \varepsilon > 0$ and $t_c = +\infty$ if $\mu^2 + \varepsilon \leq 0$.

Theorem 2.2. Let $u(t) = e^{-itH_0}u_0$. If $\mu^2 + \varepsilon > 0$, then for any $\ell \in \mathbf{Z}$,

$$WFu(t_c\ell) = \{(x,\xi) \in \dot{T}^*(\mathbf{R}^2);\ ((-1)^\ell e^{At_c\ell}x, (-1)^\ell e^{At_c\ell}\xi) \in WFu_0\} \tag{2.1}$$

and if $u_0 \in \mathcal{E}' \cap L^2(\mathbf{R}^2)$, it holds that

$$u(t) \in C^\infty(\mathbf{R}^2), \ \forall t \notin t_c\mathbf{Z}. \tag{2.2}$$

Remark 2.1. The conclusion (2.2) and (2.1) are valid for the solution $u(t) = e^{-iHt}u_0$ for a perturbed operator H satisfying an appropriate condition if $\kappa < 2$ and $\kappa < 1$, respectively

Theorem 2.3. Let $A = 0$ and $\varepsilon > 0$. Then,

$$WFu(t_c/2) = \{(x,\xi) \in \dot{T}^*(\mathbf{R}^2);\ (\xi, -x) \in WF\hat{u}_0\}.$$

Here, $\hat{u}_0$ denotes the Fourier transform of u_0: $\hat{u}_0(\xi) = \int e^{-ix\xi}u_0(x)dx$.

Now, we state our result on creation of new singularities. Suppose that the initial datum has the special form $u_\Gamma(x) = e^{i(\Gamma x, x)}$, where Γ is a nonzero real symmetric matrix. For every square matrix M, $\Lambda_{\mathbf{R}}(M)$ denotes the set of all real eigenvalues of M. We define $\Lambda_{\mathbf{R}}(M) = \emptyset$ if M has no real eigenvalues.

Theorem 2.4. *Suppose* $\mu^2 + \varepsilon = 0$. *Then* $u(t) = e^{-itH_0}u_\Gamma(x)$ *is smooth if and only if* $-(2t)^{-1} \notin \Lambda_{\mathbf{R}}(\Gamma)$.

Theorem 2.5. *Assume that $\mu^2 + \varepsilon > 0$ and set $\delta = \sqrt{\mu^2 + \varepsilon}$. Then $u(t) = e^{-itH_0}u_\Gamma$ is smooth if and only if $-2^{-1}\delta\cot(\delta t) \notin \Lambda_{\mathbf{R}}(\Gamma - A)$.*

Theorem 2.6. *Assume that $\mu^2 + \varepsilon < 0$ and set $\delta = \sqrt{-\mu^2 - \varepsilon}$. Then $u(t) = e^{-itH_0}u_\Gamma$ is smooth if and only if $-2^{-1}\delta\coth(\delta t) \notin \Lambda_{\mathbf{R}}(\Gamma - A)$.*

As a direct consequence of Theorem 2.5, every solution $u(t) = e^{-itH_0}u_\Gamma(x)$ has a singularity at a certain time if $A = 0$. Contrary to this, the solution $u(t) = e^{-itH_0}u_\Gamma$ is always smooth if $\Gamma - A$ has no real eigenvalues.

3   Propagation of Wave Packets

In this section we describe an essential idea to prove our main results. Let

$$h(x,\xi) = \sum_{|\alpha+\beta|\leq 2} a_{\alpha,\beta}x^\alpha\xi^\beta, \quad a_{\alpha,\beta} \in \mathbf{R}$$

and

$$h^w(x, D)u = (2\pi)^{-n}\int e^{ix\xi}h(\{x + y\}/2, \xi)u(y)dyd\xi, \quad u \in S(\mathbf{R}^n).$$

We denote by $\phi^t(x,\xi) = (X(t), \Xi(t))$ the integral core of the Hamilton vector field H_h of $h(x,\xi)$ through (x,ξ):

$$\dot{X} = \partial_\xi h(X, \Xi), \quad \dot{\Xi} = -\partial_x h(X, \Xi), \quad X(0) = x, \quad \Xi(0) = \xi.$$

The Wigner transformation is a mapping from $(u, v) \in S(\mathbf{R}^n)^2$ to $S(\mathbf{R}^{2n})$ defined as

$$W(u, v)(x, \xi) = \int e^{-iy\xi}u(x + \frac{y}{2})\overline{v(x - \frac{y}{2})}dy.$$

In particular, we denote $W(u, u)$ by $W[u]$. It can be naturally extended as a map from $(u, v) \in S'(\mathbf{R}^n)^2$ to $S'(\mathbf{R}^{2n})$. by making use of

$$\langle W(u, v)(x, \xi), \varphi(x, \xi)\rangle = \langle \varphi^w(x, D)u, \bar{v}\rangle, \quad \forall\varphi \in S(\mathbf{R}^{2n}).$$

The microlocal conservation law can be stated as follows.

Lemma 3.1. *Let $u_0 \in S'(\mathbf{R}^n)$. For $u(t) = e^{-ith^w(x,D)}u_0$, it holds that*

$$W[u(t)](x, \xi) = W[u(0)](\phi^{-t}(x, \xi)), \quad \forall(x, \xi) \in \mathbf{R}^{2n}.$$

This lemma is a consequence of the following simple observation.

Lemma 3.2. *Let $u_0 \in \mathcal{S}(\mathbf{R}^n)$. Then, $u(t) = e^{-ith^w(x,D)}u_0$ satisfies*

$$\{\partial_t + H_h\}W[u(t)](x,\xi) = 0, \tag{3.1}$$

The decay property of wave packet of u introduced by A. Cordoba and C. Fefferman ([CF78])

$$\mathcal{P}_\lambda[u](q,p) = \lambda^{n/4}\int e^{i\lambda(q-y)p}e^{-\lambda(q-y)^2/2}u(y)dy \text{ as } \lambda \to \infty$$

is strongly connected with the microlocal regularity of $u \in \mathcal{S}'(\mathbf{R}^n)$. Indeed,

Lemma 3.3 ([G90]). *Let u be a tempered distribution on $\mathbf{R}^n$. Then $(x_0,\xi_0) \notin WF_s(u)$ if and only if there is a neighborhood W of (x_0,ξ_0) such that*

$$\int_1^\infty \lambda^{n+2s-1}\int_W |\mathcal{P}_\lambda[u](q,p)|^2 dqdpd\lambda < \infty. \tag{3.2}$$

All the results in the previous section can be derived by making use of the following wave packet identity.

Proposition 3.1. *Let $u = e^{-itH_0}u(0)$. Then, it holds that*

$$|\mathcal{P}_\lambda[u(t)]|^2 = \pi^{-n/2}\int E_\lambda(x-q,\xi-\lambda p)W[u(0)](\phi^{-t}(x,\xi))dxd\xi. \tag{3.3}$$

Here, E is a Gaussian function defined as $E_\lambda(x,\xi) = \exp[-\lambda\{x^2 + (\xi/\lambda)^2\}]$.

Acknowledgements

The author wishes to express his gratitude to the organizers of this nice conference for their kind invitation and hospitality.

References

[CF78]   A. Cordoba and C. Fefferman, Wave packets and Fourier integral operators, Comm. P.D.Eqs., 3 (1978), 979–1005.

[CKS95]   W. Craig, T. Kappeler and W. Strauss, Microlocal dispersive smoothing for the Schrödinger equation, Comm. Pure Appl. Math. vol. XLVIII (1995), 769-860.

[D96]   S. Doi, Smoothing effects of Schrödinger evolution groups on Riemannian manifolds, Duke Math. J., 82 (1996), 679–706.

[F79]   D. Fujiwara, A construction of the fundamental solution for the Schrödinger equation, J. d'Analyse Math. 35 (1979), 41-96.

[G90]   P. Gérard, Moyennisation et regularite deux-microlocale, Ann. Sci. Ecole
           Norm. Sup. (4) 23 (1990), no. 1, 89–121.

[K96]   K. Kajitani, Analytically smoothing effect for Schrödinger equations. Dynami-
           cal systems and differential equations, Vol. I (Springfield, MO, 1996). Discrete
           Contin. Dynam. Systems 1998, Added Volume I, 350–352.

[KRY97]   L. Kapitanski, I. Rodnianski and K. Yajima, On the fundamental solution of
           a perturbed harmonic oscillator, Topol. Methods Nonlinear Anal. 9 (1997),
           77-106.

[RZ99]   L. Robbiano and C. Zuily, Microlocal analytic smoothing effect for the
           Schrodinger equation. Duke Math. J. 100 (1999), no. 1, 93–129.

[S94]   N.A. Shananin, On singularities of solutions of the Schrödinger equation for a
           free particle, Math. Notes, 55 (1994), 626–631.

[W85]   A. Weinstein, A symbol class for some Schrödinger equations on $\mathbf{R}^r$, Amer. J.
           Math., 107 (1985), 1-21.

[WZ82]   A. Weinstein and S. Zelditch, Singularities of solutions of some Schrödinger
           equations on $\mathbf{R}^n$, Bull. Amer. Math. Soc., 6 (1982), 449-452.

[W99-1]   J. Wunsch, Propagation of singularities and growth for Schrodinger operators.
           Duke Math. J. 98 (1999), no. 1, 137–186.

[W99-2]   J. Wunsch, The trace of the generalized harmonic oscillator. Ann. Inst. Fourier
           (Grenoble) 49 (1999), no. 1, 351–373.

[Y96]   K. Yajima, Smoothness and non-smoothness of the fundamental solution of
           time dependent Schrödinger equations, Comm. Math. Phys. 181 (1996), 605-
           629.

[Z83]   S. Zelditch, Reconstruction of singularities for solutions of Schrödinger equa-
           tions, Comm. Math. Phys., 90 (1983), 1-26.

Address

TAKASHI ŌKAJI, Department of Mathematics, Kyoto University, Kyoto 606-
8502, Japan

E-MAIL: okaji@kusm.kyoto-u.ac.jp

2000 Mathematics Subject Classification. Primary 35B65; Secondary 81S30

Operator Theory:
Advances and Applications, Vol. 126
© 2001 Birkhäuser Verlag Basel/Switzerland

Gevrey and Analytic Properties of the Solutions of Several Classes of Partial Differential Equations

P.R. POPIVANOV

Abstract. This paper deals with Gevrey and analytic properties of the solutions of several classes of linear partial differential operators with analytic coefficients on the torus. In many cases locally non-analytic hypoelliptic operators turn out to be analytic hypoelliptic on the torus.

1 Introduction

We remind to the reader the definition of the Gevrey class $G_s(\Omega)$ containing $C^\infty(\Omega)$ functions in the domain Ω.

Definition 1.1. The function $u \in G_s(\Omega)$, $s \geq 1$, if for each compact $K \subset\subset \Omega$ there exists a constant $C(K) > 0$ and such that for every multiindex $\alpha \in \mathbf{Z}_+^n$ the following inequality holds:

$$|D^\alpha u(x)| \leq C^{|\alpha|+1}(\alpha)^s$$

Obviously $G_1(\Omega)$ coincides with the set of real analytic functions.

Put $G_{s,0}(\Omega) = G_s \cap C_0^\infty$ for $s > 1$. Then $G_{s,0}$ is not trivial and we can define the inductive topology in $G_{s,0}$. This way the ultradistribution space G_s' is introduced as the dual space to $G_{s,0}$.

The standard definitions of local hypoellipticity in the C^∞, analytic and Gevrey categories are well-known (see [MR97], [Hör83]) and we omit it. Moreover, in the same way hypoellipticity can be defined on the n-th dimensional torus $\mathbf{T}^n$. Obviously, local hypoellipticity in the class of Schwartz distributions implies hypoellipticity on the torus. Certainly, on the torus we assume that $Pu = f$, $u \in D'(\mathbf{T}^n)$ and $f \in G_s(\mathbf{T}^n)$, $1 \leq s \leq \infty$, $G_\infty(\mathbf{T}^n) = C^\infty(\mathbf{T}^n)$. There are many operators on the torus $\mathbf{T}^n$ which are hypoelliptic there but are not hypoelliptic locally.

To begin with we shall consider several examples.

Example 1. The heat operator with the symbol $i\tau + |\xi|^2$, $\xi \in \mathbf{R}^n$, is C^∞ hypoelliptic locally and non- analytic hypoelliptic locally. It is G_s, $s \geq 2$, hypoelliptic locally but it is not G_s, $1 \leq s < 2$, hypoelliptic locally.

Example 2. (Bove, Tartakoff). Let $x = (x', x'', x''') \in \mathbf{R}^k \times \mathbf{R}^l \times \mathbf{R}^{n-k-l}$ and consider the pseudodifferential operator

$$P(x, D) = \sum_{j=1}^{k} \mu_j (D_{x_j}^2 + x_j^2 |D'''|^2) + \sum_{i=k+1}^{k+l} a_i D_{x_i}^2 + \alpha |D'''|, \qquad (1.1)$$

where the constants $\mu_j > 0$, $a_i > 0$ and $\alpha \in \mathbf{C}^1$. This is the main condition in our investigation (Grushin, Boutet de Monvel, Hörmander):

$$(A) \quad \alpha + \sum_{j=1}^{k} \mu_j (2n_j + 1) \neq 0, \forall n = (n_1, \ldots, n_k) \in \mathbf{Z}_+^k.$$

Suppose that $n - k - l = 1$, i.e. $x''' \in \mathbf{R}^1$ and the first order term is αD_n, $D''' = D_n$. Then (1.1) is a differential operator and the main condition has the form:

$$(A)_1 \quad \pm \alpha + \sum_{j=1}^{k} \mu_j (2n_j + 1) \neq 0, \forall n \in \mathbf{Z}_+^k.$$

The characteristic manifold Σ of P, is written as

$$\Sigma = \{(x, \xi) \in T^*(\Omega) \setminus 0 : x' = \xi' = 0, \xi'' = 0, |\xi'''| > 0\}$$

and is a transversal intersection of a symplectic and involute submanifolds of $T^*(\Omega) \setminus 0$.

Proposition 1.1. *(Grushin [Gru70], B. de Monvel [Mon74], Hörmander [Hör83], Bove-Tartakoff [BT96]) Assume that $(x_0, \xi_0) \in \Sigma$, $(x_0, \xi_0) = (0, \bar{x}'', \bar{x}'''; 0, 0, \bar{\xi}''' \neq 0)$, the leaf passing through the point (x_0, ξ_0) is $\Gamma_{(x_0,\xi_0)} = \{(x, \xi) \in T^*(\Omega) \setminus 0 : (x, \xi) = (0, x'', \bar{x}'''; 0, 0, \bar{\xi}''' \neq 0), x'' \in \mathbf{R}^l\}$ and the conditions (A), $(A)_1$ hold.*

Then $P(x, D)$ is locally hypoelliptic with loss of regularity equal to 1 in the scale of Sobolev spaces. Moreover, P is G_s, $s \geq 2$ hypoelliptic locally. Let W be a conical neighbourhood of the point (x_0, ξ_0) and let $1 \leq s < 2$. The conditions $(x_0, \xi_0) \notin WF_{G_s}(Pu)$, $\Gamma_{(x_0,\xi_0)} \cap (W \setminus \{(x_0, \xi_0)\}) \cap WF_{G_s}(u) = \emptyset$ imply that $(x_0, \xi_0) \notin WF_{G_s}(u)$.

In other words, the microlocalized Gevrey singularities $WF_{G_s}(u)$ propagate along the leaf $\Gamma_{(x_0,\xi_0)}$ if $1 \leq s < 2$.

In the special case $l = 0$ the operator (1.1) is analytic hypoelliptic under the condition (A). We remind the reader that if $k = 0$, i.e. the operator P has involutive characteristics, then the necessary and sufficient condition for local C^∞ hypoellipticity with loss of regularity 1 of P takes the form: $\alpha \notin \{t : t \leq 0\}$.

3. Model examples on the torus:

$$P(x,D) = \sum_{j=1}^{k} \mu_j(D_{x_j}^2 + sin^2 x_j D_n^2) + \sum_{i=k+1}^{k+l} a_i D_{x_i}^2 + \alpha D_n, \qquad (1.2)$$

$\mu_j > 0$, $a_i > 0$, $\alpha \in \mathbf{C}^1$.

Then $P(x,D)$ is locally hypoelliptic on $\mathbf{T}^n$ if the condition $(A)_1$ holds.

2 Analytic hypoelliptic differential operators on the torus

Consider now on the torus $\mathbf{T}^{n+1} = \mathbf{T}_x^n \times \mathbf{T}_t^1$ the following differential operator with analytic coefficients:

$$P(t,x,D_t,D_x) = \sum_{i,j=1}^{n+1} a_{ij}(x)D_iD_j + \sum_{i=1}^{n+1} a_i(t,x)D_i + a(t,x), \qquad (2.1)$$

where by definition $x_{n+1} = t$, $D_{n+1} = D_t$, $D_j = D_{x_j}$, $1 \le j \le n$, $a_{ij}(x) = a_{ji}(x)$ and the dual variables to $x \in \mathbf{R}^n$ are $\xi \in \mathbf{R}^n$, while the dual variables to t is τ.

The principal symbol of (2.1) is denoted by $p_2^0(x,\tau,\xi)$.

As in Hörmander [Hör83] we assume that

$(*)\quad Re\, p_2^0 \ge 0, p_2^0(x,\tau,\xi) \in \Gamma = \{z \in \mathbf{C}^1 : |Im\, z| \le \gamma Re\, z, \gamma = const > 0\}.$

$(**)\quad$ The characteristic set $\Sigma = char\, p_2^0 = \{(x,\tau,\xi) : p_2^0(x,\tau,\xi) = 0, |\xi| + |\tau| > 0\}$ is analytic submanifold of $T^*(\mathbf{T}^{n+1})$ and p_2^0 is transversally elliptic operator there.

Remarks. 1.Consider the Hessian $\tilde{Q}_{(x,\xi)}$, $(x,\xi) \in \Sigma$ of $\frac{1}{2}Re\, p_2^0$. Then $\tilde{Q} = 0$ on $T_{(x,\xi)}(\Sigma)$ and $\tilde{Q}_{(x,\xi)} \ge 0$. The operator P is said to be transversally elliptic if for each point $(x,\xi) \in \Sigma$ there exists a plane $L_{(x,\xi)}$ transversal to $T_{(x,\xi)}(\Sigma)$ and such that $\tilde{Q}|_L > 0$.

2. The subprincipal symbol p_1' of the operator (2.1) is defined as follows:

$$p_1' = p_1 + \frac{i}{2}\sum_{j=1}^{n+1} \frac{\partial^2 p_2^0}{\partial x_j \partial \xi_j}.$$

3. Put Q for the Hessian of $\frac{1}{2}p_2^0$ evaluated at the point $(x_0,\tau_0,\xi_0) \in \Sigma$ and F for the Hamilton map of Q evaluated at the same characteristic point (x_0,τ_0,ξ_0) (see [Hör83]).

According to Hörmander (see [Hör83]) the local hypoellipticity of the operator (2.1) $(*), (**)$ with loss of regularity equal to 1 in the scale of classical Sobolev spaces is equivalent to the fulfilment of condition $(***)$:

$(***)$ for each characteristic point $(x_0, \tau_0, \xi_0) \in \Sigma$, $|\tau_0| + |\xi_0| = 1$ we have:

$$p_1'(x_0, t_0, \tau_0, \xi_0) + Q(\bar{v}, v) + \sum_j (2\alpha_j + 1)\mu_j \neq 0, 0 \leq \alpha_j \in \mathbf{Z},$$

where Q is the Hessian of $\frac{1}{2}p_2^0$ evaluated at (x_0, τ_0, ξ_0), $\mu_j \in \Gamma$ are the eigenvalues of the Hamilton map of $\frac{1}{i}Q$ and v is a generalized eigenvector belonging to the eigenvalue 0 of the Hamilton map.

This is our first result.

Theorem 2.1. *Under the assumptions $(*), (**), (***)$ and $\Sigma \subset \{\xi = 0, \tau \neq 0\}$ the operator (2.1) is analytic hypoelliptic on $\mathbf{T}^{n+1}$.*

Example 4. The differential operator (1.2) is analytic hypoelliptic on $\mathbf{T}^{n+1}$ under the condition $(A)_1$.

3   Some generalizations

We shall propose here several examples of second order linear differential operators on the torus with loss of regularity in the Sobolev classes strictly greater than 1 which are analytic hypoelliptic too.

Example 5. Consider the differential operator with analytic coefficients on the torus $\mathbf{T}^2$:

$$P = D_t^2 + (\varphi(t) + i\psi(t))D_x,$$

where φ, ψ are real-valued real analytic functions on $\mathbf{T}^1$, $\varphi(t) \neq 0$, $\forall t \in \mathbf{T}^1$ and $\psi(t) \not\equiv 0$.

Thus, $\psi(t)$ has finitely many zeroes in the interval $[0, 2\pi]$: $t_1, \ldots, t_n$ with even order of vanishing $k_1, \ldots, k_n$. Put $k = max_{1 \leq j \leq n} k_j$. Obviously, k is even integer. As it was proved in [Pop75],[Men77] then the following a-priori estimate holds on each compact set $K \subset\subset \mathbf{R}^2$ and for each real s in the Sobolev spaces H^s:

$$||u||_{s + \frac{k+2}{2(k+1)}} \leq C_{s,k}(||Pu||_s + ||u||_s), \forall u \in C_0^\infty(K).$$

It is also known that the operator P is locally C^∞ hypoelliptic. Moreover, we claim that P is analytic hypoelliptic on $\mathbf{T}^2$.

Assume now that $\varphi(t) \equiv 0$ and $\psi(t) \not\equiv 0$. Then

$$P = -\partial_t^2 + \psi(t)\partial_x$$

and let $X_1 = \partial_t$, $X_2 = \psi(t)\partial_x$. The function $\psi(t)$ can have at the points t_j, $1 \leq j \leq n$ zeroes of arbitrary orders $k_j < \infty$ (even,odd). P satisfies the Hörmander's bracket condition and therefore it is C^∞ locally hypoelliptic. Moreover, P is analytic hypoelliptic on $\mathbf{T}^2$ (see [CH98]).

Remark. A similar result to this in Theorem 1 can be proved for a special non-linear perturbation of the operator (2.1). So consider the semilinear equation

$$P(t, x, D)u + f(t, x, u) = g(t, x), \tag{3.1}$$

where P is given by (2.1) and the conditions of Theorem 1 hold.

The function f depends analytically on (t, x, u) for $(t, x) \in \mathbf{T}^{n+1}$ and $u \in C^1$. L.Zanghirati showed that if g is analytic on $\mathbf{T}^{n+1}$ and $u \in C^\infty(\mathbf{T}^{n+1})$ satisfies (3.1) then u is analytic on $\mathbf{T}^{n+1}$.

On the other hand it is obviously that if $u \in H^s(\mathbf{T}^{n+1})$, $s > \frac{n+1}{2}$ satisfies (3.1) with a right hand side $g \in C^\infty(\mathbf{T}^{n+1})$ then $u \in C^\infty(\mathbf{T}^{n+1}$. Therefore, each smooth solution of (3.1) on $\mathbf{T}^{n+1}$ turns out to be analytic there for each analytic on $\mathbf{T}^{n+1}$ right hand side g.

References

[BT96]   A. Bove, D.Tartakoff. Propagation of Gevrey regularity for a class of hypoelliptic equations. Transactions of the AMS, 348, N 7, 1996, 2533-2575.

[CH98]   P. Cordaro, A.Himonas. Global analytic regularity for sums of squares of vector fields. Transactions of the AMS, vol. 350, N 12, 1998, 4993-5001.

[Gru70]   V.V. Grushin. On a class of hypoelliptic operators.Math.USSR Sbornik, vol.12 (1970),N 3, 458-476 (translated by AMS)

[Hör83]   L. Hörmander.The analysis of linear partial differential operators III, Springer-Verlag, 1983.

[Men77]   A. Menikoff. On hypoelliptic operators with double characteristics. Ann. Sc. Norm. Sup. Pisa, 4, 1977, 689-724.

[Mon74]   B. de Monvel L. Hypoelliptic operators with double characteristics and related pseudodifferential operators.Comm.PureAppl.Math. 27, 1974, 585-639.

[MR97]   M. Mascarello, L.Rodino. Partial differential equations with multiple characteristics. Akademie Verlag, 1997.

[Pop75]   P. Popivanov. Subelliptic estimates for pseudodifferential operators with double characteristics. Serdica, 1, 1975, 356-371.

Address

P. R. POPIVANOV, Institute of Mathematics, Bulgarian Academy of Sciences, Sofia 1113, Bulgaria

E-MAIL: popivano@math.bas.bg

2000 Mathematics Subject Classification. Primary 35H10; Secondary 35D10, 35B10

Operator Theory:
Advances and Applications, Vol. 126
© 2001 Birkhäuser Verlag Basel/Switzerland

Periodic Manifolds, Spectral Gaps, and Eigenvalues in Gaps

OLAF POST

Abstract. We investigate spectral properties of the Laplace operator on a class of non-compact Riemannian manifolds. We prove that for a given number N we can construct a periodic manifold such that the essential spectrum of the corresponding Laplacian has at least N open gaps. Furthermore, by perturbing the periodic metric of the manifold locally we can prove the existence of eigenvalues in a gap of the essential spectrum.

1 Introduction

There has been done many work in the analysis of periodic Schrödinger or divergence type operators. It is well-known that the spectrum of a Schrödinger-operator with periodic potential has band-gap structure under certain conditions (see e.g. [HH95]), i.e., the spectrum is the locally finite union of compact intervals and there exist an interval (a, b) not lying in the spectrum but with essential spectrum above and below the interval. Here, we want to give an example for a periodic Laplacian on a manifold *without* potential which has spectral gaps. Therefore we obtain the same qualitative results *only* by the periodic geometry. As in the Schrödinger case a decoupling procedure is responsible for the gaps. Related results can be found in [DH87] and [G97].

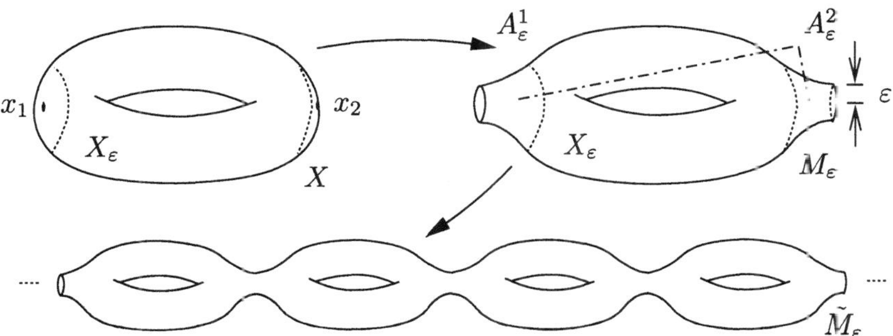

Fig. 1. Construction of the periodic manifold $\tilde{M}_\varepsilon$

We start our construction of a periodic manifold with spectral gaps from a compact Riemannian manifold X (for simplicity without boundary). We choose two different points $x_1, x_2 \in X$ and attach to x_1 resp. x_2 a cylindrical neighbourhood A_ε^1 resp. A_ε^2 (cf. Figure 1) where the "free" boundary of A_ε^i is isometric to a sphere of radius $\varepsilon > 0$. We call the resulting manifold M_ε. By glueing together $\mathbb{Z}$ copies of the period cell M_ε we obtain a $\mathbb{Z}$-periodic manifold $\tilde{M}_\varepsilon$. Our first main result is the following:

Theorem 1.1. *For each $N \in \mathbb{N}$ there exist at least N gaps in the spectrum of the Laplacian on the periodic manifold $\tilde{M}_\varepsilon$ provided ε is small enough.*

The proof basically uses Floquet Theory for which we refer to the next section. More examples of periodic manifolds with spectral gaps can be found in [P00].

Next, we locally perturb the metric of a periodic manifold $\tilde{M}$ with a spectral gap (a, b) to produce eigenvalues in the gap. Again, such effects are well studied in the case of Schrödinger or divergence type operators (see e.g. [DH86], [AADH94] or [HB00]). To simplify the notation, we only allow a conformal perturbation supported on a compact subset. More general settings (i.e., infinite range perturbations and non-conformal perturbations) can be found in [P00].

Here, the perturbation is a blow-up of some compact area, i.e., the manifold $\tilde{M}$ is perturbed by conformal factors $\rho_\tau \colon \tilde{M} \longrightarrow]0, \infty[$ starting from the constant function 1 for $\tau = 0$ and growing up to infinity only on a compact area as $\tau \to \infty$ (outside this area nothing is changed). The *Decomposition Principle* (see Theorem 4.1) assures that a spectral gap (a, b) of $\Delta_{\tilde{M}}$ remains a spectral gap in the essential spectrum of $\Delta_{\tilde{M}(\tau)}$ for all $\tau \geq 0$. Our second main result is the following:

Theorem 1.2. *Let $\lambda \in (a, b)$ be in a spectral gap. Then an infinite number of pairs (τ, u) with $\tau > 0$ and $u \neq 0$ such that $\Delta_{M(\tau)} u = \lambda u$ exist.*

The idea of the proof is quite simple (see [AADH94] or [HB00]). We show that the eigenfunctions of the full problem on $\tilde{M}$ can be approximated by eigenfunctions of an approximating problem on M^n (consisting of n copies of the period cell M), see Theorem 4.2. On the compact manifold M^n we can apply the Min-max Principle to assure the existence of eigenfunctions of the approximating problem (Theorem 4.3).

2 Periodic Manifolds and Floquet Theory

For a Riemannian manifold M (compact or not) we denote by $L_2(M)$ the usual L_2-space of square integrable functions on M with respect to the volume measure on M. The corresponding norm will be denoted by $\|\cdot\|_M$. For $u \in C_c^\infty(M)$, the space of compactly supported smooth functions, we set

$$q_M(u) := \int_M |\mathrm{d}u|^2.$$

Here du denotes the exterior derivate of u, which is a section of the cotangent bundle over M. The *Laplacian* Δ_M (for a manifold without boundary) is defined via the (closure of the) quadratic form, i.e., $q_M(u) = \langle \Delta_M u, u \rangle$ for $u \in C_c^\infty(M)$ (for details on quadratic forms see e.g. [RS80]). We therefore obtain a self-adjoint operator with spectrum lying in $[0, \infty[$.

If M is a compact manifold with (piecewise) smooth boundary $\partial M \neq \emptyset$ we can define the Laplacian with *Dirichlet* resp. *Neumann boundary conditions* in the same way. Here, we start from the (closure of the) quadratic form q_M defined on $C_c^\infty(M)$, the space of smooth functions with support *away* from the boundary, resp. on $C^\infty(M)$, the space of smooth functions up to the boundary. The corresponding operator will be denoted by Δ_M^D resp. Δ_M^N.

If M is compact the spectrum of Δ_M (with any boundary condition if $\partial M \neq \emptyset$) is purely discrete. We denote the corresponding eigenvalues by $\lambda_k(M)$ (resp. $\lambda_k^D(M)$ or $\lambda_k^N(M)$ in the Dirichlet or Neumann case) written in increasing order and repeated according to multiplicity. The *Min-max Principle* allows us to express the k-th eigenvalue of Δ_M in terms of the quadratic form q_M, i.e.,

$$\lambda_k(M) = \inf_L \sup_{u \in L, u \neq 0} \frac{q_M(u)}{\|u\|_M^2}, \tag{2.1}$$

where the infimum is taken over all k-dimensional subspaces L of the domain of the (closed) quadratic form q_M (see e.g. [D96]). Of course, the same is true for the Laplacians with boundary conditions.

A d-dimensional (non-compact) Riemannian manifold $\tilde{M}$ will be called Γ-*periodic* if $\Gamma = \mathbb{Z}^r$ acts properly discontinuously, isometrically and cocompactly, i.e., the quotient $\tilde{M}/\Gamma$ is a d-dimensional compact Riemannian manifold such that the quotient map is a local isometry. Throughout this article we study manifolds of dimension $d \geq 2$.

A closed (compact) subset M of $\tilde{M}$ is called *period cell* if M is the closure of a fundamental domain D, i.e., $M = \overline{D}$, D is open and connected, D is disjoint from any translate γD for all $\gamma \in \Gamma$, $\gamma \neq 0$, and the union over all translates γM is equal to $\tilde{M}$.

Floquet theory allows us to analyse the spectrum of the Laplacian on $\tilde{M}$ by analysing the spectra of Laplacians with quasi-periodic boundary conditions on a period cell M. In order to do this, we define θ-periodic boundary conditions. Let θ be an element of the dual group $\hat{\Gamma} = \mathrm{Hom}(\Gamma, \mathbb{T}^1)$ of $\Gamma = \mathbb{Z}^r$, which is isomorphic to the r-dimensional torus $\mathbb{T}^r = \{\theta \in \mathbb{C}^r; |\theta_i| = 1 \text{ for all } i\}$. Denote by Δ_M^θ the operator corresponding to the quadratic form q_M defined on the space of smooth functions u on M satisfying

$$u(\gamma x) = \overline{\theta(\gamma)} \, u(x)$$

for all $x \in \partial M$ and all $\gamma \in \Gamma$ such that $\gamma x \in \partial M$. Again, Δ_M^θ has purely discrete spectrum denoted by $\lambda_k^\theta(M)$. The eigenvalues depend continuously on θ. From Floquet theory we obtain

$$\operatorname{spec} \Delta_{\tilde{M}} = \bigcup_{\theta \in \hat{\Gamma}} \operatorname{spec} \Delta_M^\theta = \bigcup_{k \in \mathbb{N}} B_k(\tilde{M})$$

where $B_k = B_k(\tilde{M}) = \{\lambda_k^\theta(M); \theta \in \hat{\Gamma}\}$ is a compact interval, called *k-th band* (see e.g. [RS78], [D81]). In general, we do not know whether the intervals B_k overlap or not. But we can show the existence of gaps by proving that $\lambda_k^\theta(M)$ does not vary too much in θ.

3   Construction of a Periodic Manifold

Suppose that X is a compact Riemannian manifold of dimension $d \geq 2$ (for simplicity without boundary). We want to construct a $\mathbb{Z}^r$-periodic manifold. We choose $2r$ distinct points $x_1, \ldots, x_{2r}$. For each point x_i, denote by B_ε^i the open geodesic ball around x_i of radius $\varepsilon > 0$. Suppose further that $B_{\varepsilon_0}^i$ are pairwise disjoint, where $\varepsilon_0 > 0$ denotes the injectivity radius of X. Denote by B_ε the union of all balls B_ε^i. Let $X_\varepsilon := X \setminus B_{2\varepsilon}$ for $0 < 2\varepsilon < \varepsilon_0$ with metric inherited from X.

We now define the modified metric. For simplicity, we assume that the metric g is flat on B_{ε_0}, i.e., g is given in polar coordinates $(s, \sigma) \in]0, \varepsilon_0[\times \mathbb{S}^{d-1}$ around x_i by

$$g = \mathrm{d}s^2 + s^2 \mathrm{d}\sigma^2,$$

where $\mathrm{d}\sigma^2$ denotes the standard metric on the $(d-1)$-dimensional sphere $\mathbb{S}^{d-1}$. For a more general setting see [P00]. Let r_ε be a smooth monotone function with $r_\varepsilon(s) = \varepsilon$ in a neighbourhood of $s = 0$ and $r_\varepsilon(s) = s$ for $2\varepsilon \leq s \leq \varepsilon_0$. We denote the completion of $X \setminus \{x_1, \ldots, x_{2r}\}$ together with the modified metric

$$g_\varepsilon^i := \mathrm{d}s^2 + r_\varepsilon(s)^2 \mathrm{d}\sigma^2$$

near x^i by M_ε. Note that X_ε is embedded in M_ε and that the boundary of M_ε has $2r$ disjoint components Z_ε^i, each of them isometric to the sphere of radius ε. Let A_ε^i be the part of the manifold M_ε near x_i given in coordinates by $[0, 2\varepsilon] \times \mathbb{S}^{d-1}$. Denote by A_ε the union of all A_ε^i, $i = 1, \ldots, 2r$.

Let γM_ε be an isometric copy of M_ε with identification $x \mapsto \gamma x$ for each $\gamma \in \Gamma$. We construct a new (noncompact) manifold $\tilde{M}_\varepsilon$ by identifying $\gamma Z_\varepsilon^{2i-1}$ with $e_i \gamma Z_\varepsilon^{2i}$ for each $\gamma \in \Gamma$ and $i = 1, \ldots, r$. Here, e_i denotes the i-th generator $(0, \ldots, 1, \ldots, 0)$ of $\Gamma = \mathbb{Z}^r$. Since in a neighbourhood of Z_ε^i the manifold is isometric to a cylinder of radius ε, we can choose a smooth atlas and a smooth metric on the glued manifold $\tilde{M}_\varepsilon$. We therefore obtain a (non-compact) $\mathbb{Z}^r$-periodic manifold $\tilde{M}_\varepsilon$ and M_ε is a period cell for $\tilde{M}_\varepsilon$.

Now we are able to state the following theorem (Theorem 1.1 follows via Floquet Theory):

Theorem 3.1. *We have the convergence $\lambda_k^\theta(M_\varepsilon) \to \lambda_k(X)$ as $\varepsilon \to 0$ uniformly in $\theta \in \hat{\Gamma}$.*

Therefore, the k-th band $B_k(\tilde{M}_\varepsilon)$ reduces to the point $\{\lambda_k(X)\}$ as $\varepsilon \to 0$. Note that the convergence is *not* uniform in k (see the discussion in [CF81]). We therefore could not expect that an *infinite* number of gaps occur.

The proof of Theorem 3.1 is based on the following two lemmas. The idea is to compare the θ-periodic eigenvalues on M_ε with Dirichlet and Neumann eigenvalues on X_ε. The crucial point is, that the corresponding θ-periodic eigenfunctions on M_ε do not concentrate on A_ε, i.e., on the cylindrical ends. This will be shown in the following lemma:

Lemma 3.1. *There exists a positive function $\omega(\varepsilon)$ converging to 0 as $\varepsilon \to 0$ such that*

$$\int_{A_\varepsilon} |u|^2 \le \omega(\varepsilon) \int_{M_\varepsilon} \left(|u|^2 + |du|^2\right), \tag{3.1}$$

for all u in the domain of the quadratic form with θ-periodic boundary conditions on M_ε.

Proof. Without loss of generality, we can assume that $u \in C^\infty(M_\varepsilon)$ Suppose furthermore that $u(\varepsilon_0, \sigma) = 0$ for all $\sigma \in \mathbb{S}^{d-1}$. First we show an L_2-estimate over $A_{\varepsilon,s}^i := \{s\} \times \mathbb{S}^{d-1} \subset A_\varepsilon^i$ with its induced metric $r_\varepsilon(s)^2 d\sigma^2$.

Applying the Cauchy-Schwarz Inequality yields

$$|u(s,\sigma)|^2 = \left| \int_s^{\varepsilon_0} \partial_t u(t,\sigma)\, dt \right|^2 \le \int_s^{\varepsilon_0} r_\varepsilon(t)^{1-d} dt \int_s^{\varepsilon_0} |\partial_t u(t,\sigma)|^2 r_\varepsilon(t)^{d-1}\, dt.$$

If we integrate over $\sigma \in \mathbb{S}^{d-1}$ we obtain

$$\int_{A_{\varepsilon,s}^i} |u|^2 = \int_{\mathbb{S}^{d-1}} |u(s,\sigma)|^2 r_\varepsilon(s)^{d-1} d\sigma$$

$$\le r_\varepsilon(s)^{d-1} \int_s^{\varepsilon_0} r_\varepsilon(t)^{1-d} dt \int_{M_\varepsilon} |du|^2. \tag{3.2}$$

If $0 \le s \le 2\varepsilon$ we have $r(s)^{d-1} \le (2\varepsilon)^{d-1}$. Furthermore, the integral over t can be split into an integral over $s \le t \le 2\varepsilon$ and $2\varepsilon \le t \le \varepsilon_0$. The first integral can be estimated by ε^{2-d}, the second by $\int_{2\varepsilon}^{\varepsilon_0} t^{1-d} dt$. Therefore we have an estimate of the order $O(\varepsilon)$ if $d \ge 3$ resp. $O(\varepsilon|\ln \varepsilon|)$ if $d = 2$. Finally, if we integrate the integral on the LHS of (3.2) over $s \in [0, 2\varepsilon]$ we obtain the desired Estimate (3.1). If $u(\varepsilon_0, \sigma) \ne 0$ we choose a cut-off function. $\qquad \square$

Remark 3.1. Note that $\omega(\varepsilon)$ only depends on the geometry of X near x_i, not on u or on θ. The argument in the proof is due to [A87].

The following lemma is proven in [CF78] resp. [A87].

Lemma 3.2. *We have $\lambda_k^{\mathrm{D}}(X_\varepsilon) \to \lambda_k(X)$ resp. $\lambda_k^{\mathrm{N}}(X_\varepsilon) \to \lambda_k(X)$.*

Now we show Theorem 3.1:

Proof. From the Min-max Principle (2.1) we conclude

$$\lambda_k^{\mathrm{D}}(X_\varepsilon) \le \lambda_k^\theta(M_\varepsilon)$$

since the domains of the quadratic forms obey the opposite inclusions. In particular, $\lambda_k^\theta(M_\varepsilon)$ is bounded in θ and ε by some constant $c_k > 0$. To prove the opposite inequality we estimate

$$\frac{q_{X_\varepsilon}(u)}{\|u\|_{X_\varepsilon}^2} - \frac{q_{M_\varepsilon}(u)}{\|u\|_{M_\varepsilon}^2} \le \frac{1}{\|u\|_{X_\varepsilon}^2} \frac{q_{M_\varepsilon}(u)}{\|u\|_{M_\varepsilon}^2} \left(\|u\|_{M_\varepsilon}^2 - \|u\|_{X_\varepsilon}^2 \right)$$

$$\le \frac{\|u\|_{M_\varepsilon}^2}{\|u\|_{X_\varepsilon}^2} \lambda_k^\theta(M_\varepsilon)\, \omega(\varepsilon) \left(1 + \lambda_k^\theta(M_\varepsilon) \right) \le \omega(\varepsilon)\, \frac{c_k(1 + c_k)}{1 - \omega(\varepsilon)(1 + c_k)} =: \delta_k(\varepsilon)$$

for $u \in L$, where L denotes the space generated by the first k eigenvalues of $\Delta_{M_\varepsilon}^\theta$. Note that $\delta_k(\varepsilon) \to 0$ as $\varepsilon \to 0$ by Lemma 3.1 which we have used twice. Since $\delta_k(\varepsilon)$ is independent of $u \in L$ the Min-max Principle implies

$$\lambda_k^{\mathrm{N}}(X_\varepsilon) - \delta_k(\varepsilon) \le \lambda_k^\theta(M_\varepsilon). \tag{3.3}$$

Note that $L \restriction_{X_\varepsilon}$ is still k-dimensional (by Lemma 3.1). Together with Lemma 3.2 we have proven Theorem 3.1. $\square$

4 Eigenvalues in Gaps

In this section we discuss a simple example how to produce eigenvalues in a spectral gap by locally perturbing the metric. Suppose that $\tilde{M}_\varepsilon$ is a periodic metric as in the previous section with period cell M_ε. Let $(\Gamma^n)_n$ be an exhaustive sequence, i.e., a monotone sequence with $\bigcup_n \Gamma^n = \Gamma$. Denote by M_ε^n the union of all γM_ε with $\gamma \in \Gamma^n$. Furthermore, we assume that M_ε^n and $M_\varepsilon^n \setminus M_\varepsilon^{n_0}$ are connected.

Let $(\rho_\tau)_\tau$ be a family of smooth, strictly positive functions on $\tilde{M}_\varepsilon$ such that $\tau \mapsto \rho_\tau$ is continuous with respect to the C^1-topology. Suppose further that

$$\rho_0 = 1 \qquad\qquad \text{on } \tilde{M}_\varepsilon \tag{4.1}$$

$$\rho_\tau = 1 \qquad\qquad \text{on } \tilde{M}_\varepsilon \setminus M_\varepsilon^{n_0} \qquad\qquad \text{for all } \tau \tag{4.2}$$

$$\rho_\tau = \mathrm{e}^\tau \qquad\qquad \text{on the period cell } M_\varepsilon \qquad\qquad \text{for all } \tau \tag{4.3}$$

Finally we denote by $\tilde{M}_\varepsilon(\tau)$ the manifold $\tilde{M}_\varepsilon$ together with the metric $\rho_\tau^2 \tilde{g}_\varepsilon$ if $\tilde{g}_\varepsilon$ denotes the metric of $\tilde{M}_\varepsilon$. Similar notations are understood in the same way. Note that all domains dom $q_{M(\tau)}$ and Hilbert spaces $L_2(M(\tau))$ are the same as vector spaces if τ varies. We choose Dirichlet boundary conditions on M_ε^n in order to have the inclusion dom $q_{M_\varepsilon^n} \subset$ dom $q_{M_\varepsilon^{n'}} \subset$ dom $q_{\tilde{M}_\varepsilon}$ for the domains of the (closed) quadratic forms if $n \le n'$.

First, we guarantee that no eigenvalue of the approximating problem lies in the gap; the boundary of M_ε resp. M_ε^n is so small such that boundary conditions almost have no influence on the eigenvalues:

Lemma 4.1. *If* $\lambda_k(X) < \lambda_{k+1}(X)$ *then there exist numbers* a, b *such that* $\lambda_k(X) < a < b < \lambda_{k+1}(X)$ *and such that the interval* $I = (a, b)$ *is a common gap, i.e.,*

$$I \cap \operatorname{spec} \Delta_{\tilde{M}_\varepsilon} = \emptyset \quad \text{and} \quad I \cap \operatorname{spec} \Delta_{M_\varepsilon^n}^{\mathrm{D}} = \emptyset \tag{4.4}$$

for all $\varepsilon > 0$ *small enough.*

The lemma follows from the Dirichlet-Neumann bracketing and the Min-max Principle (see [RS78] or [P00]). Note that $\lambda_k^{\mathrm{D}}(M_\varepsilon)$, $\lambda_k^{\mathrm{N}}(M_\varepsilon) \to \lambda_k(X)$ as in Theorem 3.1 with the same error estimate (3.3).

From now on we fix $\varepsilon > 0$ and $I = (a, b)$ such that (4.4) is satisfied. We omit the index ε, e.g., $M = M_\varepsilon$ or $\tilde{M} = \tilde{M}_\varepsilon$. Furthermore, we choose $\lambda \in I$

Next, we use the Decomposition Principle (see [DL79]) to prove that the essential spectrum remains invariant under the perturbation:

Theorem 4.1. *We have* $\operatorname{ess\,spec} \Delta_{\tilde{M}} = \operatorname{ess\,spec} \Delta_{\tilde{M}(\tau)}$ *for all* $\tau \ge 0$.

In particular, $\Delta_{\tilde{M}}$ and $\Delta_{\tilde{M}(\tau)}$ have the same spectral gap. In a spectral gap of the unperturbed Laplacian, the perturbed Laplacian can only have discrete eigenvalues (possibly accumulating at the band edges). It is essential here that the perturbation is localized on a compact set.

Now we prove that eigenfunctions of the approximating problem converge to eigenfunctions of the full problem:

Theorem 4.2. *Suppose that* $\tau_n \to \tau$ *and that*

$$\Delta_{M^n(\tau_n)} u_n = \lambda u_n, \qquad \|u_n\|_{M^n} = 1.$$

Then there exists a function u *in the domain of* $\Delta_{\tilde{M}(\tau)}$ *such that* $u_n \to u$ *weakly in* $L_2(\tilde{M})$ *and strongly in* $L_{2,\mathrm{loc}}(\tilde{M})$. *Furthermore,* $u \ne 0$ *and*

$$\Delta_{\tilde{M}(\tau)} u = \lambda u \tag{4.5}$$

To prove the theorem we need the following two lemmas. The next lemma can be shown straight forward:

Lemma 4.2. *For each $\tau, \tau' \geq 0$, the (squared) norms $\|\cdot\|^2_{M(\tau)}$ and $\|\cdot\|^2_{M(\tau')}$ are equivalent. In particular, the constants depend continuously on τ and τ'. The same is true for the quadratic forms $q_{M(\tau)}$ and $q_{M(\tau')}$.*

From the last lemma and the Rellich-Kondrachov Compactness Theorem we conclude the following lemma:

Lemma 4.3. *Let u_n be the approximating eigenvalue functions of Theorem 4.2. Then there exists a subsequence of (u_n) (also denoted by (u_n)) such that $u_n \to u$ weakly in $L_2(\tilde{M})$ and strongly in $L_{2,\mathrm{loc}}(\tilde{M})$. Furthermore, Equation (4.5) is valid.*

Now we prove Theorem 4.2. We only have to show that $u \neq 0$ which is the main difficulty.

Proof. Suppose that $u = 0$. Since λ lies in a spectral gap, we have

$$\|(\Delta^{\mathrm{D}}_{M^n} - \lambda)u_n\|_{M^n} \geq (b - a)\|u_n\|_{M^n} \geq \mathrm{const} > 0. \tag{4.6}$$

by the spectral calculus. On the other hand, we estimate

$$\|(\Delta^{\mathrm{D}}_{M^n} - \lambda)u_n\|_{M^n}$$
$$\leq \|(\Delta^{\mathrm{D}}_{M^n} - \Delta^{\mathrm{D}}_{M^n(\tau_n)})u_n\|_{M^n \setminus M^{n_0}} + \|\Delta^{\mathrm{D}}_{M^n} u_n\|_{M^{n_0}} + \lambda\|u_n\|_{M^{n_0}} \tag{4.7}$$

for $n \geq n_0$ where the first term on the RHS is equal to 0 (note that the perturbation of the metric is localized on M^{n_0}). The second term can be estimated to

$$\|\Delta^{\mathrm{D}}_{M^n} u_n\|_{M^{n_0}} \leq \mathrm{const}\big(\|u_n\|_{M^{n_1}} + \|\Delta^{\mathrm{D}}_{M^n(\tau_n)} u_n\|_{M^{n_1}}\big) \leq \mathrm{const}'\|u_n\|_{M^{n_1}}$$

for some appropriate $n_1 > n_0$ and all $n \geq n_1$ by regularity theory. Since (u_n) converges strongly to $u = 0$ in $L_{2,\mathrm{loc}}(\tilde{M})$, the LHS of (4.7) converges to 0 which contradicts (4.6). $\qquad\square$

In order to show the existence of eigenfunctions of the approximating problem we define the *eigenvalue counting function*

$$\mathcal{N}_{\tau_0, \tau}(Q(\cdot), \lambda) := \sum_{\tau_0 \leq \tau' \leq \tau} \dim \ker(Q(\tau') - \lambda).$$

This function counts the number of eigenvalues λ (with multiplicity) of the family $(Q(\tau'))_{\tau_0 \leq \tau' \leq \tau}$. Note the difference to the eigenvalue counting function of a single operator $Q \geq 0$ counting the number of eigenvalues below λ, i.e.,

$$\dim_\lambda(Q) := \sum_{0 < \lambda' < \lambda} \dim \ker(Q - \lambda').$$

The next lemma follows from the fact that the eigenvalue branches $\tau \mapsto \lambda_k^{\mathrm{D}}(M^n(\tau))$ are continuous and that the number of eigenvalue branches coming from above is a lower bound for the number how often the eigenvalue branches cross the level λ. Note that the eigenvalue branches could oscillate several times around λ.

Lemma 4.4. *We have*

$$\mathcal{N}_{\tau_0,\tau}(\Delta_{M^n(\cdot)},\lambda) \geq \dim_\lambda(\Delta_{M^n}^{\mathrm{D}}(\tau)) - \dim_\lambda(\Delta_{M^n}^{\mathrm{D}}(\tau_0)).$$

The proof of the following lemma is essentially the same as the proof of Lemma 4.1:

Lemma 4.5. *For $n \geq n_0$ we have*

$$\dim_\lambda(\Delta_{M^n}^{\mathrm{D}}(\tau)) - \dim_\lambda(\Delta_{M^n}^{\mathrm{D}}(\tau_0)) = \dim_\lambda(\Delta_{M^{n_0}}^{\mathrm{D}}(\tau)) - \dim_\lambda(\Delta_{M^{n_0}}^{\mathrm{D}}(\tau_0)).$$

Finally we prove the existence of approximating eigenfunctions. Together with Theorem 4.1 and Theorem 4.2 we conclude Theorem 1.2.

Theorem 4.3. *There exist an infinite number of sequences (τ_n) and (u_n) such that $\tau_n \to \hat{\tau}$ as $n \to \infty$ and such that u_n is an eigenfunction of the Dirichlet-Laplacian on $M^n(\tau_n)$ with eigenvalue λ.*

Proof. The Min-max Principle yields

$$0 \leq \lambda_k^{\mathrm{D}}(M^{n_0}(\tau)) \leq \lambda_k^{\mathrm{D}}(M(\tau)) = \mathrm{e}^{-2\tau}\lambda_k^{\mathrm{D}}(M) \to 0$$

and therefore $\dim_\lambda(\Delta_{M^{n_0}(\tau)}^{\mathrm{D}}) \to \infty$ as $\tau \to \infty$. From the last two lemmas we conclude $\mathcal{N}_{\tau_0,\tau}(\Delta_{M^n(\cdot)}^{\mathrm{D}},\lambda) \to \infty$ uniformly in $n \in \mathbb{N}$ as $\tau \to \infty$. If the counting number increases by 1 at the parameter τ we can choose a sequence (τ_n), $\tau_0 \leq \tau_n \leq \tau$ converging to some number $\hat{\tau}$. To this sequence corresponds a sequence of eigenvalues (u_n). In the next step we let τ_0 be the old value of τ. We raise τ until the counting number increases again by 1 and so forth. $\qquad\square$

Acknowledgements

I would like to thank the organizers of the conference for their kind invitation. In addition I am indebted to my thesis advisor Rainer Hempel for his permanent support. Furthermore I would like to thank Colette Anné for the helpful discussion concerning her article.

References

[AADH94] S. Alama, M. Avellaneda, P.A. Deift, and R. Hempel, *On the existence of eigenvalues of a divergence-form operator $A+\lambda B$ in a gap of $\sigma(A)$*, Asymptotic Anal. **8** (1994), 311–344.

[A87]   C. Anné, *Spectre du Laplacien et écrasement d'anses*, Ann. Sci. Éc. Norm. Super., IV. Sér. **20** (1987), 271–280.

[CF78]   I. Chavel and E. A. Feldman, *Spectra of domains in compact manifolds*, Journal of Functional Analysis **30** (1978), 198 – 222.

[CF81]   I. Chavel and E. A. Feldman, *Spectra of manifolds with small handles*, Comment. Math. Helvetici **56** (1981), 83–102.

[DH87]   E. B. Davies and Evans M. II Harrell, *Conformally flat Riemannian metrics, Schrödinger operators, and semiclassical approximation*, J. Differ. Equations **66** (1987), 165–188.

[D96]   E.B. Davies, *Spectral theory and differential operators*, Cambridge Univerity Press, Cambridge, 1996.

[DH86]   P. A. Deift and R. Hempel, *On the existence of eigenvalues of the Schroedinger operator $H - \lambda W$ in a gap of $\sigma(H)$.*, Commun. Math. Phys. **103** (1986), 461–490.

[DL79]   H. Donnelly and P. Li, *Pure point spectrum and negative curvature for non-compact manifolds*, Duke Math. J. **46** (1979), 497–503.

[D81]   H. Donnelly, *On $L^2-Betti$ numbers for Abelian groups*, Can. Math. Bull. **24** (1981), 91–95.

[G97]   E. L. Green, *Spectral theory of Laplace-Beltrami operators with periodic metrics*, J. Differ. Equations **133** (1997), no. 1, 15–29.

[HB00]   R. Hempel and A. Besch, *Magnetic barriers of compact support and eigenvalues in spectral gaps*, Preprint (2000).

[HH95]   R. Hempel and I. Herbst, *Strong magnetic fields, Dirichlet boundaries, and spectral gaps.*, Commun. Math. Phys. **169** (1995), no. 2, 237–259.

[P00]   O. Post, *Periodic manifolds, spectral gaps, and eigenvalues in gaps*, Ph.D. thesis, Technische Universität Braunschweig, 2000.

[RS78]   M. Reed and B. Simon, *Methods of modern mathematical physics. IV: Analysis of operators*, Academic Press, New York, 1978.

[RS80]   M. Reed and B. Simon, *Methods of modern mathematical physics. I: Functional analysis. Rev. and enl. ed.*, Academic Press, New York, 1980.

Address

OLAF POST, Institut für Reine und Angewandte Mathematik, Rheinisch-Westfälische Technische Hochschule Aachen, Templergraben 55, 52062 Aachen, Germany

E-MAIL: post@iram.rwth-aachen.de

2000 Mathematics Subject Classification. Primary 58J50, 35P05; Secondary 35A15, 57M10

Operator Theory:
Advances and Applications, Vol. 126
© 2001 Birkhäuser Verlag Basel/Switzerland

Stability of Inverse Operators of Boundary Value Problems in Non-Smooth Expanding Domains

V. S. Rabinovich

Abstract. We study the problem of uniform boundedness of inverse operators of boundary value problems in expanding domains with conical points on the boundary. We give applications to the problem of approximation of solutions of boundary value problems in large domains by solutions of equations on $\mathbf{R}^n$. Moreover we establish a connection between boundary value problems with a parameter in bounded domains and boundary value problems in expanding domains what allows us to give necessary and sufficient conditions for the invertibility of boundary value problems for large values of the parameter in domains with conical points on the boundary.

1 Introduction

Let $\mathcal{D}$ be an open bounded domain in $\mathbf{R}^n$ with a smooth boundary except for a finite set $sing\ \partial\mathcal{D} = \{v_1, ..., v_N\}$ of conic points. That is, there is a neighborhood of each singular point v_j such that in this neighborhood $\mathcal{D}$ coincides with a cone Γ_{v_j} with centre at the point v_j cutting out a smooth domain on the unit sphere with centre at v_j. Let

$$\mathcal{D}_r = \left\{ x \in \mathbf{R}^n : \frac{x}{r} \in \mathcal{D}, r > 0 \right\}.$$

We consider a boundary value problem in $\mathcal{D}_r$

$$Au = f, x \in D_r; \tag{1.1}$$
$$B_1 u = f_1, ..., B_m u = f_m, x \in \partial D_r \backslash sing\ \partial D_r$$

where

$$A = \sum_{|\alpha| \leq 2m} a_\alpha(x) D^\alpha, \quad B_j = \sum_{|\alpha| \leq m_j} b_{\alpha j}(x) D^\alpha, j = 1, ..., m$$

are differential operators with coefficients in the space $C_b^\infty(\mathbf{R}^n)$, where $C_b^\infty(\mathbf{R}^n)$ is the space of infinitely differentiable functions on $\mathbf{R}^n$ bounded with all their derivatives. We associate with the boundary value problem (1.1) an operator $\mathfrak{A}_r : X_r \to Y_r$ where X_r, Y_r are functional spaces suitable for the problem, which will be defined later, and consider the question of stability of $\mathfrak{A}_r^{-1} : Y_r \to X_r$ for large r, i.e. the existence of $r_0 > 0$ such that the operators $\mathfrak{A}_r : X_r \to Y_r$ are invertible for $r \geq r_0$ and

$$\sup_{r \geq r_0} \left\| \mathfrak{A}_r^{-1} \right\|_{Y_r \to X_r} < \infty. \tag{1.2}$$

Such problems arise in the investigation of approximation of solutions of boundary value problems in bounded domains by solution of equations $Au = f$ in $\mathbb{R}^n$. Note that this question was discussed in the book of Maz'ya, Nazarov, Plamenevskii [MNP91], see also papers [Sh73a], [Sh73b], [Do76], [Do79] under the assumption on stabilization of coefficients of differential operators at infinity. Our approach is essentially different from the approaches of the cited papers and allows us to obtain the criteria of stability $\mathfrak{A}_r^{-1} : Y_r \to X_r$ for large r without the assumption of stabilization. This approach is based on an investigation of local invertibility at infinity of a special class of boundary value problems in cones with edges for which we apply the limit operators method which has been developed by the author for boundary value problems in the papers [Ra99], [Ra93]. We also consider elliptic boundary value problems with a parameter in domains with conic points on the boundary. Note that the boundary value problems with a parameter in domains with smooth boundaries were considered in the well-known paper of Agranovich, Vishik [AV64], and for domains with conic points on the boundary in the books of Kozlov, Mazya, Rossmann [KMR], B.-W. Schulze [Sc94]. Our approach gives the necessary and sufficient conditions for invertibility of the problems for large values of parameters as a corollary of stability of the inverse operators for boundary value problems for expanding domains.

2   Local Invertibility at Infinity of a Class of Boundary Value Problems

Let $\mathbf{V}_{\mathcal{D}}$ be the cone in $\mathbf{R}^{n+1}$ generated by $\mathcal{D} \subset \mathbf{R}^n$:

$$\mathbf{V}_{\mathcal{D}} = \{(x, t) \in \mathbf{R}^n \times \mathbf{R}_+ : x \in \mathcal{D}_t\}.$$

We consider a boundary value problem in $\mathbf{V}_{\mathcal{D}}$

$$Au = f, \ (x, t) \in \mathbf{V}_{\mathcal{D}}; \ B_1 u = f_1, ..., B_m u = f_m, \tag{2.1}$$
$$(x, t) \in \partial \mathbf{V}_{\mathcal{D}} \backslash \cup_{j=1} L_{\nu_j}$$

where $L_{\nu_j} = \left\{(x, t) \in \mathbf{R}^{n+1} : x = \nu_j t, \ t \in \mathbf{R}_+\right\}$ are the edges generated by the singular points ν_j, and

$$A = \sum_{|\alpha| \leq 2m} a_\alpha(x, t) D_x^\alpha, \ B_j = \sum_{|\alpha| \leq m_j} b_{\alpha j}(x, t) D_x^\alpha, \ j = 1, ..., m$$

are differential operators with respect to x. We suppose that the coefficients belong to the class $SC_b^\infty(\mathbf{V}_{\mathcal{D}})$ of functions $a(x, t)$ defined on $\mathbf{V}_{\mathcal{D}}$ and satisfying the following conditions:

1. $a(x, t) \in C^\infty(\mathbf{V}_{\mathcal{D}})$,

2.

$$\sup_{\mathbf{V}_{\mathcal{D}}} \left| \partial_x^\alpha \partial_t^j a(x,t) \right| \le C_{\alpha j} \text{ for all multi-indices } (\alpha, j). \tag{2.2}$$

and

$$\lim_{\mathbf{V}_{\mathcal{D}} \ni (x,t) \to \infty} \partial_t a(x,t) = 0. \tag{2.3}$$

Let $s \in \mathbb{N}$, $\gamma \in \mathbf{R}, v \in sing\ \partial\mathcal{D}$, Γ_v be a cone with center at v which coincides with $\mathcal{D}$ in a neighborhood of v. We denote by $V^{s,\gamma}(\Gamma_v)$ a space with norm (see for instance [Ko67], [KMR])

$$\|u\|_{V^{s,\gamma}(\Gamma_v)} = \left(\sum_{|\alpha| \le s} \int_{\Gamma_v} |x - v|^{2(|\alpha|+\gamma)} |D^\alpha u|^2\, dx \right)^{\frac{1}{2}}.$$

The space $V^{s,\gamma}(\Gamma_v)$ for non- integer s is defined by interpolation and for negative s by means of duality. If $\mathcal{D}$ is the domain in $\mathbf{R}^n$ with the finite set $sing\ \partial\mathcal{D} = \{v_1, ..., v_N\}$ of conic points, then the space $V^{s,(\gamma)}(\mathcal{D})$ coincides in a neighborhood of each singular point v_j with $V^{s,\gamma_j}(\Gamma_{v_j})$, $j = 1, ..., N$ and it is the usual Sobolev space outside the union of neighborhoods of the points $\{v_1, ..., v_N\}$. Let $\{\varphi_j\}_{j=1}^N$ be a system of C_0^∞ -functions such that $supp\ \varphi_j$ lies in a small neighborhood of v_j, and $\varphi_0 = 1 - \sum_{j=1}^N \varphi_j$. The norm in $V^{s,(\gamma)}(\mathcal{D})$ is defined by the formula

$$\|u\|_{V^{s,(\gamma)}(\mathcal{D})} = \left(\|\varphi_0 u\|^2_{H^s(\mathcal{D})} + \sum_{j=1}^N \|\varphi_j u\|^2_{V^{s,\gamma_j}(\Gamma_{v_j})} \right)^{1/2}.$$

The space $V^{s-1/2,(\gamma)}(\partial\mathcal{D})$ is defined as the space of traces on the boundary of elements in $V^{s,(\gamma)}(\mathcal{D})$, $s > 1/2$. We denote by $\mathcal{V}^{s,(\gamma)}(\mathbf{V}_{\mathcal{D}})$ the space of distribution on $\mathbf{V}_{\mathcal{D}}$ with norm

$$\|u\|_{\mathcal{V}^{s,(\gamma)}(\mathbf{V}_{\mathcal{D}})} = \left(\int_{\mathbf{R}_+} \|u(x,t)\|^2_{V^{s,(\gamma)}(\mathcal{D}_t)}\, dt \right)^{1/2},$$

and associate with the boundary value problem (2.1) the operator $\mathfrak{A} : X \to Y$ where

$$X = \mathcal{V}^{s,(\gamma)}(\mathbf{V}_{\mathcal{D}}), Y = \mathcal{V}^{s-2m,(\gamma-2m)}(\mathbf{V}_{\mathcal{D}}) \oplus_{k=1}^m \mathcal{V}^{s-m_k-1/2,(\gamma-m_j)}(\partial \mathbf{V}_{\mathcal{D}}).$$

We will say that the operator $\mathfrak{A} : X \to Y$ is locally invertible at infinity if there are constants $R > 0, C > 0$ such that the following estimates hold

$$\|\mathfrak{A}\chi_R f\|_Y \ge C\, \|\chi_R f\|_X\ , \|\chi_R \mathfrak{A}^* \psi\|_{X_*} \ge C\, \|\chi_R \psi\|_{Y_*}$$

where $\chi \in C_0^\infty(\mathbf{R}^{n+1})$ and $\chi(x) = 1$ if $|x| \le 1$ and $\chi(x) = 0$ if $|x| \ge 2$, and $\chi_R(x) = \chi(x/R)$. The criteria of local invertibility at infinity of the operator

$\mathfrak{A} : X \to Y$ is given in terms of limit operators, which we will introduce below. Let ω be the point in $\mathbf{R}^n_x$. Then we denote by L_ω the ray in $\mathbf{R}^{n+1}_{(x,t)}$ passing through the origin and at the point $(\omega, 1)$ and we denote by η_ω the infinitely distant point corresponding to the ray L_ω. We define neighborhoods of the point η_ω as the sets

$$U_{\eta_\omega} = \left\{ (x,t) \in \mathbf{R}^{n+1} : t > R \text{ and } x \in W^\omega_t \right\}$$

where W^ω is a neighborhood of $\omega \in \mathbf{R}^n_x$. We say that the sequence $\mathbb{Z}^{n+1} \ni h_m \to \eta_\omega$ if for each neighborhood U_{η_ω} of the point η_ω there exists m_0 such that $h_m \in U_{\eta_\omega}$ if $m \geq m_0$. In the same way we define the convergence $(x,t) \to \eta_\omega$. Let $a(x,t) \in C^\infty_b(\mathbf{R}^{n+1})$ and $\mathbb{Z}^{n+1} \ni h_m \to \eta_\omega$. Then there exists a subsequence $h_{m_k} \to \eta_\omega$ such that there exists a limit in the C^∞–topology

$$\lim_{k \to \infty} a((x,t) + h_{m_k}) = a^h(x,t)$$

where $a_h(x,t) \in C^\infty_b(\mathbf{R}^{n+1})$. Let

$$\mathcal{C} = \sum_{|\alpha| \leq p} c_\alpha(x,t) D^\alpha_x$$

where $c_\alpha(x,t) \in C^\infty_b(\mathbf{R}^{n+1})$. We denote by $\sigma_{\eta_\omega}(\mathcal{C})$ the set of differential operators

$$\mathcal{C}^h = \sum_{|\alpha| \leq p} c^h_\alpha(x,t) D^\alpha_x,$$

where

$$\lim_{m \to \infty} c_\alpha((x,t) + h_m) = c^h_\alpha(x,t), \ h_m \to \eta_\omega \text{ for all } \alpha : |\alpha| \leq p$$

and the limits exist in $C^\infty(\mathbf{R}^{n+1})$ Note that, if the coefficients $c_\alpha(x,t)$ are slowly varying at infinity as $t \to \infty$ then the functions $c^h_\alpha(x,t) \equiv c^h_\alpha(x)$ do not depend on t. Let $\omega \in \partial\mathcal{D} \backslash sing \ \partial\mathcal{D}$. We denote by $\mathbb{H}_\omega$ the half-space bounded by the tangent space to $\partial\mathbf{V}_\mathcal{D}$ passing through the ray L_ω defined by the interior normal vector to the boundary at the point ω. We have

$$\mathbb{H}_\omega = \left\{ (x,t) \in \mathbf{R}^{n+1} : x \in \mathbf{H}_{\omega,t} = \mathbf{H}^0_\omega + t\omega \right\}$$

where $\mathbf{H}_\omega$ is a half-space bounded by the tangent space to $\mathcal{D}$ at the point ω, and $\mathbf{H}^0_\omega = \mathbf{H}_\omega - \omega$. If $\omega = v_j \in sing \ \partial\mathcal{D}$ then we denote by $\mathbb{K}_{v_j}$ the wedge

$$\mathbb{K}_{v_j} = \left\{ (x,t) \in \mathbf{R}^{n+1} : x \in \Gamma_{v_j,t} = \Gamma^0_{v_j} + tv_j, t \in \mathbf{R} \right\},$$

where $\Gamma^0_{v_j} = \Gamma_{v_j} - v_j$ is the shift of Γ_{v_j} at the origin. We denote by $\mathcal{V}^{s,\gamma}(\mathbb{K}_{v_j})$ the space with the following norm:

$$\|u\|_{\mathcal{V}^{s,\gamma}(\mathbb{K}_{v_j})} = \left(\int_{\mathbf{R}} \|u(x,t)\|^2_{V^{s,\gamma}(\Gamma_{v_j,t})} \, dt \right)^{1/2} .$$

Note that the space $\mathcal{V}^{s,\gamma}(\mathbb{K}_{v_j})$ is invariant with respect to shifts by the vectors $h = \alpha(v_j, 1), \alpha \in \mathbf{R}$ which are parallel to the edge of the wedge $\mathbb{K}_{v_j}$. Indeed,

$$\|u(x - \alpha v_j, t - \alpha\|_{\mathcal{V}^{s,\gamma}(\mathbb{K}_{v_j})} = \left(\int_{\mathbf{R}} \|u(x, t - \alpha)\|^2_{\mathcal{V}^{s,\gamma}(\Gamma_{v_j, t-\alpha})} \, dt \right)^{1/2}$$
$$= \|u\|_{\mathcal{V}^{s,\gamma}(\mathbb{K}_{v_j})}.$$

Theorem 2.1. *The operator*

$$\mathfrak{A} : \mathcal{V}^{s,(\gamma)}(\mathbf{V}_{\mathcal{D}}) \to \mathcal{V}^{s-2m,(\gamma-2m)}(\mathbf{V}_{\mathcal{D}}) \oplus_{k=1}^m \mathcal{V}^{s-m_k-1/2,(\gamma-m_k)}(\partial\mathbf{V}_{\mathcal{D}})$$

is locally invertible at infinity if and only if : 1) for each infinitely distant point η_ω where $\omega \in int\,\mathcal{D}$ all limit operators $A^h \in \sigma_{\eta_\omega}(A)$ are invertible from $H^s(\mathbf{R}^{n+1})$ into $H^{s-2m}(\mathbf{R}^{n+1})$; 2) for each infinitely distant point η_ω where $\omega \in \partial\mathcal{D} \setminus sing\,\partial\mathcal{D}$ all operators $\mathfrak{A}^h_{\mathbb{H}_\omega}$ of boundary value problems in the half-space $\mathbb{H}_\omega$

$$\mathfrak{A}^h_{\mathbb{H}_\omega} : H^s(\mathbb{H}_\omega) \to H^{s-2m}(\mathbb{H}_\omega) \oplus_{j=1}^m H^{s-m_j-1/2}(\partial\mathbb{H}_\omega),$$

defined by the interior operator $A^h \in \sigma_{\eta_\omega}(A)$ and the boundary operators $B^h_j \in \sigma_{\eta_\omega}(B_j)$, are invertible; 3) for each infinitely distant point η_{v_j} where $v_j \in sing\,\partial\mathcal{D}$ all operators $\mathfrak{A}_{\mathbb{K}_{v_j}}$ of boundary value problems in the wedge $\mathbb{K}_{v_j}$

$$\mathfrak{A}_{\mathbb{K}_{v_j}} : \mathcal{V}^{s,\gamma}(\mathbb{K}_{v_j}) \to \mathcal{V}^{s-2m,\gamma-2m}(\mathbb{K}_{v_j}) \oplus_{j=1}^m \mathcal{V}^{s-m_j-1/2,\gamma-m_j}(\partial\mathbb{K}_{v_j})$$

defined by the interior operator $A^h \in \sigma_{\eta_{v_j}}(A)$ and the boundary operators $B^h_k \in \sigma_{\eta_{v_j}}(B_k)$, are invertible.

Proof. The idea of the proof is following. As follows from local principle the operator $\mathfrak{A}$ is locally invertible at infinity if and only if this operator is locally invertible in each infinitely distant point η_ω where $\omega \in \bar{\mathcal{D}}$. This means that there exists a neighborhood U_{η_ω} of the point η_ω such that

$$\left\|\mathfrak{A}\psi_{U_{\eta_\omega}} f\right\|_Y \geq C \left\|\psi_{U_{\eta_\omega}} f\right\|_X, \quad \left\|\psi_{U_{\eta_\omega}} \mathfrak{A}^*\varphi\right\|_{X^*} \geq C \left\|\psi_{U_{\eta_\omega}} \varphi\right\|_{Y^*},$$

where $\psi_{U_{\eta_\omega}}$ is a cut-off function of the neighborhood of U_{η_ω}. The results of [Ra99], [Ra93] imply: 1) Let $\omega \in int\mathcal{D}$. The operator $\mathfrak{A}$ is locally invertible at the point η_ω if and only if the condition 1) of Theorem 2.1 holds. 2) Let $\omega \in \partial\mathcal{D} \setminus sing\,\partial\mathcal{D}$. The operator $\mathfrak{A}$ is locally invertible at the point η_ω if and only if the condition 2) of Theorem 2.1 holds. 3) Let $v_j \in sing\,\partial\mathcal{D}$. The operator $\mathfrak{A}$ is locally invertible at the point η_{v_j} if and only if the condition 3) of Theorem 2.1 holds $\qquad \square$

3 Stability of a Sequence of Inverse Operators

Let us associate to the operator

$$\mathfrak{A} : \mathcal{V}^{s,(\gamma)}(\mathbf{V}_{\mathcal{D}}) \to \mathcal{V}^{s-2m,(\gamma-2m)}(\mathbf{V}_{\mathcal{D}}) \oplus_{k=1}^m \mathcal{V}^{s-m_k-1/2,(\gamma-m_k)}(\partial\mathbf{V}_{\mathcal{D}})$$

the family of operators

$$\mathfrak{B}_t : V^{s,(\gamma)}(\mathcal{D}_t) \to V^{s-2m,(\gamma-2m)}(\mathcal{D}_t) \oplus_{j=1}^m V^{s-m_j-1/2,(\gamma-m_j)}(\partial\mathcal{D}_t)$$

where $\mathfrak{B}_t$ is the operator of the boundary value problem in the domain $\mathcal{D}_t$

$$A_t u(x) = f(x), x \in \mathcal{D}_t, \tag{3.1}$$
$$B_{1t} u(x) = f_1(x), ..., B_{mt} u(x) = f_m(x), x \in \partial\mathcal{D}_t,$$

and

$$A_t = \sum_{|\alpha| \leq 2m} a_\alpha(x,t) D_x^\alpha, \; B_{jt} = \sum_{|\alpha| \leq m_j} b_{\alpha j}(x,t) D_x^\alpha, \; j = 1, ..., m$$

are differential operators on $\mathcal{D}_t$ with coefficients depending on the parameter $t \in \mathbf{R}_+$.

Definition 3.1. We say that the sequence $\mathfrak{B}_t^{-1}$ is stable for large t if there exists $T > 0$ such that for $t > T$ the operators $\mathfrak{B}_t^{-1}$ are invertible and

$$\sup_{t>T} \left\| \mathfrak{B}_t^{-1} \right\| < \infty.$$

The following theorem, which is a straight corollary of Theorem 2.1, gives the necessary and sufficient conditions for the family of operators $\mathfrak{B}_t^{-1}$ to be stable.

Theorem 3.1. *The family $\mathfrak{B}_t^{-1}$ is stable if and only if: 1) for each infinitely distant point η_ω where $\omega \in \mathrm{int}\mathcal{D}$, all operators $A^h : H^s(\mathbf{R}^n) \to H^{s-2m}(\mathbf{R}^n)$ are invertible, where*

$$A^h = \sum_{|\alpha| \leq 2m} a_\alpha^h(x) D_x^\alpha \in \sigma_{\eta_\omega}(A),$$

2) for each infinitely distant point η_ω where $\omega \in \partial\mathcal{D} \setminus \mathrm{sing}\; \partial\mathcal{D}$, all operators

$$\mathfrak{A}_{\mathbf{H}_\omega^0}^h : H^s(\mathbf{H}_\omega^0) \to H^{s-2m}(\mathbf{H}_\omega^0) \oplus_{j=1}^m H^{s-m_j-1/2}(\partial\mathbf{H}_\omega^0)$$

are invertible, where $\mathfrak{A}_{\mathbf{H}_\omega^0}^h$ is the operator of boundary value problems for the half-space $\mathbf{H}_\omega^0 \subset \mathbf{R}^n$ defined by the interior operator

$$A^h = \sum_{|\alpha| \leq 2m} a_\alpha^h(x) D_x^\alpha \in \sigma_{\eta_\omega}(A),$$

and the boundary operators

$$B_j^h = \sum_{|\alpha| \leq m_j} b_{\alpha j}^h(x) D_x^\alpha \in \sigma_{\eta_\omega}(B_j), \; j = 1, ..., m;$$

3) for each infinitely distant point η_{ν_j} where $\nu_j \in \text{sing } \partial D$ the operators $\mathfrak{A}^h_{\Gamma_{\nu_j}}$ of boundary value problems in the cone $\Gamma_{\nu_j} \subset \mathbf{R}^n$

$$\mathfrak{A}^h_{\Gamma_{\nu_j}} : V^{s,\gamma}(\Gamma_{\nu_j}) \to V^{s-2m,\gamma-2m}(\Gamma_{\nu_j}) \oplus_{j=1}^m V^{s-m_j-1/2,\gamma-m_j}(\partial\Gamma_{\nu_j})$$

defined by the interior operator $A^h \in \sigma_{\eta_{\nu_j}}(A)$ and the boundary operators $B^h_k \in \sigma_{\eta_{\nu_j}}(B_j)$, are invertible.

Let us consider the case when the coefficients of the operators A and B_j do not depend on the parameter t. Denote by η'_ω the infinitely distant point corresponding to the ray in $\mathbf{R}^n$ outgoing from the origin and passing through the point $\omega \neq 0$. We denote by $\sigma_{\eta'_\omega}(A)$ the set of limit operators of $A = \sum_{|\alpha|\leq 2m} a_\alpha(x)D^\alpha$, where $a_\alpha(x) \in C_b^\infty(\mathbf{R}^n)$.   We will consider the following cases: 1) Let $0 \in \text{int}D$. Then $\cup_{t>0}\mathcal{D}_t = \mathbf{R}^n$. In this case the Theorem 3.1 admits the following realization.

Theorem 3.2. *The set $\mathfrak{A}_r^{-1} : X_r \to Y_r$ is stable if and only if : 1) The operator $A : H^s(\mathbf{R}^n) \to H^{s-2m}(\mathbf{R}^n)$ is invertible; 2) For each point $\omega \in \partial D \backslash \text{sing } \partial D$ the operators of the boundary value problems*

$$\mathfrak{A}^h_{\mathbf{H}^0_v} : H^s(\mathbf{H}^0_v) \to H^{s-2m}(\mathbf{H}^0_v) \oplus_{j=1}^m H^{s-m_j-1/2}(\partial\mathbf{H}^0_v),$$

defined by the interior operator $A^h \in \sigma_{\eta'_\omega}(A)$, and the boundary operators $B^h_j \in \sigma_{\eta'_\omega}(B_j)$, $j = 1, ..., m$, are invertible. 3) For each point $v_j \in \text{sing } \partial D$

$$\mathfrak{A}^h_{\Gamma^0_{v_j}} : V^{s,\gamma_j}(\Gamma^0_{v_j}) \to V^{s-2m,\gamma_j-2m}(\Gamma^0_{v_j}) \oplus_{k=1}^m V^{s-m_k,\gamma_j-m_k}(\partial\Gamma^0_{v_j}),$$

the operators of boundary value problems defined by interior operators $A^h \in \sigma_{\eta'_{v_j}}(A)$, and the boundary operators $B^h_k \in \sigma_{\eta'_{v_j}}(B_k)$, $k = 1, ..., m$, are invertible.

Let us consider the application of Theorem 3.2 to the problem of the approximation of solutions of boundary value problems in expanding domains by solution of an equation on $\mathbf{R}^n$. We consider the following boundary value problem

$$Au_r = f_r, \ x \in \mathcal{D}_r, B_j u(x) = 0, \ x \in \partial\mathcal{D}_r, \ j = 1, 2, ..., m \tag{3.2}$$

where A and B have been introduced earlier, and f_r is the restriction of a function $f \in H^{s-2m}(\mathbf{R}^n)$ to $\mathcal{D}_r$. It is easy to see that $f_r \in W^{s-2m,(\gamma-2m)}(\mathcal{D}_r)$, if $\gamma_j \geq 2m, j = 1, ..., N$. Theorem 3.2 in this case has the following important corollary.

Corollary 3.1. *Let the conditions of Theorem 3.2 be fulfilled. Then there exists a r_0 such that for $r \geq r_0$ the problem (3.2) has a unique solution u_r. Moreover,*

$$\lim_{r\to 0} \|\hat{u}_r - u_r\|_{W^{s,(\gamma)}(\mathcal{D}_r)} = 0,$$

where $\hat{u}_r$ is the restriction to $\mathcal{D}_r$ of unique solution $u \in H^{s-2m}(\mathbf{R}^n)$ of the equation

$$Au = f, \ f \in H^{s-2m}(\mathbf{R}^n).$$

This corollary gives conditions for approximation of solutions of boundary value problems in large domains $\mathcal{D}_r$ by solutions of differential equations on $\mathbf{R}^n$ and conversely. 2) Let $0 \in \partial\mathcal{D}\backslash sing \ \partial\mathcal{D}$. Then

$$\cup_{t>0}\mathcal{D}_t = \mathbf{H}_{\mu^0} = \left\{ x \in \mathbf{R}^n : (x, \mu^0) = x_1\mu_1^0 + ... + x_n\mu_n^0 > 0 \right\}$$

is a half-space in $\mathbf{R}^n$, defined by the vector of interior normal to the boundary $\partial\mathcal{D}$ at the point 0.

Theorem 3.3. *The set* $\mathfrak{A}_r^{-1} : X_r \to Y_r$ *is stable if and only if : 1) The operator*

$$\mathfrak{A}_{\mathbf{H}_{\mu^0}} : H^s(\mathbf{H}_{\mu^0}) \to H^{s-2m}(\mathbf{H}_{\mu^0}) \oplus_{k=1}^m H^{s-m_k-1/2}(\partial\mathbf{H}_{\mu^0})$$

of boundary value problem in the half-space $\mathbf{H}_{\mu^0}$ *defined by the operators* $A, B_1, ...,$ B_m, *is invertible; 2) for each infinitely distant point* η'_ω *where* $\omega \in \acute{\mathcal{D}}$ *all operators* $A^h : H^s(\mathbf{R}^n) \to H^{s-2m}(\mathbf{R}^n)$ *are invertible, where*

$$A^h = \sum_{|\alpha| \leq 2m} a_\alpha^h(x)D_x^\alpha \in \sigma_{\eta'_\omega}(A),$$

3) For each point $\omega \in \partial\mathcal{D}\backslash sing \ \partial\mathcal{D} \ \ \omega \neq 0$, *the operators of the boundary value problem*

$$\mathfrak{A}_{\mathbf{H}_v^0}^h : H^s(\mathbf{H}_v^0) \to H^{s-2m}(\mathbf{H}_v^0) \oplus_{j=1}^m H^{s-m_j-1/2}(\partial\mathbf{H}_v^0),$$

defined by the interior operator $A^h \in \sigma_{\eta'_\omega}(A)$, *and the boundary operators* $B_j^h \in \sigma_{\eta'_\omega}(B_j)$, $j = 1, ..., m$ *are invertible. 3) For each point* $v_j \in sing \ \partial\mathcal{D}$,

$$\mathfrak{A}_{\mathbf{\Gamma}_{v_j}^0}^h : V^{s,\gamma_j}(\Gamma_{v_j}^0) \to V^{s-2m,\gamma_j-2m}(\Gamma_{v_j}^0) \oplus_{k=1}^m V^{s-m_k,\gamma_j-m_k}(\partial\Gamma_{v_j}^0)$$

the operators of boundary value problems defined by the interior operator $A^h \in \sigma_{\eta'_{v_j}}(A)$, *and the boundary operators* $B_k^h \in \sigma_{\eta'_{v_j}}(B_k)$, $k = 1, ..., m$, *are invertible.*

Corollary 3.2. *Let the conditions of Theorem 3.3 be fulfilled. Then there exists a* r_0 *such that for* $r \geq r_0$ *the problem (3.2) has a unique solution* u_r. *Moreover,*

$$\lim_{r \to 0} \|\hat{u}_r - u_r\|_{V^{s,(\gamma)}(\mathcal{D}_r)} = 0,$$

where $\hat{u}_r$ *is the restriction to* $\mathcal{D}_r$ *of a unique solution* $u \in H^{s-2m}(\mathbf{H}_{\mu^0})$ *of the boundary value problem in the half-space* $\mathbf{H}_{\mu_0}$

$$Au = f, \ x \in \mathbf{H}_{\mu^0},$$
$$B_1 u = 0, ..., \ B_m u = 0, x \in \partial\mathbf{H}_{\mu^0}$$

This corollary gives conditions for the approximation of solutions of boundary value problems in large domains $\mathcal{D}_r$ by solutions of boundary value problems in half-space. 3) Let $0 \in \partial\mathcal{D}\backslash sing \ \partial\mathcal{D}$. Then $\cup_{t>0}\mathcal{D}_t = \mathbf{\Gamma}_0$ is a cone with center at the point 0.

Theorem 3.4. *The set $\mathfrak{A}_r^{-1} : X_r \to Y_r$ is stable if and only if : 1) The operator*

$$\mathfrak{A}_{\Gamma_0} : V^{s,\gamma_0}(\Gamma_0) \to V^{s-2m,\gamma_0-2m}(\Gamma_0) \oplus_{k=1}^{m} V^{s-m_k,\gamma_0-m_k}(\partial\Gamma_0)$$

of the boundary value problem in the cone Γ_0 defined by the operators $A, B_1, ..., B_m$ is invertible; 2) for each infinitely distant point η'_ω where $\omega \in \operatorname{int} D$, all operators $A^h : H^s(\mathbf{R}^n) \to H^{s-2m}(\mathbf{R}^n)$, are invertible, where

$$A^h = \sum_{|\alpha| \leq 2m} a_\alpha^h(x) D_x^\alpha \in \sigma_{\eta'_\omega}(A),$$

3) For each point $\omega \in \partial D \backslash \operatorname{sing} \partial D$ the operators of the boundary value problem

$$\mathfrak{A}_{\mathbf{H}_v^0}^h : H^s(\mathbf{H}_v^0) \to H^{s-2m}(\mathbf{H}_v^0) \oplus_{j=1}^{m} H^{s-m_j-1/2}(\partial\mathbf{H}_v^0),$$

defined by interior operator $A^h \in \sigma_{\eta'_\omega}(A)$, and the boundary operators $B_j^h \in \sigma_{\eta'_\omega}(B_j)$, $j = 1, ..., m$, are invertible. 4) For each point $v_j \in \operatorname{sing} \partial D \backslash 0$,

$$\mathfrak{A}_{\Gamma_{v_j}^0}^h : V^{s,\gamma_j}(\Gamma_{v_j}^0) \to V^{s-2m,\gamma_j-2m}(\Gamma_{v_j}^0) \oplus_{k=1}^{m} V^{s-m_k,\gamma_j-m_k}(\partial\Gamma_{v_j}^0)$$

the boundary value problem defined by interior operator $A^h \in \sigma_{\gamma'_{v_j}}(A)$, and the boundary operators $B_k^h \in \sigma_{\eta'_{v_j}}(B_k)$, $k = 1, ..., m$, are invertible.

Corollary 3.3. *Let the conditions of Theorem 3.4 be fulfilled. Then there exists a r_0 such that for $r \geq r_0$ the problem (3.2) has a unique solution u_r. Moreover,*

$$\lim_{r \to 0} \|\hat{u}_r - u_r\|_{V^{s,(\gamma)}(\mathcal{D}_r)} = 0,$$

where $\hat{u}_r$ is the restriction to $\mathcal{D}_r$ of a unique solution $u \in V^{s-2m,\gamma_0-2m}(\Gamma_0)$ of the boundary value problem in the cone Γ_0

$$Au = f , \ x \in \ ,$$
$$B_1 u = 0, ..., \ B_m u = 0, x \in \ \partial\Gamma_0 \backslash 0$$

This corollary gives conditions for the approximation of solutions of boundary value problems in a large domains $\mathcal{D}_r$ by solutions of the boundary value problem on the cone Γ_0.

Example 3.1. For simplicity of illustration we will consider only the case of a domain $\mathcal{D}(\ni 0)$ with a smooth boundary ∂D.

Let us consider the Dirichlet problem for the Schrödinger operator

$$(-\Delta + q(x)) u_r = f_r, \ x \in \mathcal{D}_r; \ u_r \mid_{\partial\mathcal{D}_r} = 0. \tag{3.3}$$

Denote by $\mathfrak{A}_r : H^s(\mathcal{D}_r) \to H^{s-2}(\mathcal{D}_r), s > 1/2$ the operator of the problem (3.3).

We suppose that the potential $q(x)$ is slowly varying at infinity, that is $q(x) \in C_b^\infty(\mathbf{R}^n)$ and

$$\lim_{x \to \infty} \partial_{x_j} q(x) = 0, \; j = 1, ..., n.$$

In this case the limits $\lim_{k \to \infty} q(x + h_k), h_k \to \infty$, (if they exists) do not depend on x. We also suppose that that the potential $q(x)$ is sectorial, that is, if

$$I(q) = \{z \in \mathbf{C} : \mathbf{z} = q(x), x \in \mathbf{R}^n, \}$$

then

$$\overline{I(q)} \subset \Gamma_\varepsilon = \{z \in \mathbf{C} : |z| > \varepsilon\} \cap \{z \in \mathbf{C} : \pi + \varepsilon < \arg z < 2\pi - \varepsilon\},$$

where $\varepsilon > 0$ is small.

Applying the calculus of pseudodifferential operators with slowly varying symbols [Gr70], it is easy to show that the operator $A = -\Delta + q(x) : H^s \to H^{s-2}$ is Fredholm with index zero. Moreover sectoriality of the potential provides that the kernel of A has the dimension zero. Consequently, A is invertible, that is the condition 1) of the Theorem 3.2 is fulfilled.

Let us check the condition 2) of the Theorem 3.2 . In our case the operators $\mathfrak{A}_{\mathbf{H}_v^0}^h : H^s(\mathbf{H}_v^0) \to H^{s-2}(\mathbf{H}_v^0)$ are the operators of the Dirichlet problems for the half-space $\mathbf{H}_v^0$, defined by the operators $-\Delta + q_h$, where q_h are partial limits of $q(x)$ when x tending to the infinitely distant point η_ω'. As follows from sectoriality of $q(x)$ all $q_h \in \Gamma_\varepsilon$. The standard elliptic theory [Es] provides the uniform invertibility of all operators $\mathfrak{A}_{\mathbf{H}_v^0}^h$.

Thus all conditions of the Theorem 3.2 are fulfilled, consequently the sequence $\mathfrak{A}_r^{-1} : H^{s-2}(\mathcal{D}_r) \to H^s(\mathcal{D}_r), s > 1/2$ is stable.

Let $f \in H^{s-2}(\mathbf{R}^n), s > 1/2$ and f_r is a restriction of f to $\mathcal{D}_r$. Then there exists r_0 such that for $r \geq r_0$ the problem 3.3 has a unique solution u_r. Moreover,

$$\lim_{r \to 0} \|\hat{u}_r - u_r\|_{H^s(\mathcal{D}_r)} = 0,$$

where $\hat{u}_r$ is the restriction to $\mathcal{D}_r$ of a unique solution $u \in H^{s-2m}(\mathbf{R}^n)$ of the equation

$$(-\Delta + q(x)) u = f, \; f \in H^{s-2}(\mathbf{R}^n).$$

4 Boundary Value Problems with a Parameter

In the domain $\mathcal{D} \subset \mathbf{R}^n$ we consider the boundary value problem with a parameter $\zeta = q + i\lambda > 0$, where $q > 0$ and $\lambda (\in \mathbf{R})$ is fixed:

$$A(\zeta)u = f, \; x \in \mathcal{D}; \; B_1(\zeta)u = f_1, ..., B_m(\zeta)u = f_m, \tag{4.1}$$
$$x \in \mathcal{D} \setminus (sing \; \partial\mathcal{D}),$$

where

$$A(\zeta) = \sum_{|\alpha| \leq 2m} a_\alpha(x) \left(\frac{1}{\zeta} D_x \right)^\alpha, \quad B_k(\zeta) = \sum_{|\alpha| \leq m_k} b_{\alpha k}(x) \left(\frac{1}{\zeta} D_x \right)^\alpha$$

with $a_\alpha(x), b_{\alpha k}(x) \in C^\infty(\overline{\mathcal{D}})$. Let us introduce the functional spaces with norm depending on a parameter in which we will consider the boundary value problem (4.1). Denote by $H^s(\mathcal{D}, q), s \in \mathbf{N}$, the space with norm

$$\|u\|_{H^s(\mathcal{D},q)} = \left(\sum_{|\alpha| \leq s} \int_{\mathcal{D}} \left| \left(\frac{1}{q} D_x \right)^\alpha u \right|^2 dx \right)^{1/2}$$

Let $\{\varphi_j\}_{j=1}^N$ be a system of C_0^∞ -functions such that $supp\ \varphi_j$ lies in a small neighborhood of ν_j, $\varphi_0 = 1 - \sum_{j=1}^N \varphi_j$. We define the space $V_q^{s,(\gamma)}(\mathcal{D})$ as the space with norm

$$\|u\|_{V_q^{s,(\gamma)}(\mathcal{D})} = \left(\|\varphi_0^q u\|_{H_q^s(\mathcal{D})}^2 + \sum_{j=0}^N \|\varphi_j^q u\|_{V_q^{s,\gamma_j}(\Gamma_{v_j})}^2 \right)^{1/2},$$

where $\varphi_j^q(x) = \varphi_j(qx), j = 0, 1, ..., N$, and $V_q^{s,\gamma_j}(\Gamma_{v_j})$ has the norm

$$\|u\|_{V_q^{s,\gamma_j}(\Gamma_{v_j})} = q^\gamma \left(\sum_{|\alpha| \leq s} \int_{\Gamma_{v_j}} |x - v_j|^{2(|\alpha|+\gamma)} |D^\alpha u|^2 dx \right)^{\frac{1}{2}}.$$

The space $V_q^{s,(\gamma)}(\mathcal{D})$ for positive but non-integer s is defined by interpolation and for negative s by means of duality. As earlier we introduce the space $V_q^{s-1/2,(\gamma)}(\mathcal{D})$, $s > 1/2$ as the space of traces on the boundary $\partial\mathcal{D}$ of elements in $V_q^{s,(\gamma)}(\mathcal{D})$.

We assosiate with the boundary value problem the operator

$$\mathcal{L}_\zeta : X_q = V_q^{s,(\gamma)}(\mathcal{D}) \to V_q^{s-2m,(\gamma-2m)}(\mathcal{D}) \oplus_{k=1}^m V_q^{s-m_k-1/2,(\gamma-m_k)}(\partial\mathcal{D}) = Y_q.$$

Let

$$(U_q f)(y) = q^{-n/2} f(y/q), q > 0.$$

The operator U_q is unitary from $V_q^{s,(\gamma)}(\mathcal{D})$ into $V^{s,(\gamma)}(\mathcal{D}_q)$, and from $V_q^{s,(\gamma)}(\partial\mathcal{D})$ into $V^{s,(\gamma)}(\partial\mathcal{D}_q)$. It is easy to check that the operator

$$U_q^{-1} \mathcal{L}_\zeta U_q : V^{s,(\gamma)}(\mathcal{D}_q) \to V^{s-2m,(\gamma-2m)}(\mathcal{D}_q) \oplus_{k=1}^m V^{s-m_k-1/2,(\gamma-m_k)}(\partial\mathcal{D}_q)$$

is the operator of the boundary value problem defined by the operators

$$\hat{A}(q) = \sum_{|\alpha|\leq 2m} a_\alpha(x/q)\left(\frac{q}{q+i\gamma}\right)^\alpha D_x^\alpha,$$

$$\hat{B}_k(q) = \sum_{|\alpha|\leq m_k} b_{\alpha k}(x/q)\left(\frac{q}{q+i\gamma}\right)^\alpha D_x^\alpha.$$

The coefficients

$$a_\alpha(x/q)\left(\frac{q}{q+i\gamma}\right)^\alpha, b_{\alpha k}(x/q)\left(\frac{q}{q+i\gamma}\right)^\alpha \in SC_b^\infty(\mathbf{V}_{\mathcal{D}})$$

satisfy the conditions 2.2, 2.3 with respect to (x,q). Moreover, their limits

$$\lim_{(x,q)\to\eta_\omega} a_\alpha(x/q)\left(\frac{q}{q+i\gamma}\right)^\alpha = a_\alpha(\omega), \omega\in\overline{\mathcal{D}}$$

$$\lim_{(x,q)\to\eta_\omega} b_{\alpha k}(x/q)\left(\frac{q}{q+i\gamma}\right)^\alpha = b_{\alpha k}(\omega), \omega\in\overline{\mathcal{D}}$$

do not depend on x. Hence there is only one limit operator $\hat{A}^\omega\in\sigma_{\eta_\omega}(A)$, with constant coefficients

$$\hat{A}^\omega = \sum_{|\alpha|\leq 2m} a_\alpha^h(\omega)D_x^\alpha,\ \omega\in\overline{\mathcal{D}}.$$

The same situation holds for the operators $\hat{B}_k(q)$. For each $\hat{B}_k(q)$ and a point $\omega\in\partial\mathcal{D}$ there is only one limit operator

$$\hat{B}_k^\omega = \sum_{|\alpha|\leq m_k} b_{\alpha k}(\omega)D_x^\alpha,\ \omega\in\partial\mathcal{D}.$$

Hence, the invertibility of the family of operators $\mathcal{L}_q$ depending on the parameter $q>0$ for large values of q follows from Theorem 3.1.

Theorem 4.1. *The following statements are equivalent: 1) There exists $q_0 > 0$ such that the operators*

$$\mathcal{L}_\zeta : V_q^{s,(\gamma)}(\mathcal{D}) \to V_q^{s-2m,(\gamma-2m)}(\mathcal{D}) \oplus_{k=1}^m V_q^{s-m_k-1/2,(\gamma-m_k)}(\partial\mathcal{D})$$

are invertible if $q\geq q_0$ and

$$\sup_{q\geq q_0}\left\|\mathcal{L}_\zeta^{-1}\right\|_{Y_q\to X_q} < \infty;$$

2) (i) for each point $\omega\in\overline{\mathcal{D}}$ exists a $C_\omega > 0$ such that

$$\left|\sum_{|\alpha|\leq 2m} a_\alpha(\omega)\xi^\alpha\right| \geq C_\omega\left(1+|\xi|\right)^{2m},$$

ii) for each $\omega \in \partial D \setminus sing \ \partial D$ all operators $\mathfrak{L}_{\mathbf{H}^0_\omega}$

$$\mathfrak{L}_{\mathbf{H}^0_\omega} : H^s(\mathbf{H}^0_\omega) \to H^{s-2m}(\mathbf{H}^0_\omega) \oplus_{k=1}^m H^{s-m_k-1/2}(\partial \mathbf{H}^0_\omega),$$

of boundary value problems in the half-spaces $\mathbf{H}^0_\omega$, defined by the interior operator $\hat{A}^\omega$ and the boundary operators $\hat{B}^\omega_k, k = 1, ..., m$, are invertible, ii) for each singular point $\nu_j \in sing \ \partial D$ all operators $\mathfrak{L}_{\mathbf{K}_{v_j}}$ of boundary value problems in the cone $\mathbf{K}_{v_j}$

$$\mathfrak{L}_{\mathbf{K}_{v_j}} : V^{s,\gamma}(\mathbf{K}_{v_j}) \to \mathcal{V}^{s-2m,\gamma-2m}(\mathbf{K}_{v_j}) \oplus_{j=1}^m \mathcal{V}^{s-m_j-1/2,\gamma-m_j}(\partial \mathbf{K}_{v_j})$$

defined by the interior operator $\hat{A}^\omega$ and the boundary operators $\bar{B}^\omega_k, k = 1, ..., m$, are invertible;

Remark 4.1. The necessary and sufficient conditions for invertibility of the operator $\mathfrak{L}_{\mathbf{H}^0_\omega}$ well-known (see, for instance [AV64], [Es], [RS82]).

Remark 4.2. The problem of invertibility of the operator $\mathfrak{L}_{\mathbf{K}_{v_j}}$ is more complicated. The necessary conditions for invertibility of $\mathfrak{L}_{\mathbf{K}_{v_j}}$ in terms of eigenvalues of an operator pencil associated with the main homogeneous part of $\mathfrak{L}_{\mathbf{K}_{v_j}}$ are given in (see [Ko67], [KMR] and references given there). Some sufficient conditions of invertibility of the operator $\mathfrak{L}_{\mathbf{K}_{v_j}}$ are given in the book [KMR].

Acknowledgements

The work is supported by the CONACYT project 32424-E and RFFI project 98-01-01-023

References

[AV64]   Agranovich, M.S., Vishik, M.I., Elliptic problems with parameter and parabolic problems of general type. Uspekhi Mat.Nauk, 19 (1964) 3, 53-161

[Ko67]   Kondratiev, V.A., Boundary value problems in domains with conical and angular points, Trudy Mosk. Math. Soc. 16 (1967), 209-292

[KMR]   Kozlov, V.A., Maz'ya, V.G., Rossmann, J., Elliptic Boundary Value Problems in Domains with Point Singularities. Mathematical Surveys and Monographs, vol.52, AMS

[MNP91]   Maz'ya, V.G., Nazarov, S.A., Plamenevskii, B.A., Asymptotische Theorie elliptisher Randwertaufgaben in singulär gestörten Gebieten, Vol. 1,2, Akademie-Verlag, Berlin, 1991

[Sc94]   B.-W. Schulze, Pseudo-Differential Boundary Value Problems, Conical Singularities, and Asymptotics, Akademie-Verlag, Berlin, 1994.

[KR91]   Kozlov, V.A., Rossmann, J. , On the behavior of the spectrum of parameter-depending operators under small variations of the domain and application to operator pencils generated by elliptic boundary value problems in a cone, Math. Nachr. 153 (1991) 123-129

[Do76]   Doctorskii R.Ya., On the approximation of solutions of well-posed problems in unbounded domains for elliptic pseudodifferential operators, Soviet Math. Dokl. Vol.17 (1976), No.4, 1185-1188

[Do79]   Doctorskii R.Ya., Some sufficient conditions of invertibility and Noether properties of operators of boundary value problems. Sibirskii Mathem. Jornal, v.20, No.6, 1979, 1249-1260

[Sh73a]   Shimon L., On approximation of solutions of boundary problems in non-bounded domains, Differencialnie Uravnenia (in Russian), 1973 V. 9, No.8, 1482-1492

[Sh73b]   Shimon L., Matem. sbornik (in Russian), On approximation of solutions of boundary problems in domains with unbounded boundary, 1973, v.91, No.4, 488-500

[Ra99]   Rabinovich V.S. , Criterion for Local Invertibility of Pseudodifferential Operators with Operator Symbols and Some Applications. In the book: Proceedings of the St.Petersburg Math. Society, Vol.5, 1998. Amer.Math. Soc. Translations, Ser.2, Vol.193, 1999, 239-258.

[Ra93]   Rabinovich V.S., Fredholmness of the boundary value problems on non compact manifolds and limit operators. Russian Akad. Sciences. Doklady, Math., 46, 1993, no.153-58.

[Es]   G.I.Eskin, Boundary Value Problems for Elliptic Pseudodifferential Equations, Transl. of Math. Monogr., AMS, vol.52

[RS82]   S.Rempel, B-W. Schulze, Index Theory of Elliptic Boundary Problems, Akademie-Verlag, Berlin, 1982

[Gr70]   V.V. Grushin, *Pseudodifferential operators on $\mathbf{R}^n$ with bounded symbols*. Functionalniy Analiz i ego Prilojenia, **4**, No. 3 (1970), 37-50

Address

V. RABINOVICH, Departamento de Telecomunicaciones, ESIME del I.P.N., Unidad Zacatenco, Av. I.P.N., s/n Ed. 1, 07738 México, D.F., México

E-MAIL: rabinov@maya.esimez.ipn.mx

2000 Mathematics Subject Classification. Primary 35J40

Operator Theory:
Advances and Applications, Vol. 126
© 2001 Birkhäuser Verlag Basel/Switzerland

Local Algebra of a Non-Symmetric Corner

V. Rabinovich, B.-W. Schulze, and N. Tarkhanov

Abstract. The paper is aimed at describing a local algebra of boundary value problems near a corner point of the boundary. We show a symbol which controls the local solvability of the problem at the corner.

1 Introduction

Boundary value problems in domains with point singularities and smooth edges on the boundary are studied by many authors. The classical work here is that by Kondrat'ev [Kon67].

Domains with edges on the boundary intersecting at non-zero angles belong to the next step in the hierarchy of manifolds with singularities that consists of manifolds with corners. Stability under products is of primary importance, enough alone to justify the detailed study of manifolds with corners, because of the Schwartz kernel theorem.

In the literature there are several approaches to the analysis on manifolds with corners, cf. [MP77], [Mel87], [Sch92], [RST99]. As a necessary intermediate tool one requires parameter-dependent versions of calculi for lower order singularities.

So far one restricted the discussion to domains which behave symmetrically in a neighbourhood of the singularity. In the study of such problems one uses the parameter-dependent theory of [AV64].

This paper starts the study of boundary value problems in domains of anisotropic structure close to a corner on the boundary. Thus, we deal with a non-standard theory of boundary value problems depending on a parameter, cf. [Gru71,Gru72].

Our work is motivated by a recent question of V.A. Kondrat'ev, "What about the Dirichlet problem in the domain $\mathcal{D} = \{(x,y,z) \in \mathbb{R}^3 : z^2 \geq x^2 + y^4\}$?" Note that the origin is a singular point on the boundary of $\mathcal{D}$. It is actually a conical point, however, it goes beyond the class of conical points treated in the original paper [Kon67].

2 Typical Symbols

More generally, let $f(x)$ be a quasihomogeneous function of weights $(p_1, \ldots, p_n)$ and degree d in $\mathbb{R}^n$, i.e.,

$$f(r^{p_1} x_1, \ldots, r^{p_n} x_n) = r^d f(x)$$

for all $x \in \mathbb{R}^n \setminus \{0\}$ and $r > 0$. As but one example of such a function we show
$f(x) = (x_1^{2/p_1} + \ldots + x_n^{2/p_n})^{d/2}$.

Consider the domain $\mathcal{D}$ in $\mathbb{R}^{n+1}$ consisting of all $w = (w_1, \ldots, w_n, w_{n+1})$ with
$w_{n+1} \geq f(w_1, \ldots, w_n)$. The origin $w = 0$ is a singular point of $\partial\mathcal{D}$, namely a
corner in the hierarchy of singularities. Our goal is to describe a local algebra of
pseudodifferential operators in $\mathcal{D}$ close to $w = 0$.

As usual we apply a blow-up procedure at $w = 0$. To this end, we introduce
the generalised polar coordinates $w_j = r^{p_j} x_j$, for $j = 1, \ldots, n$, and $w_{n+1} = r^d$
(rescaling). Under this change of variables $\mathcal{D}$ transforms into the semicylinder
$\mathbb{R}_+ \times B$, where $B = \{x \in \mathbb{R}^n : f(x) \leq 1\}$.

The local geometry of $\mathcal{D}$ near $w = 0$ is now encoded in the induced Riemannian
metric on $\mathbb{R}_+ \times B$, which degenerates at $r = 0$ provided some p_j is positive.

Let us look at the pull-backs of differential operators under the change $w = \pi(r, x)$.
An easy calculation shows that

$$\begin{cases} D_{w_1} = r^{-p_1} D_{x_1}, \\ \quad \cdots\cdots\cdots \\ D_{w_n} = r^{-p_n} D_{x_n}, \\ D_{w_{n+1}} = \dfrac{1}{d} r^{-d} (r D_r - p_1 x_1 D_{x_1} - \ldots - p_n x_n D_{x_n}) \end{cases}$$

hence a differential operator $A = \sum_{|\alpha| \leq m} A_\alpha(w) D_w^\alpha$ with smooth coefficients near
$w = 0$ transforms to

$$\pi^* A = \frac{1}{r^{pm}} \sum_{j + |\beta| \leq m} A_{j,\beta}(r, x) \, (r^{p-p_1} D_{x_1})^{\beta_1} \ldots (r^{p-p_n} D_{x_n})^{\beta_n} (r^{p-d+1} D_r)^j$$

where $p = \max\{p_1, \ldots, p_n\}$, and the coefficients $A_{j,\beta}(r, x)$ still inherit "good"
properties of $A_\alpha(w)$ provided $p \geq d$.

We might say that the blow-up works well unless we actually have a regular case.

Degenerate differential operators were intensively studied by Grushin at the be-
ginning of the '70s, cf. [Gru71,Gru72]. This study was extended by Levendorskii in
the '80s and '90s, [Lev93]. Still the partial differential equations which stem from
the singularities of the underlying manifold fall out the classes already treated.

As usual $\pi^* A$ is completed by boundary conditions on $\mathbb{R}_+ \times \partial B$, which corresponds
to imposing boundary conditions away from the singular point on $\partial\mathcal{D}$. We will
restrict our attention to $\pi^* A$ itself, because the boundary operators transform
similarly.

3   Quantisation

The standard way is to microlocalise the singular derivative $r^{p-d-1}D_r$ and invert
the corresponding family of boundary value problems on B. To this end we make
the change of variables $t = \delta(r)$, where

$$\delta(r) = \int_{r_0}^{r} \frac{d\theta}{\theta^{p-d+1}}$$

$$= \frac{1}{d-p}\left(r^{d-p} - r_0^{d-p}\right)$$

for small $r > 0$. We may neglect the constant and take $\delta(r) = r^{d-p}/(d-p)$ for
small r. If $d = p$ we consider $\delta(r) = \log r$.

Under this change of variables the operator $\pi^* A$ in turn transforms to a differential
operator near the base $t = \delta(0)$ of the cylinder $(\delta(0), \infty) \times B$,

$$\frac{1}{t^{\frac{pm}{d-p}}} \sum_{j+|\beta|\leq m} A_{j,\beta}(t^{\frac{1}{d-p}}, x)\,(t^{\frac{p-p_1}{d-p}} D_{x_1})^{\beta_1} \ldots (t^{\frac{p-p_n}{d-p}} D_{x_n})^{\beta_n} D_t^j.$$

If $d - p > 0$ then $\delta(0) = 0$, i.e., $t = \delta(r)$ is a diffeomorphism of $\mathbb{R}_+$ onto $\mathbb{R}_+$. Hence
the operator needs boundary conditions on the base $t = 0$, which corresponds to
the "regular" case of usual Sobolev spaces and boundary conditions up to and
including $w = 0$, cf. [VE67].

If $d - p \leq 0$ then δ takes $r = 0$ to $t = -\infty$. In this case relevant function tools are
weighted Sobolev spaces at the cylindrical end. Namely,

$$\|U\|^2_{H^{s,\gamma,\mu}(\mathbb{R}\times B)} = \int_{\mathbb{R}} e^{-2\gamma t} t^{2\mu} \sum_{j+|\beta|\leq s} \int_B |(t^{q_1} D_{x_1})^{\beta_1} \ldots (t^{q_n} D_{x_n})^{\beta_n} D_t^j U|^2 dt dx$$

where

$$q_j = \frac{p - p_j}{d - p}$$

for $j = 1,\ldots,n$. The pull-backs of these spaces under $t = \delta(r)$ yield a scale of
weighted Sobolev near the base $r = 0$ of $\mathbb{R}_+ \times B$,

$$\|u\|^2_{\mathcal{H}^{s,\gamma,\mu}(\mathbb{R}_+\times B)} = \int_{\mathbb{R}} e^{-2\gamma\delta} \delta^{2\mu} \sum_{j+|\beta|\leq s} \int_B |(r^{p-p_1} D_{x_1})^{\beta_1} \ldots (r^{p-p_n} D_{x_n})^{\beta_n} \left(\frac{1}{\delta'} D_r\right)^j u|^2 d\delta dx.$$

The weight function $e^{-\gamma\delta(r)}$ is very essential in the analysis near cuspidal points.
The weight function $(\delta(r))^\mu$ is dominated by the former, and the exponent μ does
not affect the Fredholm property, cf. [RST00,RST98,RST99].

It is now clear that the local algebra at $w = 0$ is simply the algebra of pseudodifferential operators with operator-valued symbols near $t = -\infty$, pulled back by $r \mapsto t = \delta(r)$. The elements of this latter are Fourier integral operators on the semiaxis

$$\mathcal{A}u\,(r) = \frac{1}{2\pi} \int_{\mathbb{R} - \imath\gamma} dz \int_{\mathbb{R}_+} e^{\imath(\delta(r) - \delta(r'))z}\, a(r, z)\, u(r')\, d\delta(r'),$$

where $a(r, z)$ takes the values in the algebra of boundary value problems on B. In our case

$$a(r, z) = \frac{1}{r^{pm}} \sum_{j + |\beta| \leq m} A_{j,\beta}(r, x)\, (r^{p - p_1} D_{x_1})^{\beta_1} \ldots (r^{p - p_n} D_{x_n})^{\beta_n} z^j$$

completed by a family of boundary conditions on ∂B. The factor $1/r^{pm}$ can be traced back to the weighted Sobolev spaces, and so we omit it. Then $\mathcal{A} = \mathrm{op}\,(a)$ acts by

$$\mathcal{A}:\ H^{s,\gamma,\mu}(\mathbb{R}_+ \times B) \to \begin{array}{c} H^{s-m,\gamma,\mu}(\mathbb{R}_+ \times B) \\ \oplus \\ \oplus_i H^{s-m_i-1/2,\gamma,\mu}(\mathbb{R}_+ \times \partial B), \end{array}$$

m_i standing for the orders of boundary operators.

The local diffeomorphism $t = \delta(r)$ gives an explicit link of the "smoothness" of the coefficients of $\mathcal{A}$ near $r = 0$ and the behaviour of the symbols of pseudodifferential operators when $t \to -\infty$.

4   Local Invertibility

Recall that $\mathcal{A}$ is said to be *locally invertible* at $w = 0$ if there is a cut-off function $\chi(r)$ and bounded operators $\mathcal{B}'$, $\mathcal{B}''$ acting in the opposite direction, such that $\mathcal{B}'\mathcal{A}\chi = \chi$ and $\chi\,\mathcal{A}\mathcal{B}'' = \chi$.

Then the main result is that the local invertibility of $\mathcal{A}$ at $w = 0$ is equivalent to the invertibility of the symbol $\sigma(\mathcal{A})$ of $\mathcal{A}$ at $w = 0$, the latter acting in weighted Sobolev spaces on B with parameter $r \ll 1$,

$$\|u\|^2_{H^s_r(B)} = \int_B \sum_{|\beta| \leq s} |(r^{p-p_1} D_{x_1})^{\beta_1} \ldots (r^{p-p_n} D_{x_n})^{\beta_n} u|^2 dx.$$

Theorem 4.1. *Suppose s is an integer $\geq m$. Then $\mathcal{A}$ is locally invertible at $w = 0$ if and only if*

$$\sigma(\mathcal{A})(r, z):\ H^s_r(B) \to \begin{array}{c} H^{s-m}_r(B) \\ \oplus \\ \oplus_i H^{s-m_i-1/2}_r(\partial B), \end{array} \tag{4.1}$$

is invertible for all $r \ll 1$ and $z \in \mathbb{R} - \imath\gamma$, and the inverse is bounded uniformly in r and z.

Changing the variables by $x_j = r^{p-p_j} y_j$, $j = 1, \ldots, n$, we can move the parameter r from $\sigma(\mathcal{A})(r, z)$ and Sobolev spaces to the domain. Namely, the condition (4.1) just amounts to the fact that

$$\left(\sum_{j+|\beta| \leq m} A_{j,\beta}(0,0)\, D_y^\beta\, z^j \;+\; \text{boundary family} \right) : H^s(B_r) \to \begin{array}{c} H^{s-m}(B_r) \\ \oplus \\ \oplus_i H_r^{s-m_i-1/2}(\partial B_r), \end{array}$$

has a bounded inverse uniformly in $r \ll 1$ and $z \in \mathbb{R} - \imath\gamma$, where

$$B_r = \{ y \in \mathbb{R}^n : \ f(r^p y) \leq r^d \}.$$

Example 4.1. Let $\mathcal{D} = \{(x, y, z) \in \mathbb{R}^3 : \ z^2 \geq x^2 + y^4\}$. Then $p_1 = 1$, $p_2 = 1/2$ and $d = 1$. In this case $B_r = \{(x,y) \in \mathbb{R}^2 : \ x^2 + r^2 y^4 \leq 1\}$ expands to the strip $[-1, 1] \times \mathbb{R}$ along the y-axis.

References

[AV64]   M. S. Agranovich and M. I. Vishik, *Elliptic problems with parameter and parabolic problems of general type*, Uspekhi Mat. Nauk **19** (1964), no. 3, 53–160.

[Gru71]   V. V. Grushin, *On a class of elliptic pseudodifferential operators degenerate on a submanifold*, Math. USSR Sbornik **13** (1971), no. 2, 155–183.

[Gru72]   V. V. Grushin, *Hypoelliptic differential equations and pseudodifferential operators with operator-valued symbols*, Math. USSR Sbornik **17** (1972), no. 2, 497–517.

[Kon67]   V. A. Kondrat'ev, *Boundary value problems for elliptic equations in domains with conical points*, Trudy Mosk. Mat. Obshch. **16** (1967), 209–292.

[Lev93]   S. Levendorskii, *Degenerate Elliptic Equations*, Kluwer Academic Publishers, Dordrecht NL, 1993.

[Mel87]   R. B. Melrose, *Pseudodifferential Operators on Manifolds with Corners*, Manuscript MIT, Boston, 1987.

[MP77]   V. G. Maz'ya and B. A. Plamenevskii, *Elliptic boundary value problems on manifolds with singularities*, Problems of Mathematical Analysis, Vol. 6, Univ. of Leningrad, 1977, pp. 85–142.

[RST00]   V. Rabinovich, B.-W. Schulze, and N. Tarkhanov, *A calculus of boundary value problems in domains with non-Lipschitz singular points*, Math. Nachr. **215** (2000), 115–160.

[RST98]   V. Rabinovich, B.-W. Schulze, and N. Tarkhanov, *Boundary Value Problems in Cuspidal Wedges*, Preprint 24, Univ. Potsdam, October 1998. 66 pp.

[RST99]   V. Rabinovich, B.-W. Schulze, and N. Tarkhanov, *Boundary Value Problems in Domains with Corners*, Preprint 99/19, Univ. Potsdam, September 1999, 32 pp.

[Sch92]   B.-W. Schulze, *The Mellin pseudodifferential calculus on manifolds with corners*, In: Symposium "Analysis on Manifolds with Singularities," Breitenbrunn, 1990, Teubner-Verlag, Leipzig, 1992, pp. 208–289.

[VE67]   M. I. Vishik and G. I. Eskin, *Elliptic equations in convolution in a bounded domain and their applications*, Uspekhi Mat. Nauk **22** (1967), 15–76.

Addresses

VLADIMIR RABINOVICH, Departamento de Telecomunicaciones, ESIME del
I.P.N., Unidad Zacatenco, Av. I.P.N., s/n Ed. 1, 07738 México, D.F.,
México

E-MAIL: rabinov@maya.esimez.ipn.mx

BERT-WOLFGANG SCHULZE, Universität Potsdam, Institut für Mathematik, Postfach 60 15 53, 14415 Potsdam, Germany

E-MAIL: schulze@math.uni-potsdam.de

NIKOLAI TARKHANOV, Universität Potsdam, Institut für Mathematik, Postfach 60 15 53, 14415 Potsdam, Germany

E-MAIL: tarkhan@math.uni-potsdam.de

2000 Mathematics Subject Classification. Primary 35S05; Secondary 35S15, 46E40

Operator Theory:
Advances and Applications, Vol. 126
© 2001 Birkhäuser Verlag Basel/Switzerland

The Integrated Density of States for a Random Schrödinger Operator in Strong Magnetic Fields. II. Asymptotics near Higher Landau Levels

GEORGI D. RAIKOV

Abstract. We consider the three-dimensional Schrödinger operator with strong constant magnetic field and random electric potential. We investigate the asymptotic behaviour of its integrated density of states near the qth Landau level, for any fixed $q > 1$.

1 Introduction

In this paper we consider the three-dimensional Schrödinger operator with constant magnetic field and random scalar potential, and analyze the asymptotic behaviour of its integrated density of states (IDOS) as the intensity b of the magnetic field tends to infinity. The paper should be regarded as a continuation of [KR00] where we studied the asymptotics as $b \to \infty$ of the IDOS near the *first* Landau level, while here we consider the same type of asymptotics near the qth Landau level, $q > 1$. Here we recall briefly the basic definitions from [KR00].
Let $\mathbf{b} := (0, 0, b)$, $b > 0$, $\mathbf{x} = (x, y, z) \in \mathbb{R}^3$. Introduce the unperturbed self-adjoint Schrödinger operator

$$H_0(b) := \left(i\nabla + \frac{\mathbf{b} \wedge \mathbf{x}}{2}\right)^2 \equiv \left(i\frac{\partial}{\partial x} - \frac{by}{2}\right)^2 + \left(i\frac{\partial}{\partial y} + \frac{bx}{2}\right)^2 - \frac{\partial^2}{\partial z^2}, \qquad (1.1)$$

defined originally on $C_0^\infty(\mathbb{R}^3)$, and then closed in $L^2(\mathbb{R}^3)$. We have

$$\sigma(H_0(b)) = [b, +\infty), b > 0, \qquad (1.2)$$

where $\sigma(H_0(b))$ denotes the spectrum of the operator $H_0(b)$ (see e.g. [AHS78]). Let $(\Omega, \mathcal{F}, \mathbb{P})$ be a probability space, and $V_\omega(\mathbf{x})$, $\omega \in \Omega$, $\mathbf{x} \in \mathbb{R}^3$, be a real random field. We assume that V_ω is $\mathbb{G}^3$-ergodic with $\mathbb{G} = \mathbb{Z}$ or $\mathbb{G} = \mathbb{R}$ (see [Kir89, Section 3.1]). In other words, there exists an ergodic group of measure preserving automorphisms $T_\mathbf{k} : \Omega \to \Omega$, $\mathbf{k} \in \mathbb{G}^3$, such that $V_\omega(\mathbf{x} + \mathbf{k}) = V_{T_\mathbf{k}\omega}(\mathbf{x})$ for $\mathbf{x} \in \mathbb{R}^3$ and $\omega \in \Omega$. We recall that ergodicity of a group G of automorphisms of Ω means that the G-invariance of a given set $\mathcal{A} \in \mathcal{F}$ implies either $\mathbb{P}(\mathcal{A}) = 1$ or $\mathbb{P}(\mathcal{A}) = 0$.
For $\mathbf{x} \in \mathbb{R}^3$ we write $\mathbf{x} = (X, z)$ with $X \in \mathbb{R}^2$, $z \in \mathbb{R}$. Hence, z is the variable along the magnetic field $\mathbf{b} = (0, 0, b)$, while X runs over the plane perpendicular to $\mathbf{b}$. We suppose that V_ω is $\mathbb{G}$-ergodic with $\mathbb{G} = \mathbb{Z}$ or $\mathbb{G} = \mathbb{R}$ in the direction of the

magnetic field, i.e. that the subgroup $\{\mathcal{T}_{\mathbf{k}}|\mathbf{k} = (0,0,k),\ k \in \mathbb{G}\}$ is ergodic. Further, we assume that the realizations of V_ω are almost surely uniformly bounded, i.e.

$$c_0 := \mathrm{ess} - \sup_{\omega \in \Omega}\ \sup_{\mathbf{x} \in \mathbb{R}^3} |V_\omega(\mathbf{x})| < \infty. \tag{1.3}$$

Finally, we suppose that the realizations of V_ω are almost surely continuous.

Let T be a selfadjoint operator in a Hilbert space. Denote by $P_{\mathcal{I}}(T)$ its spectral projection corresponding to the interval $\mathcal{I} \subset \mathbb{R}$. Set $N(\lambda; T) := \mathrm{rank}\, P_{(-\infty,\lambda)}(T)$, $\lambda \in \mathbb{R}$. If $T = T^*$ is compact, put $n_\pm(s; T) := \mathrm{rank}\, P_{(s,+\infty)}(\pm T)$, $s > 0$. Finally, if T is a linear compact operator in Hilbert space which is not necessarily self-adjoint, set $n_*(s; T) := \mathrm{rank}\, P_{(s^2,+\infty)}(T^*T)$, $s > 0$.

On $D(H_0(b))$ define the perturbed Schrödinger operator $H(b,\omega) := H_0(b) + V_\omega$. On the Sobolev space $\mathrm{H}^2\left(\left(-\frac{R}{2}, \frac{R}{2}\right)^3\right)$ with Dirichlet boundary conditions, define the operator $H_{0,R}^D(b) := \left(i\nabla + \frac{\mathbf{b} \wedge \mathbf{x}}{2}\right)^2$. Then there exists a non-random non-decreasing function $\mathcal{D}_b : \mathbb{R} \to \mathbb{R}_+$ such that almost surely

$$\lim_{R \to \infty} R^{-3}\, N(\mu; H_{0,R}^D(b) + V_\omega) = \mathcal{D}_b(\mu), \tag{1.4}$$

provided that $\mu \in \mathbb{R}$ is a continuity point of $\mathcal{D}_b$ (see [Nak00], [HLMW00]). The function $\mathcal{D}_b(\mu)$, $\mu \in \mathbb{R}$, is called the IDOS for the operator $H(b,\omega)$.
In this paper we consider the asymptotic behaviour as $b \to \infty$ of $\mathcal{D}_b(\lambda_2 + (2q-1)b) - \mathcal{D}_b(\lambda_1 + (2q-1)b)$, the parameters $\lambda_1, \lambda_2 \in \mathbb{R}$, $\lambda_1 < \lambda_2$, and $q \in \mathbb{N}_* := \{1, 2, \ldots\}$, being fixed. Recall that the numbers $\{(2q-1)b\}_{q=1}^\infty$ are called Landau levels.

2   Statement of Main Result

Let $h_{0,R} := -\frac{d^2}{dz^2}$ be the self-adjoint operator defined on $\mathrm{H}^2\left(\left(-\frac{R}{2}, \frac{R}{2}\right)\right)$ with Dirichlet boundary conditions.

Proposition 2.1. ([Kir89, Chapter7], [PF92, Chapter III]) *Let* $\mathbb{G} = \mathbb{Z}$ *or* $\mathbb{G} = \mathbb{R}$. *Let* $f_\omega(z)$, $\omega \in \Omega$, $z \in \mathbb{R}$, *be a real* $\mathbb{G}$-*ergodic random field whose realizations are almost surely uniformly bounded and continuous. Then for each* $\lambda \in \mathbb{R}$ *the limit*

$$\varrho(\lambda; f) := \lim_{R \to \infty} R^{-1}\, N(\lambda; h_{0,R} + f_\omega) \tag{2.1}$$

exists almost surely. Moreover, the function $\varrho(\lambda; f)$ *is non-random, and continuous with respect to* $\lambda \in \mathbb{R}$.

Our assumptions concerning V_ω guarantee that the random field $f_\omega = V_\omega(X, .)$ depending on the parameter $X \in \mathbb{R}^2$, satisfies the hypotheses of Proposition 2.1.

Moreover, if $\mathbb{G} = \mathbb{Z}$, then the function $\varrho(\lambda; V(X, .))$ is periodic with respect to $X \in \mathbb{R}^2$, while in the case $\mathbb{G} = \mathbb{R}$ the quantity $\varrho(\lambda; V(X, .))$ is independent of $X \in \mathbb{R}^2$ (see [KR00]). For $\lambda \in \mathbb{R}$ set

$$k(\lambda) = k(\lambda; V) := \begin{cases} \int_{(-\frac{1}{2}, \frac{1}{2})^2} \varrho(\lambda, V(X, .)) \, dX & \text{if} \quad \mathbb{G} = \mathbb{Z} \\ \varrho(\lambda, V(0, .)) & \text{if} \quad \mathbb{G} = \mathbb{R}. \end{cases}$$

Obviously, $k(\lambda)$ is non-decreasing and continuous with respect to λ.

Theorem 2.1. *Let $\mathbb{G} = \mathbb{Z}$ or $\mathbb{G} = \mathbb{R}$. Let V_ω be a real $\mathbb{G}^3$-ergodic random field whose realizations are almost surely uniformly bounded and continuous. Assume in addition that V_ω is $\mathbb{G}$-ergodic in the direction of the magnetic field. Then for each $q \in \mathbb{N}_*$, and $\lambda_1, \lambda_2 \in \mathbb{R}$, $\lambda_1 < \lambda_2$, we have*

$$\lim_{b \to \infty} b^{-1} \left(\mathcal{D}_b(\lambda_2 + (2q - 1)b) - \mathcal{D}_b(\lambda_1 + (2q - 1)b) \right) = \frac{1}{2\pi} \left(k(\lambda_2) - k(\lambda_1) \right). \quad (2.2)$$

Theorem 2.1 contains the asymptotics as $b \to \infty$ of the IDOS $\mathcal{D}_b$ near the qth Landau level, $q \geq 1$. Since (2.2) has been proved in [KR00] for $q = 1$, we shall prove it here for $q > 1$. The methods applied in this paper are similar to the ones used in [KR00], and are based on the Birman-Schwinger principle (see [Bir66, Lemma 1.1]), a suitable version of the Kac-Murdock-Szegö theorem (see [Rai99, Lemma 3.2]), and the Birkhoff-Khintchine ergodic theorem. However, the analysis near higher Landau levels is more complicated since the first Landau level coincides with lower bound of the spectrum of $H_0(b)$ (see (1.2)), while the higher Landau levels $(2q - 1)b$, $q > 1$, are internal points of $\sigma(H_0(b))$. The proof of Theorem 2.1 can be found in Section 4, while Section 3 contains preliminary estimates.

3 Preliminary Estimates

3.1. Let $H_{0,R}^N(b)$ be the self-adjoint operator generated in $L^2\left(\left(-\frac{R}{2}, \frac{R}{2}\right)^3\right)$ by the closed quadratic form $\int_{(-\frac{R}{2}, \frac{R}{2})^3} \left| i\nabla u + \frac{\mathbf{b} \wedge \mathbf{x}}{2} u \right|^2 \, d\mathbf{x}$, $u \in \mathrm{H}^1\left(\left(-\frac{R}{2}, \frac{R}{2}\right)^3\right)$.

Lemma 3.1. ([Nak00, Theorem 1], [HLMW00, Theorem 3.1]) *Let $\mathbb{G} = \mathbb{Z}$ or $\mathbb{G} = \mathbb{R}$. Assume that V_ω is a real $\mathbb{G}^3$-ergodic random field whose realizations are almost surely uniformly bounded and continuous. Let $\mu \in \mathbb{R}$ be a continuity point of $\mathcal{D}_b$. Then almost surely*

$$\mathcal{D}_b(\mu) = \lim_{R \to \infty} R^{-3} \, N(\mu; H_{0,R}^N(b) + V_\omega). \quad (3.1)$$

Set $\chi_R(\mathbf{x}) := \mathbf{1}_{(-\frac{R}{2}, \frac{R}{2})^3}(\mathbf{x})$, $\mathbf{x} \in \mathbb{R}^3$. On $D(H_0(b))$ introduce the operator $H_0(b) + (V_\omega - \mu)\chi_R$ with $b > 0$, $\omega \in \Omega$, $\mu \in \mathbb{R}$, $R > 0$.

Proposition 3.1. *Under the hypotheses of Lemma 3.1 almost surely*

$$\mathcal{D}_b(\mu) = \lim_{R\to\infty} R^{-3} N(0; H_0(b) + (V_\omega - \mu)\chi_R). \tag{3.2}$$

Proof. By the minimax principle,

$$N(\mu; H_{0,R}^D + V_\omega) = N(0; H_{0,R}^D + V_\omega - \mu) \leq N(0; H_0(b) + (V_\omega - \mu)\chi_R). \tag{3.3}$$

Set $\mathcal{O}_R := \mathbb{R}^3 \setminus \left[-\frac{R}{2}, \frac{R}{2}\right]^3$. Denote by $\tilde{H}_{0,R}^N$ the self-adjoint operator generated by the quadratic form $\int_{\mathcal{O}_R} \left|i\nabla u + \frac{\mathbf{b}\wedge\mathbf{x}}{2}u\right|^2 d\mathbf{x}$, defined initially for $u \in C_0^\infty\left(\overline{\mathcal{O}_R}\right)$, and then closed in $L^2(\mathcal{O}_R)$. The minimax principle implies

$$N(0; H_0(b) + (V_\omega - \mu)\chi_R) \leq N(0; H_{0,R}^N + V_\omega - \mu) + N(0; \tilde{H}_{0,R}^N). \tag{3.4}$$

Since $\tilde{H}_{0,R}^N \geq 0$, we have $N(0; \tilde{H}_{0,R}^N) = 0$. Hence, (3.4) can be re-written as

$$N(0; H_0(b) + (V_\omega - \mu)\chi_R) \leq N(0; H_{0,R}^N + V_\omega - \mu) = N(\mu; H_{0,R}^N + V_\omega). \tag{3.5}$$

Combining (3.3) and (3.5) with (1.4) and (3.1), we get (3.2). $\Diamond$

Remark. Proposition 3.1 is very similar to [KR00, Proposition 4.1]. However, the proof presented here is much simpler because now we dispose of Lemma 3.1.

Introduce the compact Birman-Schwinger-type operators

$$T(\mu) = T_{b,\omega,R}(\mu) := H_0(b)^{-1/2}(V_\omega - \mu)\chi_R H_0(b)^{-1/2}, \ \mu \in \mathbb{R}, \tag{3.6}$$

$$\tilde{T} = \tilde{T}_{b,R} := H_0(b)^{-1/2}\chi_R H_0(b)^{-1/2}, \tag{3.7}$$

so that we have $T(\mu) = T(0) - \mu\tilde{T}$.

Corollary 3.1. *Under the assumptions of Lemma 3.1 almost surely*

$$\mathcal{D}_b(\mu) = \lim_{R\to\infty} R^{-3} n_-(1; T_{b,\omega,R}(\mu)). \tag{3.8}$$

Proof. It suffices to recall (3.2), and to apply the Birman-Schwinger principle. $\Diamond$

3.2. Let $\mathcal{H}_0(b) := \left(i\frac{\partial}{\partial x} - \frac{by}{2}\right)^2 + \left(i\frac{\partial}{\partial y} + \frac{bx}{2}\right)^2$ be the selfadjoint operator defined originally on $C_0^\infty(\mathbb{R}^2)$, and then closed in $L^2(\mathbb{R}^2)$. The spectrum of $\mathcal{H}_0(b)$ coincides with the set of the Landau levels, i.e. $\sigma(\mathcal{H}_0(b)) = \bigcup_{q=1}^\infty\{(2q-1)b\}$. Fix $q \geq 1$. Denote by $p_q = p_{q,b} : L^2(\mathbb{R}^2) \to L^2(\mathbb{R}^2)$ the orthogonal projection onto the eigenspace of $\mathcal{H}_0(b)$ associated with the qth Landau level $(2q-1)b$. In other words, $p_q w = w$ implies $w \in D(\mathcal{H}_0(b))$ and $\mathcal{H}_0(b)w = (2q-1)bw$. It is well-known that

$$(p_q w)(x,y) = \int_{\mathbb{R}^2} \mathcal{P}_q(x,y;x',y')w(x',y')\,dx'dy', \quad w \in L^2(\mathbb{R}^2),$$

with

$$\mathcal{P}_q(x, y; x', y') :=$$

$$\frac{b}{2\pi} e^{-\frac{b}{4}\left[(x-x')^2+(y-y')^2+2i(xy'-yx')\right]} L_{q-1}\left(\frac{(x-x')^2 + (y-y')^2}{2}\right) \tag{3.9}$$

where $L_s(\xi) := \frac{1}{s!} e^\xi \frac{d^s}{d\xi^s}\left(\xi^s e^{-\xi}\right)$, $s \geq 0$, $\xi \in \mathbb{R}$, is the Laguerre polynomial of order s. Note that we have

$$\mathcal{P}_q(x, y; x, y) = \frac{b}{2\pi}, \quad (x, y) \in \mathbb{R}^2, \ \forall q \in \mathbb{N}_*. \tag{3.10}$$

Define the orthogonal projection $P_q = P_{q,b} : L^2(\mathbb{R}^3) \to L^2(\mathbb{R}^3)$ by $P_q := \int_{\mathbb{R}}^{\oplus} p_q \, dz$, i.e.

$$(P_q u)(x, y, z) = \int_{\mathbb{R}^2} \mathcal{P}_q(x, y; x', y') u(x', y', z) \, dx' dy', \quad u \in L^2(\mathbb{R}^3).$$

Then P_q commutes with $H_0(b)$ and $\frac{\partial}{\partial z}$, and we have

$$H_0(b) P_q u = \left(-\frac{\partial^2}{\partial z^2} + (2q - 1)b\right) P_q u, \quad u \in D(H_0(b)), \tag{3.11}$$

(see (1.1)). For $\gamma > 0$ define the operator

$$r(\gamma) := \left(-\frac{\partial^2}{\partial z^2} + \gamma\right)^{-1/2}, \tag{3.12}$$

bounded and selfadjoint in $L^2(\mathbb{R}^3)$. Evidently,

$$(r(\gamma)^2 u)(x, y, z) = \frac{1}{2\sqrt{\gamma}} \int_{\mathbb{R}} e^{-\sqrt{\gamma}|z-z'|} u(x, y, z') \, dz', \quad u \in L^2(\mathbb{R}^3). \tag{3.13}$$

Moreover, the operators P_q and $r(\gamma)$ commute.

3.3. In this subsection we estimate a quantity which yields the main asymptotic term as $b \to \infty$ of $\mathcal{D}_b(\lambda_2 + (2q - 1)b) - \mathcal{D}_b(\lambda_1 + (2q - 1)b)$.

Proposition 3.2. *Let the hypotheses of Theorem 2.1 hold. Then for every $\lambda \in \mathbb{R}$, $q \in \mathbb{N}_*$, $s > 0$, and $\gamma > 0$ almost surely*

$$\lim_{b\to\infty} \lim_{R\to\infty} \frac{1}{bR^3} n_-(s; r(\gamma) P_q(V_\omega - \lambda - \gamma)\chi_R P_q r(\gamma)) = \frac{1}{2\pi} k\left(\frac{\lambda + \gamma}{s} - \gamma; \frac{1}{s} V\right). \tag{3.14}$$

Idea of the proof. The proof is analogous to the one of [KR00, Corollary 4.3] which corresponds to $q = 1$ and $\gamma = 1$. By (3.9), (3.10), and (3.13), the extension to general q and γ can be carried out in a quite straightforward manner. $\Diamond$

Let $\phi : [b_0, +\infty) \to \mathbb{R}_+$, $b_0 > 0$, be a strictly decreasing function such that $\phi(b_0) < 1$ and $\lim_{b\to\infty} b\phi(b) = 0$.

Corollary 3.2. *Let the hypotheses of Theorem 2.1 hold. Then for every* $\lambda \in \mathbb{R}$, $q \in \mathbb{N}_*$, *almost surely*

$$\lim_{b \to \infty} \lim_{R \to \infty} b^{-1} R^{-3} n_-(1 \pm \phi(b); P_q T(\lambda + (2q-1)b) P_q) = \frac{1}{2\pi} k(\lambda; V), \qquad (3.15)$$

the operator $T(\mu)$ *being defined in (3.6).*

Proof. First of all note that

$$P_q T(\lambda + (2q-1)b) P_q = r((2q-1)b) P_q (V_\omega - \lambda - (2q-1)b) \chi_R P_q r((2q-1)b)$$

(see (3.6), (3.11) and (3.12)). Then the Birman-Schwinger principle entails

$$n_-(1 \pm \phi(b); P_q T(\lambda + (2q-1)b) P_q)) =$$

$$N\left(0; -\frac{\partial^2}{\partial z^2} + (2q-1)b + \varphi_\pm(b) P_q (V_\omega - \lambda - (2q-1)b) \chi_R P_q\right), \qquad (3.16)$$

where $\varphi_\pm(b) := 1/(1 \pm \phi(b))$. Note that $\lim_{b \to \infty} \varphi_\pm(b) = 1$, $\lim_{b \to \infty} b(1 - \varphi_\pm(b)) = 0$. Let us re-arrange the terms appearing at the right-hand side of (3.16). We have

$$-\frac{\partial^2}{\partial z^2} + (2q-1)b + \varphi_\pm(b) P_q (V_\omega - \lambda - (2q-1)b) \chi_R P_q = -\frac{\partial^2}{\partial z^2} + \gamma + E_{q,b} + W_{q,b}^\pm, \quad (3.17)$$

where $\gamma > 0$ is a fixed number, $E_{q,b} := ((2q-1)b - \gamma)(1 - P_q \chi_R P_q)$, and $W_{q,b}^\pm := P_q \{V_\omega - \lambda - \gamma - (1 - \varphi_\pm(b))(V_\omega - \lambda - (2q-1)b)\} \chi_R P_q$. Fix $\delta > 0$, and assume that b is so large that we have

$$P_q(V_\omega - \lambda - \gamma - \delta)\chi_R P_q \leq W_{q,b}^\pm \leq P_q(V_\omega - \lambda - \gamma + \delta)\chi_R P_q. \qquad (3.18)$$

Assume as well that $b > \gamma/(2q-1)$ so that the operator $E_{q,b}$ is non-negative. By the minimax principle, (3.16) – (3.18) imply

$$N\left(0; -\frac{\partial^2}{\partial z^2} + \gamma + E_{q,b} + P_q(V_\omega - \lambda - \gamma + \delta)\chi_R P_q\right) \leq$$

$$n_-(1 \pm \phi(b); P_q T(\lambda + (2q-1)b) P_q) \leq N(0; -\frac{\partial^2}{\partial z^2} + \gamma + P_q(V_\omega - \lambda - \gamma - \delta)\chi_R P_q).$$
$$(3.19)$$

By the Birman-Schwinger principle,

$$N\left(0; -\frac{\partial^2}{\partial z^2} + \gamma + P_q(V_\omega - \lambda - \gamma - \delta)\chi_R P_q\right) =$$

$$n_-(1; r(\gamma) P_q(V_\omega - \lambda - \gamma - \delta)\chi_R P_q r(\gamma)). \qquad (3.20)$$

It follows from (3.14) with $s = 1$ that

$$\lim_{b\to\infty}\lim_{R\to\infty}\frac{1}{bR^3}n_-(1;r(\gamma)P_q(V_\omega - \lambda - \gamma - \delta)\chi_R P_q r(\gamma)) = \frac{1}{2\pi}k(\lambda + \delta; V). \quad (3.21)$$

The second inequality in (3.19), and (3.20) - (3.21) imply that for any $\hat{c} > 0$

$$\limsup_{b\to\infty}\limsup_{R\to\infty}\frac{1}{bR^3}n_-(1 \pm \phi(b); P_q T(\lambda + (2q - 1)b)P_q) \leq \frac{1}{2\pi}k(\lambda + \delta; V). \quad (3.22)$$

Now assume that $\gamma > 0$ is so large that almost surely $V_\omega(\mathbf{x}) - \lambda - \gamma + \delta \leq 0$ for every $\mathbf{x} \in \mathbb{R}^3$. Then we have $P_q(V_\omega - \lambda - \gamma + \delta)\chi_R P_q = -S^*S$ with $S := U\chi_R P_q$, and $U := \sqrt{\gamma + \lambda - V_\omega - \delta}$. Introduce the operator $\tilde{r}(\gamma) := \left(-\frac{\partial^2}{\partial z^2} + \gamma + E_{q,b}\right)^{-1/2}$, bounded and self-adjoint in $L^2(\mathbb{R}^3)$. By the Birman-Schwinger principle,

$$N\left(0; -\frac{\partial^2}{\partial z^2} + \gamma + E_{q,b} + P_q(V_\omega - \lambda - \gamma + \delta)\chi_R P_q\right) = n_+(1;\tilde{r}(\gamma)S^*S\tilde{r}(\gamma)) =$$

$$n_*(1; S\tilde{r}(\gamma)) = n_*(1; \tilde{r}(\gamma)S^*) = n_+(1; S\tilde{r}(\gamma)^2 S^*). \quad (3.23)$$

Applying the resolvent identity $\tilde{r}(\gamma)^2 = r(\gamma)^2 - r(\gamma)^2 E_{q,b}\tilde{r}(\gamma)^2$, we get

$$n_+(1; S\tilde{r}(\gamma)^2 S^*) \geq n_+(1 + \varepsilon; Sr(\gamma)^2 S^*) - n_+(\varepsilon; Sr(\gamma)^2 E_{q,b}\tilde{r}(\gamma)^2 S^*), \ \forall \varepsilon > 0. \quad (3.24)$$

Let us estimate the first term at the right-hand side of (3.24). We have

$$n_+(1 + \varepsilon; Sr(\gamma)^2 S^*) = n_+(1 + \varepsilon; r(\gamma)S^* Sr(\gamma)) =$$

$$n_-(1 + \varepsilon; r(\gamma)P_q(V_\omega - \lambda - \gamma + \delta)\chi_R P_q r(\gamma)), \ \forall \varepsilon > 0. \quad (3.25)$$

By (3.14) with $s = 1 + \varepsilon$,

$$\lim_{b\to\infty}\lim_{R\to\infty} b^{-1}R^{-3}n_-(1 + \varepsilon; r(\gamma)P_q(V_\omega - \lambda - \gamma + \delta)\chi_R P_q r(\gamma))$$

$$= \frac{1}{2\pi}k\left(\frac{\lambda + \gamma - \delta}{1 + \varepsilon} - \gamma; \frac{V}{1 + \varepsilon}\right). \quad (3.26)$$

Let us now estimate the second term at the right-hand side of (3.24). We have

$$Sr(\gamma)^2 E_{q,b}\tilde{r}(\gamma)^2 S^* = ((2q - 1)b - \gamma)U\chi_R P_q r(\gamma)^2(1 - P_q\chi_R P_q)\tilde{r}(\gamma)^2 S^*.$$

The operators $((2q - 1)b - \gamma)U$ and $\tilde{r}(\gamma)^2 S^*$ are uniformly bounded with respect to R. Therefore,

$$n_+(\varepsilon; Sr(\gamma)^2 E_{q,b}\tilde{r}(\gamma)^2 S^*) \leq n_*(\eta; \chi_R P_q r(\gamma)^2(1 - P_q\chi_R P_q)) \quad (3.27)$$

where $\eta > 0$ is independent of R. Further,

$$n_*(\eta; \chi_R P_q r(\gamma)^2(1 - P_q\chi_R P_q)) \leq \eta^{-2}\|\chi_R P_q r(\gamma)^2(1 - P_q\chi_R P_q)\|_{HS}^2, \quad (3.28)$$

$\|\cdot\|_{HS}$ being the Hilbert-Schmidt norm. A straightforward calculation yields

$$\lim_{R\to\infty} R^{-3}\|\chi_R P_q r(\gamma)^2(1 - P_q\chi_R P_q)\|_{HS}^2 = 0. \tag{3.29}$$

Putting together (3.27)–(3.29), we obtain

$$\lim_{R\to\infty} R^{-3}n_+(\varepsilon; Sr(\gamma)^2 E_{q,b}\tilde{r}(\gamma)^2 S^*) = 0, \ \forall\varepsilon > 0. \tag{3.30}$$

Combining the first inequality in (3.19) with (3.23)–(3.26) and (3.30), we get

$$\liminf_{b\to\infty} \liminf_{R\to\infty} b^{-1}R^{-3}n_-(1 \pm \phi(b); P_q T(\lambda + (2q - 1)b)P_q) \geq$$

$$\frac{1}{2\pi}k\left(\frac{\lambda + \gamma - \delta}{1 + \varepsilon} - \gamma; \frac{V}{1 + \varepsilon}\right). \tag{3.31}$$

Letting $\varepsilon \downarrow 0$ in (3.31), and then $\delta \downarrow 0$ in (3.22) and (3.31), and taking into account the boundedness of V_ω and the continuity of k, we arrive at (3.15). $\Diamond$

3.4. In this subsection we estimate certain quantities which do not contribute to the main asymptotic term as $b \to \infty$ of $\mathcal{D}_b(\lambda_2 + (2q - 1)b) - \mathcal{D}_b(\lambda_1 + (2q - 1)b)$. Introduce the orthogonal projections $P_q^- := \sum_{s=1}^{q-1} P_s$, $q > 1$, $\tilde{P}_q^- := \sum_{s=1}^q P_s$, $q \geq 1$, and $P_q^+ := \sum_{s=q+1}^{\infty} P_s$, $q \geq 1$.

Proposition 3.3. *Assume that (1.3) holds. Let $\mu_-, \mu_+ \in \mathbb{R}$, $\mu_- < \mu_+$, $q > 1$. Then almost surely*

$$\lim_{b\to\infty} \limsup_{R\to\infty} b^{-1}R^{-3}(n_-(1 - \phi(b); P_q^- T(\mu_+ + (2q - 1)b)\chi_R P_q^-)$$

$$-n_-(1 + \phi(b); P_q^- T(\mu_- + (2q - 1)b)\chi_R P_q^-) = 0. \tag{3.32}$$

Proof. Assume that b is so large that $(2q - 1)b + \mu_\pm \pm c_0 > 0$, c_0 being defined in (1.3). Set $\psi_\pm(b) := (1 \mp \phi(b))/((2q - 1)b + \mu_\pm \pm c_0)$. Evidently,

$$n_-(1 - \phi(b); P_q^- T(\mu_+ + (2q - 1)b)\chi_R P_q^-) \leq n_+(\psi_+(b); P_q^-\tilde{T}P_q^-), \tag{3.33}$$

$$n_-(1 + \phi(b); P_q^- T(\mu_- + (2q - 1)b)\chi_R P_q^-) \geq n_+(\psi_-(b); P_q^-\tilde{T}P_q^-), \tag{3.34}$$

$\tilde{T}$ being defined in (3.7). Obviously, $P_q^-\tilde{T}P_q^-$ is a trace-class operator, and

$$\lim_{R\to\infty} R^{-3}\mathrm{Tr}\,(P_q^-\tilde{T}P_q^-)^l = \frac{b}{(2\pi)^2}\sum_{s=1}^{q-1}\int_{\mathbb{R}} \frac{d\zeta}{(\zeta^2 + (2s - 1)b)^l}, \ l \in \mathbb{N}_*.$$

Applying the Kac-Murdock-Szegö theorem (see [Rai99, Lemma 3.2]), we get

$$\lim_{R\to\infty} R^{-3}n_+(\varepsilon; P_q^-\tilde{T}P_q^-) = \frac{b}{(2\pi)^2}\sum_{s=1}^{q-1} \mathrm{meas}\ \{\zeta \in \mathbb{R}|(\zeta^2 + (2s - 1)b)^{-1} > \varepsilon\} =$$

$$\frac{b}{2\pi^2}\sum_{s=1}^{q-1}\left(\varepsilon^{-1}-(2s-1)b\right)_+^{1/2}, \ \forall \varepsilon > 0, \forall b > 0. \tag{3.35}$$

Combining (3.33)–(3.35), we get

$$\limsup_{R\to\infty} R^{-3}\big(n_-(1-\phi(b); P_q^- T(\mu_+ + (2q-1)b)\chi_R P_q^-)$$

$$-n_-(1+\phi(b); P_q^- T(\mu_- + (2q-1)b)\chi_R P_q^-)) \le$$

$$\frac{b}{2\pi^2}\sum_{s=1}^{q-1}\left\{(\psi_+(b)^{-1}-(2s-1)b)_+^{1/2}-(\psi_-(b)^{-1}-(2s-1)b)_+^{1/2}\right\}. \tag{3.36}$$

Rationalizing, we find that for each $s = 1,\dots,q-1$, we have

$$\lim_{b\to\infty}\left\{(\psi_+(b)^{-1}-(2s-1)b)_+^{1/2}-(\psi_-(b)^{-1}-(2s-1)b)_+^{1/2}\right\}=0. \tag{3.37}$$

Now, (3.32) follows from (3.36) and (3.37). $\Diamond$

Proposition 3.4. *Assume that (1.3) holds. Fix $\lambda \in \mathbb{R}$. Then here exists $b_* > 0$ independent of R, such that $b > b_*$ implies*

$$n_-(1 \pm \phi(b)); P_q^+ T(\lambda + (2q-1)b)\chi_R P_q^+) = 0, \ \forall R > 0. \tag{3.38}$$

Proof. It suffices to note that $\lim_{b\to\infty}(1\pm\phi(b)) = 1$, and $\limsup_{b\to\infty}\|P_q^+ T(\lambda + (2q-1)b)\chi_R P_q^+\| \le \lim_{b\to\infty}\frac{c_0+|\lambda|+(2q-1)b}{(2q+1)b} = \frac{2q-1}{2q+1} < 1.$ $\Diamond$

Proposition 3.5. *For each $\varepsilon > 0$ and $q < 1$ we have*

$$\lim_{R\to\infty} R^{-3} n_\pm(\varepsilon; 2\mathrm{Re}\,(P_q^- \tilde{T} P_q + \tilde{P}_q^- \tilde{T} P_q^+)) = 0. \tag{3.39}$$

Proof. Write the estimates $n_\pm(\varepsilon; 2\mathrm{Re}(P_q^- \tilde{T} P_q + \tilde{P}_q^- \tilde{T} P_q^+)) \le 2n_*(\varepsilon/2; P_q^- \tilde{T} P_q) + 2n_*(\varepsilon/2; \tilde{P}_q^- \tilde{T} P_q^+) \le 8\varepsilon^{-2}\left(\|P_q^- \tilde{T} P_q\|_{HS}^2 + \|\tilde{P}_q^- \tilde{T} P_q^+\|_{HS}^2\right)$, and verify by direct calculation that $\lim_{R\to\infty} R^{-3}\|P_q^- \tilde{T} P_q\|_{HS}^2 = \lim_{R\to\infty} R^{-3}\|\tilde{P}_q^- \tilde{T} P_q^+\|_{HS}^2 = 0.$ $\Diamond$

4 Proof of Theorem 2.1

Fix $q > 1$, $\lambda_1, \lambda_2 \in \mathbb{R}$, $\lambda_1 < \lambda_2$. In order to prove (2.2), it suffices to show that for each sequence $\{b_j\}_{j\ge 1}$ such that $b_j \to \infty$ as $j \to \infty$, we have

$$\lim_{j\to\infty} b_j^{-1}\left(\mathcal{D}_{b_j}(\lambda_2 + (2q-1)b_j) - \mathcal{D}_{b_j}(\lambda_1 + (2q-1)b_j)\right) = \frac{1}{2\pi}\left(k(\lambda_2) - k(\lambda_1)\right). \tag{4.1}$$

Fix four sequences $\{\lambda_{l,m}^{\pm}\}_{m\geq 1}$, $l = 1,2$, such that $\lambda_{l,m}^{-} < \lambda_l < \lambda_{l,m}^{+}$, $m \geq 1$, $\lim_{m\to\infty} \lambda_{l,m}^{\pm} = \lambda_l$, and $\lambda_{l,m}^{\pm} + (2q-1)b_j$ are continuity points of $\mathcal{D}_{b_j}$ for all $m \geq 1$ and $j \geq 1$, $l = 1,2$. Then by Corollary 3.1 we have

$$\limsup_{j\to\infty} b_j^{-1}\left(\mathcal{D}_{b_j}(\lambda_2 + (2q-1)b_j) - \mathcal{D}_{b_j}(\lambda_1 + (2q-1)b_j)\right) \leq$$

$$\limsup_{j\to\infty}\lim_{R\to\infty} b_j^{-1}R^{-3}(n_-(1;T(\lambda_{2,m}^{+} + (2q-1)b_j)) - n_-(1;T(\lambda_{1,m}^{-} + (2q-1)b_j))),$$

$$(4.2)$$

$$\liminf_{j\to\infty} b_j^{-1}\left(\mathcal{D}_{b_j}(\lambda_2 + (2q-1)b_j) - \mathcal{D}_{b_j}(\lambda_1 + (2q-1)b_j)\right) \geq$$

$$\liminf_{j\to\infty}\lim_{R\to\infty} b_j^{-1}R^{-3}(n_-(1;T(\lambda_{2,m}^{-} + (2q-1)b_j)) - n_-(1;T(\lambda_{1,m}^{+} + (2q-1)b_j))).$$

$$(4.3)$$

Further, note that the elementary operator inequalities

$$T(\lambda + (2q-1)b) \geq P_q T(\lambda + (2q-1)b + 2c_0\delta)P_q +$$

$$P_q^- T(\lambda + (2q-1)b + c_0(1+\delta^{-1}))P_q^- + P_q^+ T(\lambda + (2q-1)b + c_0(1+\delta^{-1}))P_q^+ -$$

$$2(\lambda + (2q-1)b)\mathrm{Re}(P_q^+\tilde{T}\tilde{P}_q^- + P_q^-\tilde{T}P_q),$$

$$T(\lambda + (2q-1)b) \leq P_q T(\lambda + (2q-1)b - 2c_0\delta)P_q +$$

$$P_q^- T(\lambda + (2q-1)b - c_0(1+\delta^{-1}))P_q^- + P_q^+ T(\lambda + (2q-1)b - c_0(1+\delta^{-1}))P_q^+ -$$

$$2(\lambda + (2q-1)b)\mathrm{Re}(P_q^+\tilde{T}\tilde{P}_q^- + P_q^-\tilde{T}P_q),$$

are valid for each $\lambda \in \mathbb{R}$, $b > 0$, and $\delta > 0$. Therefore,

$$n_-(1;T(\lambda + (2q-1)b)) \leq n_-(1 - \phi(b); P_q T(\lambda + (2q-1)b + 2c_0\delta)P_q) +$$

$$n_-(1 - \phi(b); P_q^- T(\lambda + (2q-1)b + c_0(1+\delta^{-1}))P_q^-) +$$

$$n_-(1 - \phi(b); P_q^+ T(\lambda + (2q-1)b + c_0(1+\delta^{-1}))P_q^+) +$$

$$n_+(\phi(b); 2(\lambda + (2q-1)b)\mathrm{Re}(P_q^+\tilde{T}\tilde{P}_q^- + P_q^-\tilde{T}P_q)),$$

and

$$n_-(1;T(\lambda + (2q-1)b)) \geq n_-(1 + \phi(b); P_q T(\lambda + (2q-1)b - 2c_0\delta)P_q) +$$

$$n_-(1 + \phi(b); P_q^- T(\lambda + (2q-1)b - c_0(1+\delta^{-1}))P_q^-) +$$

$$n_-(1 + \phi(b); P_q^+ T(\lambda + (2q-1)b - c_0(1+\delta^{-1}))P_q^+) -$$

$$n_-(\phi(b); 2(\lambda + (2q-1)b)\mathrm{Re}(P_q^+\tilde{T}\tilde{P}_q^- + P_q^-\tilde{T}P_q)),$$

the numerical function ϕ being introduced before Corollary 3.2. Hence, we get

$$\limsup_{j\to\infty}\lim_{R\to\infty} b_j^{-1}R^{-3}(n_-(1;T(\lambda_{2,m}^{+} + (2q-1)b_j)) - n_-(1;T(\lambda_{1,m}^{-} + (2q-1)b_j))) \leq$$

$$\limsup_{j\to\infty}\limsup_{R\to\infty} b_j^{-1}R^{-3}(n_-(1-\phi(b);P_qT(\lambda_{2,m}^+ + (2q-1)b_j - 2c_0\delta)P_q)$$

$$-n_-(1+\phi(b);P_qT(\lambda_{1,m}^- + (2q-1)b_j - 2c_0\delta)P_q))+$$

$$\limsup_{j\to\infty}\limsup_{R\to\infty} b_j^{-1}R^{-3}(n_-(1-\phi(b);P_q^-T(\lambda_{2,m}^+ + (2q-1)b_j + c_0(1+\delta^{-1}))P_q^-)$$

$$-n_-(1+\phi(b);P_q^-T(\lambda_{1,m}^- + (2q-1)b_j - c_0(1+\delta^{-1}))P_q^-))-$$

$$\limsup_{j\to\infty}\limsup_{R\to\infty} b_j^{-1}R^{-3}n_-(1-\phi(b);P_q^+T(\lambda_{2,m}^+ + (2q-1)b_j + c_0(1+\delta^{-1})P_q^+)+$$

$$\limsup_{j\to\infty}\limsup_{R\to\infty} b_j^{-1}R^{-3}(n_+(\phi(b);2(\lambda_{2,m}^+ + (2q-1)b_j)\mathrm{Re}(P_q^+\tilde{T}\tilde{P}_c^- + P_q^-\tilde{T}P_q))$$

$$+n_-(\phi(b);2(\lambda_{1,m}^- + (2q-1)b)\mathrm{Re}(P_q^+\tilde{T}\tilde{P}_q^- + P_q^-\tilde{T}P_q)), \tag{4.4}$$

and

$$\liminf_{j\to\infty}\lim_{R\to\infty} b_j^{-1}R^{-3}(n_-(1;T(\lambda_{2,m}^- + (2q-1)b_j)) - n_-(1;T(\lambda_{1,m}^+ + (2q-1)b_j))) \geq$$

$$\liminf_{j\to\infty}\liminf_{R\to\infty} b_j^{-1}R^{-3}(n_-(1+\phi(b);P_qT(\lambda_{2,m}^- + (2q-1)b_j - 2c_0\delta)P_q)$$

$$-n_-(1-\phi(b);P_qT(\lambda_{1,m}^+ + (2q-1)b_j + 2c_0\delta)P_q))-$$

$$\liminf_{j\to\infty}\liminf_{R\to\infty} b_j^{-1}R^{-3}(n_-(1-\phi(b);P_q^-T(\lambda_{1,m}^+ + (2q-1)b_j + c_0(1+\delta^{-1}))P_q^-)$$

$$-n_-(1+\phi(b);P_q^-T(\lambda_{2,m}^- + (2q-1)b_j - c_0(1+\delta^{-1}))P_q^-))-$$

$$\liminf_{j\to\infty}\liminf_{R\to\infty} b_j^{-1}R^{-3}n_-(1-\phi(b);P_q^+T(\lambda_{1,m}^+ + (2q-1)b_j + c_0(1+\delta^{-1}))P_q^+)-$$

$$\liminf_{j\to\infty}\liminf_{R\to\infty} b_j^{-1}R^{-3}(n_+(\phi(b);2(\lambda_{1,m}^+ + (2q-1)b_j)\mathrm{Re}(P_q^+\tilde{T}\tilde{P}_q^- + P_q^-\tilde{T}P_q))$$

$$+n_-(\phi(b);2(\lambda_{2,m}^- + (2q-1)b)\mathrm{Re}(P_q^+\tilde{T}\tilde{P}_q^- + P_q^-\tilde{T}P_q)). \tag{4.5}$$

Employing Corollary 3.2, we find that the first term at the right-hand side of (4.4) is equal to $\frac{1}{2\pi}\left(k(\lambda_{2,m}^+ + 2c_0\delta)) - k(\lambda_{1,m}^- - 2c_0\delta)\right)$, while the first term at the side of (4.5) is equal to $\frac{1}{2\pi}\left(k(\lambda_{2,m}^- - 2c_0\delta)) - k(\lambda_{1,m}^+ + 2c_0\delta)\right)$. Assume that $\delta > 0$ is small enough, and apply Proposition 3.3 to the second terms at the right-hand sides of (4.4) and (4.5) in order to check that these terms vanish. Similarly, utilize Proposition 3.4 (respectively, Proposition 3.5) in order to verify that the third (respectively, fourth) terms at the right-hand sides of (4.4) and (4.5) vanish. Putting together (4.2) – (4.5), we get

$$\limsup_{j\to\infty} b_j^{-1}\left(\mathcal{D}_{b_j}(\lambda_2 + (2q-1)b_j) - \mathcal{D}_{b_j}(\lambda_1 + (2q-1)b_j)\right) \leq$$

$$\frac{1}{2\pi}\left(k(\lambda_{2,m}^+ + 2c_0\delta)) - k(\lambda_{1,m}^- - 2c_0\delta)\right), \tag{4.6}$$

$$\liminf_{j\to\infty} b_j^{-1}\left(\mathcal{D}_{b_j}(\lambda_2 + (2q-1)b_j) - \mathcal{D}_{b_j}(\lambda_1 + (2q-1)b_j)\right) \geq$$

$$\frac{1}{2\pi}\left(k(\lambda_{2,m}^- - 2c_0\delta)) - k(\lambda_{1,m}^+ + 2c_0\delta)\right). \tag{4.7}$$

Letting $m \to \infty$, and $\delta \downarrow 0$ in (4.6)-(4.7), we obtain (4.1).

Acknowledgements

The financial support of the Organizing Committee of the Conference on Partial Differential Equations, Clausthal, July 24 -28, 2000, and of Grant MM 706/97 of the Bulgarian Science Foundation, is gratefully acknowledged. The author thanks the referee whose valuable remarks contributed to the improvement of the exposition.

References

[AHS78]   J.AVRON, I.HERBST, B.SIMON, *Schrödinger operators with magnetic fields. I. General interactions*, Duke. Math. J. **45** (1978), 847-883.

[Bir66]   M.Š.BIRMAN, *On the spectrum of singular boundary value problems*, Mat. Sbornik **55** (1961), 125-174 (Russian); Engl. transl. in Amer. Math. Soc. Transl., (2) **53** (1966), 23-80.

[HLMW00]   T.HUPFER, H.LESCHKE, P.MÜLLER, S.WARZEL, *The integrated density of states and its absolute continuity for magnetic Schrödinger operators with unbounded random potentials*, math-ph Preprint 0010013 available at http://xxx.uni-augsburg.de/abs/math-ph

[Kir89]   W.KIRSCH, *Random Schrödinger operators: a course*. In: Schrödinger operators, Proc. Nord. Summer Sch. Math., Sandbjerg Slot, Soenderborg/Denmark 1988, Lect. Notes Phys. **345**, (1989) 264-370.

[KR00]   W.KIRSCH, G.D.RAIKOV, *Strong magnetic field asymptotics of the integrated density of states for a random 3D Schrödinger operator*, mp_arc Preprint 99-345 available at http://www.ma.utexas.edu/mp_arc (to appear in Annales Henri Poincaré)

[Nak00]   S.NAKAMURA, *A remark on the Dirichlet-Neumann decoupling and the integrated density of states*, mp_arc Preprint 00-198 available at http://www.ma. utexas.edu/mp_arc (to appear in J.Funct.Anal.)

[PF92]   L.PASTUR, A.FIGOTIN, *Spectra of Random and Almost-Periodic Operators*. Grundlehren der Mathematischen Wissenschaften **297**. Springer-Verlag, Berlin etc. (1992).

[Rai99]   G.D.RAIKOV, *Eigenvalue asymptotics for the Pauli operator in strong nonconstant magnetic fields*, Ann.Inst.Fourier, **49**, (1999), 1603-1636.

Address

GEORGI D. RAIKOV, Institute of Mathematics, Bulgarian Academy of Sciences, Acad.G.Bonchev Str., bl.8, 1113 Sofia, Bulgaria

E-MAIL: gdraikov@yahoo.com

2000 Mathematics Subject Classification. Primary 82B44; Secondary 81Q10, 47B80

Operator Theory:
Advances and Applications, Vol. 126
© 2001 Birkhäuser Verlag Basel/Switzerland

Partial Differential Operators with Multiple Symplectic Characteristics

Luigi Rodino, Maria Mascarello, and Nguyen Minh Tri

Abstract. In the first section of the paper we shortly review classes of operators with multiple symplectic characteristics for which hypoellipticity and solvability were studied by Grushin, Parenti, Rodino, Mascarello, and others. In the second section we present a new result concerning the second order model operator of Grushin. As well known, hypoellipticity for it depends on discrete conditions on a parameter; here we give explicit formulas for fundamental solutions using hypergeometric functions.

1 Reviewing Classes of Operators with Symplectic characteristics

Operators with symplectic characteristics, in a broad sense, are differential or pseudo-differential operators whose principal symbol vanishes on a non-involutive set Σ. In turn, non-involutive smooth manifolds Σ can be defined as those submanifolds of the cotangent bundle which are transformed, via a symplectic diffeomorphism φ, into the flat sets $\varphi(\Sigma) = \{x_1 = \cdots = x_p = 0, \xi_1 = \cdots = \xi_p = 0\} \subset (\mathbb{R}^p_x \times \mathbb{R}^{n-p}_y) \times (\mathbb{R}^p_\xi \times \mathbb{R}^{n-p}_\eta \backslash 0), 0 < p < n$. As relevant examples in the case Σ flat, $p = 1, n = 2$, we may quote the Mizohata operator $X = \partial/\partial x + ix^k \partial/\partial y$, with k positive integer and $i = \sqrt{-1}$, and the Grushin operator

$$G = \frac{\partial^2}{\partial x^2} + x^{2k} \frac{\partial^2}{\partial y^2} + i\lambda x^{k-1} \frac{\partial}{\partial y}, \quad \lambda \in \mathbb{C}. \tag{1.1}$$

Grushin [Gr71] treated (1.1) as a particular case of the operator with multiple characteristics

$$P = \sum c_{\alpha\beta\gamma} x^\alpha D_x^\beta D_y^\gamma + W, \tag{1.2}$$

where now $x = (x_1, \ldots, x_p), y = (y_1, \ldots, y_{n-p})$ and we use standard multi-index notation; in the sum $|\alpha|/k + |\beta| = m - j(k+1)/k$ with $0 \le j \le mk/(k+1)$, $\gamma = m + (|\alpha| - |\beta| - km)/(k+1)$ and order $W < m/(k+1)$. The coefficients $c_{\alpha\beta\gamma}$ are complex constants, or smooth functions of (x, y), or else more generally zero order classical pseudo-differential operators, cf. Parenti-Rodino [PR80].

Returning to the case of a general non-involutive characteristic manifold Σ, one may further generalize (1.2) by considering as in Mascarello-Rodino [MR81]

$$P = \sum U_2^\alpha c_{\alpha\beta\gamma} D_y^\gamma U_1^\beta + W, \tag{1.3}$$

where U_1, U_2 are pseudo-differential operators with symbol $u_1 = \xi \circ \varphi^{-1}, u_2 = x \circ \varphi^{-1}$ respectively.

The study of hypoellipticity and solvability of (1.2), (1.3) can be related to the analysis of some operators with polynomial coefficients in $\mathbb{R}^p$. In particular, considering G in (1.1) with k odd integer, one deduces that G is hypoelliptic if and only if (see Mascarello-Rodino [MR97] and the references there):

$$\lambda \neq \pm[2N(k+1)+k], \lambda \neq \pm[2N(k+1)+k+2], \quad N = 0, 1, 2, \ldots. \tag{1.4}$$

The aim of the present paper is to construct an explicit fundamental solution of G under the assumption (1.4); the expression is new, as far as we know, cf. the recent related contributions of Beals [Be99], Beals-Greiner-Gaveau [BGG99], Tri [Tr99] and, concerning a similar hyperbolic problem, Yagdjian [Ya97], section 2.1.

In conclusion, we take the opportunity to correct a mistake in Mascarello-Rodino [MR81]. Namely, the authors stated there that for a pseudo-differential operator P with symbol $p(x, \xi) \sim \sum_{j=0}^{\infty} p_{m-j}(x, \xi)$, the assumption (1.4) is equivalent to the estimates

$$|p_{m-j}(x, \xi)|/|\xi|^{m-j} \leq C(d_{\Sigma_1}(x, \xi) + d_{\Sigma_2}^k(x, \xi))^{m-j(k+1)/k} \tag{1.5}$$

for $0 \leq j \leq mk/(k+1)$, where $\Sigma_j, j = 1, 2$, is given by $u_j = 0$ and d_{Σ_j} is the distance from $(x, \xi/|\xi|)$ to Σ_j for $|\xi| \geq 1, \Sigma = \Sigma_1 \cap \Sigma_2$. Actually, as observed by Mughetti [Mu99], the equivalence holds when Σ is flat, i.e. for P in (1.2), or else if $m = 1$ or $m = 2$ by taking as p_{m-1} the sub-principal symbol, but it may fail in general for non-flat Σ and $m \geq 3$.

2   Fundamental Solution of the Model of Grushin

Let us consider the operator G, defined as in (1.1), with k odd positive integer and the parameter $\lambda \in \mathbb{C}$ satisfying the hypoellipticity condition (1.4). The following quantities will be important for us:

$$R = (x^{k+1} + u^{k+1})^2 + (k+1)^2(y-v)^2, \quad p = 4x^{k+1}u^{k+1}/R,$$
$$A_+ = x^{k+1} + u^{k+1} + i(k+1)(y-v), \quad A_- = x^{k+1} + u^{k+1} - i(k+1)(y-v),$$
$$M = A_+^{-\frac{k+\lambda}{2k+2}} A_-^{-\frac{k-\lambda}{2k+2}},$$

here we take $z_1^{z_2} = e^{z_2 \ln z_1}$ for $z_1, z_2 \in \mathbb{C}$ and if $z_1 = re^{i\phi}, -\pi < \phi \leq \pi$ then $\ln z_1 = \ln r + i\phi$. Let us rewrite $G = X_2 X_1 + i(\lambda + k)x^{k-1}\frac{\partial}{\partial y}$ where $X_1 = \frac{\partial}{\partial x} - ix^k \frac{\partial}{\partial y}$, $X_2 = \frac{\partial}{\partial x} + ix^k \frac{\partial}{\partial y}$.

We will find the uniform fundamental solution $E(x, y, u, v)$ of G in the following form

$$E(x, y, u, v) = H(p)M. \tag{2.1}$$

Note that $X_1 A_+ = 2(k+1)x^k, X_1 A_- = 0, X_2 A_+ = 0, X_2 A_- = 2(k+1)x^k$. After some computations we arrive at

$$GE = 16(k+1)^2 u^{2k+2} x^{2k} \left[(u^{k+1} - x^{k+1})^2 + (k+1)^2(y-v)^2\right] MR^{-3} H''(p)+$$
$$+4(k+1)\left[k(x^{2k+2} + u^{2k+2} + (k+1)^2(y-v)^2) - (6k+4)x^{k+1}u^{k+1}\right] R^{-2} H'(p)+$$
$$+(\lambda^2 - k^2)MR^{-1}H(p).$$

Therefore if $H(p)$ satisfies the following hypergeometric equation (see Bateman-Erdelyi [BE53], p. 56)

$$p(1-p)H''(p) + [c - (1+a+b)p]H'(p) - abH(p) = 0 \tag{2.2}$$

with $a = \frac{k+\lambda}{2k+2}, b = \frac{k-\lambda}{2k+2}, c = \frac{k}{k+1}$, then formally we will have

$$GE = 0.$$

The general solution of (2.2) is

$$
\begin{aligned}
H(p) &= C_1 F\left(\tfrac{k+\lambda}{2k+2}, \tfrac{k-\lambda}{2k+2}, \tfrac{k}{k+1}, p\right) + \\
&+ C_2 p^{\frac{1}{k+1}} F\left(\tfrac{k+2+\lambda}{2k+2}, \tfrac{k+2-\lambda}{2k+2}, \tfrac{k+2}{k+1}, p\right),
\end{aligned}
\tag{2.3}
$$

where C_1, C_2 are some complex constants and for $|z| < 1$ we define as standard

$$F(a,b,c,z) = \sum_{n=0}^{\infty} \frac{(a)_n (b)_n}{(c)_n} \frac{z^n}{n!},$$

with $(t)_n = t(t+1) \cdots (t+n-1)$, cf. [BE53], p. 74. We note that $0 \leq p \leq 1$ since k is odd. Let us fix in (2.3):

$$C_1 = -\frac{\Gamma\left(\frac{k+\lambda}{2k+2}\right)\Gamma\left(\frac{k-\lambda}{2k+2}\right)}{2^{\frac{2k+3}{k+1}}\pi\Gamma\left(\frac{k}{k+1}\right)}, \quad C_2 = -\frac{\Gamma\left(\frac{k+2+\lambda}{2k+2}\right)\Gamma\left(\frac{k+2-\lambda}{2k+2}\right)}{2^{\frac{2k+3}{k+1}}\pi\Gamma\left(\frac{k+2}{k+1}\right)}. \tag{2.4}$$

It is easy to see that $E(x, y, u, v)$ in (2.1) has then singularity only for $x = u, y = v$. Moreover, (1.4) guarantees that $H(p)$ has a logarithm growth at $(x, y) = (u, v)$.

Theorem 2.1. *Let E be defined by (2.1), (2.3), (2.4). Then*

$$GE(x, y, u, v) = \delta(x - u, y - v).$$

Proof. We begin by fixing $(u, v) \in \mathbb{R}^2$. First assume that $u \neq 0$. Then $A_+^\alpha, A_-^\beta \in C^\infty$ for every α and β. Let us introduce the following polar coordinates

$$x = u + r\cos\phi, y = v + r\sin\phi.$$

It is easy to see that $E(\cdot, \cdot, u, v) \in L^q_{loc}(\mathbb{R}^2)$ for every $1 \le q < \infty$. Let $B_\varepsilon(u, v) = \{(x, y) | r < \varepsilon\}$ and $\mathbb{R}^2_\varepsilon(u, v) = \mathbb{R}^2 \backslash B_\varepsilon(u, v) = \{(x, y) \in \mathbb{R}^2 | r \ge \varepsilon\}$. By applying Green's formula we have

$$\int_{\mathbb{R}^2_\varepsilon(u,v)} f(x, y) \, {}^tGw(x, y) \, dxdy = \int_{\mathbb{R}^2_\varepsilon(u,v)} w(x, y) Gf(x, y) \, dxdy +$$
$$+ \int_{r=\varepsilon} f(x, y) \left\{ (\nu_1 - ix^k \nu_2) X_2 w - i(\lambda + k) x^{k-1} \nu_2 w \right\} ds - \qquad (2.5)$$
$$- \int_{r=\varepsilon} w(x, y) \left\{ (\nu_1 + ix^k \nu_2) X_1 f(x, y) \right\} ds =: I_1 + I_2 + I_3$$

for every $w(x, y) \in C_0^\infty(\mathbb{R}^2), f(x, y) \in C^\infty(\mathbb{R}^2 \backslash (u, v))$, where $\nu = (\nu_1, \nu_2)$ is the unit outward normal to $\partial \mathbb{R}^2_\varepsilon(u, v)$. Replace $f(x, y)$ in (2.5) by $E(x, y, u, v)$. The first integral I_1 in the right side of (2.5) vanishes. We now compute the second and the third integral in the right side of (2.5). When (x, y) tends to (u, v) it is easy to check that

$$M = (2u^{k+1} + o(1))^{-\frac{k+\lambda}{2k+2}} (2u^{k+1} + o(1))^{-\frac{k-\lambda}{2k+2}} = 2^{-\frac{k}{k+1}} u^{-k} + o(1),$$
$$x^k = u^k + ku^{k-1} r \cos\phi + o(r), \quad R = 4u^{2k+2} + o(1),$$
$$\nu_1|_{\partial B_\varepsilon(u,v)} = -\cos\phi, \quad ix^k \nu_2|_{\partial B_\varepsilon(u,v)} = -iu^k \sin\phi + o(1),$$
$$1 - p = (k+1)^2 (u^{2k} \cos^2\phi + \sin^2\phi) u^{-2k-2}/4 + o(r^2),$$
$$X_1 p = (k+1)^2 u^{-k-2}(-u^k \cos\phi + i\sin\phi) r/2 + o(r).$$

Moreover we have (see [BE53], p. 75)

$$F'\left(\frac{k+\lambda}{2k+2}, \frac{k-\lambda}{2k+2}, \frac{k}{k+1}, p\right) = \frac{\Gamma\left(\frac{k}{k+1}\right)}{\Gamma\left(\frac{k+\lambda}{2k+2}\right) \Gamma\left(\frac{k-\lambda}{2k+2}\right)} (1 - p) + O(1)$$

$$F'\left(\frac{k+2+\lambda}{2k+2}, \frac{k+2-\lambda}{2k+2}, \frac{k+2}{k+1}, p\right) = \frac{\Gamma\left(\frac{k+2}{k+1}\right)}{\Gamma\left(\frac{k+2+\lambda}{2k+2}\right) \Gamma\left(\frac{k+2-\lambda}{2k+2}\right)} (1 - p) + O(1)$$

Hence

$$X_1 E = X_1(H(p)M) = H(p) X_1 M + M X_1 p H'(p) =$$
$$= 2^{\frac{k+2}{k+1}} (-u^k \cos\phi + i\sin\phi)(u^{2k} \cos^2\phi + \sin^2\phi)^{-1} r^{-1} + o(r^{-1}).$$

It follows immediately that

$$I_3 \to \frac{w(u, v)}{2\pi} \int_0^{2\pi} \frac{u^k \, d\phi}{u^{2k} \cos^2\phi + \sin^2\phi} = w(u, v),$$

and $I_2 \to 0$ as $\varepsilon \to 0$.

We deduce that

$$(GE, w(x, y)) = (E, {}^tGw(x, y)) = \lim_{\varepsilon \to 0} \int_{r \ge \varepsilon} E \, {}^tGw(x, y) \, dxdy = w(u, v).$$

Taking the limit when $u \to 0$, we complete the proof of the theorem. $\qquad \square$

References

[BE53]   H. Bateman and A. Erdelyi. Higher Transcendental Functions, Vol. I. *McGraw-Hill*, New York, 1953.

[Be99]   R. Beals. A Note on Fundamental Solutions. *Comm. Part. Diff. Equat.*, 24:369-376, 1999.

[BGG99]   R. Beals, P. Greiner and B. Gaveau. Green's functions for some highly degenerate elliptic operators. *J. Funct. Anal.*, 165:407-429, 1999.

[Gr71]   V. V. Grushin. On a Class of Elliptic Pseudo Differential Operators Degenerate on a Submanifold. *Math. USSR Sbornik*, 13:155-183, 1971.

[MR81]   M. Mascarello and L. Rodino. A Class of Pseudo Differential Operators with Multiple Non-Involutive Characteristics. *Ann. Scuola Norm. Sup. Pisa*, Cl. Sc., Ser. IV, 8:575-603, 1981.

[MR97]   M. Mascarello and L. Rodino. Partial Differential Equations with Multiple Characteristics. *Akademie Verlag-Wiley*, Berlin, 1997.

[Mu99]   M. Mughetti. A Problem of Transversal Anisotropic Ellipticity. *Preprint Dip. Mat. Univ. Bologna Italy*, 1999.

[PR80]   C. Parenti and L. Rodino. Parametrices for a Class of Pseudo Differential Operators I, II. *Annali Mat. Pura Appl.*, 125:221-254, 125:255-278, 1980.

[Tr99]   N. M. Tri. Remark on Non-Uniform Fundamental and Non-Smooth Solutions of Some Classes of Differential Operators with Double Characteristics. *J. Math. Sci. Univ. Tokyo*, 6:437-452, 1999.

[Ya97]   K. Yagdjian. The Cauchy Problem for Hyperbolic Operators. Multiple Characteristics. Micro-Local Approach. *Akademie Verlag-Wiley*, Berlin, 1997.

Addresses

LUIGI RODINO, Dipartimento di Matematica, Università di Torino, Via Carlo Alberto 10, 10123 Torino, Italy

MARIA MASCARELLO, Dipartimento di Matematica, Politecnico di Torino, Corso Duca degli Abruzzi 24, 10129 Torino, Italy

NGUYEN MINH TRI, Institute of Mathematics, P.O. Box 631, Boho 10000, Hanoi, Vietnam

2000 Mathematics Subject Classification. Primary 35H05; Secondary 35S05

Operator Theory:
Advances and Applications, Vol. 126
© 2001 Birkhäuser Verlag Basel/Switzerland

On the Homotopy Classification of Elliptic Boundary Value Problems

ANTON SAVIN, BERT-WOLFGANG SCHULZE, AND BORIS STERNIN

Abstract. The present paper deals with the homotopy classification problem of boundary value problems for elliptic operators. We start with classical boundary value problems. The ellipticity condition allows us to reduce classical problems to the Dirichlet problem for the Laplace operator and also to obtain the homotopy classification. We then study general case of operators, that do not necessarily satisfy the Atiyah–Bott condition. The boundary value problem reduces then to the so-called spectral boundary value problem.

1   Classical boundary value problems

1. A *classical boundary value problem* for an elliptic differential operator D on a manifold M with boundary ∂M is a system of equations of the form

$$\begin{cases} Du = f, \\ Bj_{\partial M}^{m-1}u = g, \end{cases} \quad u \in C^\infty(M, E), f \in C^\infty(M, F), g \in C^\infty(\partial M, G),$$

where m is the order of D and $j_{\partial M}^{m-1}$ is jet of order $m-1$ restricted to the boundary in normal direction:

$$j_{\partial M}^{m-1}u = \left(u, -i\frac{\partial}{\partial t}u, ..., \left(-i\frac{\partial}{\partial t}\right)^{m-1}u \right)\Bigg|_{\partial M} \in C^\infty\left(\partial M, E^m|_{\partial M}\right).$$

The boundary condition is defined by a matrix pseudodifferential operator

$$B : C^\infty\left(\partial M, E^m|_{\partial M}\right) \longrightarrow C^\infty(\partial M, G)$$

on ∂M. The boundary value problem is denoted by (D, B). It defines a bounded operator in Sobolev spaces:

$$(D, B) : H^s(M, E) \longrightarrow H^{s-m}(M, F) \oplus H^\delta(\partial M, G),$$

for $s - m > -1/2$ and a suitable δ (e.g., see [Hör85]).

The class of Fredholm boundary value problems in these Sobolev spaces is determined by the well-known Shapiro-Lopatinskii condition. It is formulated as follows. At each point (x, ξ) of the cosphere bundle $\pi : S^*(\partial M) \to M$, consider the initial data subspace for solutions of the equation

$$\sigma(D)\left(x, 0, \xi, -i\frac{d}{dt}\right)u(t) = 0,$$

which are bounded as $t \to +\infty$. These subspaces define a vector subbundle $L_+(D)$ in $\pi^* E^m$. The boundary value problem (D, B) has the Fredholm property if and only if the restriction of the principal symbol of B to the subbundle $L_+(D)$ defines an isomorphism

$$\sigma(B) : L_+(D) \longrightarrow \pi^* G. \tag{1.1}$$

2. The topological content of the Shapiro-Lopatinskii condition (1.1) was uncovered by Atiyah and Bott. They showed [AB64] that for the existence of classical boundary value problems for a given elliptic operator D it is necessary that the following element vanishes

$$i^* [\sigma(D)] \in K_c^1(T^* \partial M), \quad i : T^* M|_{\partial M} \subset T^* M. \tag{1.2}$$

This element $i^* [\sigma(D)]$ is called the *Atiyah–Bott obstruction*. It can be shown that the vanishing of (1.2) implies that in a neighbourhood of the boundary the symbol is (stably) homotopic to a symbol, that does not depend on the cotangent variables. There arises a natural question. Is it possible to make a similar simplification of the boundary value problem?

It turns out that the desired simplification is possible, if one considers boundary value problems for operators of a more general type, namely, for operators of the form

$$\sum_{0 \le k \le m} D_k(t) \left(-i \frac{\partial}{\partial t}\right)^{m-k} \tag{1.3}$$

in a neighbourhood of the boundary. Here $D_k(t)$ denotes a smooth family of pseudodifferential operators of order k on the boundary, while $D_0(t)$ is a vector bundle homomorphism. The boundary value problems for elliptic operators of this form are posed in the same way as for usual differential operators (see [Hör85]). Within this class of operators, it is possible to obtain a homotopy classification for classical boundary value problems.

3. In order to classify boundary value problems, it is necessary to be able to compare boundary value problems for operators of different orders. This is done by means of the special first order operator

$$D_+ = -\frac{\partial}{\partial t} + \sqrt{-\Delta_{\partial M}},$$

where $\Delta_{\partial M}$ is the Laplacian on ∂M. This operator extends as $\sqrt{-\Delta_M}$ far from the boundary. The operator D_+ has the Fredholm property (without boundary conditions) and has index zero.

The Dirichlet problem for the Laplacian will be referred to as a *trivial* boundary value problem.

Elliptic boundary value problems $\mathcal{D}_1$ and $\mathcal{D}_2$ are said to be *stably homotopic*, if – after their orders have been made equal by composition with (suitable powers of)

D_+ and the possible addition to each of them of them of some trivial boundary value problem – they become homotopic.

It turns out that an arbitrary boundary value problem is stably homotopic to an operator (1.3) of order zero. Note that these operators *do not require* boundary conditions. This homotopy admits a good geometric description in the case of boundary value problems for first order operators

$$D = \frac{\partial}{\partial t} + A. \tag{1.4}$$

Indeed, it is easy to see that for $(x, \xi) \in S^* (\partial M)$ the spectral projection of the symbol $\sigma (A)$, which projects on the subspace corresponding to the eigenvalues with positive imaginary part, defines the bundle $L_+ (D)$. By virtue of the ellipticity of the boundary value problem, the principal symbol of B is an isomorphism $L_+ (D)$ and the bundle $\pi^* G$, which is a pull-back from the base. On the other hand, one knows that isomorphic vector bundles are homotopic, if they are realized as *subbundles*. In our case such a homotopy deforms (1.4) to the form

$$\left(\frac{\partial}{\partial t} - \sqrt{-\Delta_{\partial M}} \right) \oplus \left(\frac{\partial}{\partial t} + \sqrt{-\Delta_{\partial M}} \right).$$

In addition, the boundary value problem (D, B) is homotopic to a direct sum of two boundary value problems: the first component coincides with D_+ (it has no boundary conditions), while the second one has Dirichlet boundary condition.

Operators of order greater than one are reduced to first order operators by homotopies constructed by Hörmander [Hör85].

These reductions of boundary value problems make it possible to obtain the homotopy classification, i.e. an isomorphism

$$\chi : \mathrm{Ell}\,(M) \xrightarrow{\simeq} K_c \left(T^* \left(M \backslash \partial M \right) \right), \tag{1.5}$$

where $\mathrm{Ell}\,(M)$ is the group of stable homotopy classes of boundary value problems. Here, χ maps a boundary value problem into the difference construction of the principal symbol of the corresponding zero order operator.

As a corollary to the homotopy classification (1.5), one immediately obtains the index formula for boundary value problems

$$\mathrm{ind}\,(D, B) = p_! \left(\chi\,[D, B] \right), \quad p_! : K_c \left(T^* \left(M \backslash \partial M \right) \right) \longrightarrow \mathbb{Z}.$$

It should be noted that the homotopy classification of boundary value problems was first obtained in [BdM71] in the algebra of pseudodifferential boundary value problems. The homotopy classification (1.5) does not use this algebra.

2 Boundary value problems for general elliptic operators

1. We have seen that an elliptic operator may have no classical boundary value problems, e.g., the Atiyah–Bott obstruction (1.2) does not vanish for the Dirac and Hirzebruch operators. A class of well-posed (Fredholm) boundary value problems for general elliptic operators, i.e. for operators which may violate the Atiyah–Bott condition, was introduced in [SSS98], (see also [Cal76], [BBW93], and [Sch01]).

This extension of the class of boundary value problems is based on the idea of symmetrization of the Shapiro–Lopatinskii condition (1.1) by placing in its right-hand side some vector bundle over the cospheres $S^*(\partial M)$, which (in general) is not a pull-back from the base ∂M.

More precisely, we consider boundary value problems of the form

$$\begin{cases} Du = f, \\ Bj_{\partial M}^{m-1}u = g, \end{cases} \qquad g \in \mathrm{Im}\, P \subset C^\infty(\partial M, G). \tag{2.1}$$

In contrast to the classical boundary value problems, here the boundary value g is an element of a *subspace*, which is defined as the range of a *pseudodifferential projection* P on ∂M. An example of (2.1) is given by the boundary value problem for the Cauchy–Riemann operator in the disk. In this case, the boundary value is an element of the Hardy (sub)space of boundary values of holomorphic functions.

2. Let us now turn to the homotopy classification problem for elliptic equations (2.1). First of all, it can be shown that a general boundary problem (2.1) is homotopic to the so-called spectral boundary value problem for a first order operator. By a *spectral boundary value problem* we mean the problem of the form (cf. [APS75])

$$\begin{cases} Du = f, \\ P_+\, u|_{\partial M} = g, \end{cases} \qquad g \in \mathrm{Im}\, P_+ \subset C^\infty(\partial M, E|_{\partial M}); \tag{2.2}$$

here, we assume that the elliptic D has the form

$$D = \frac{\partial}{\partial t} + A$$

in a neighbourhood of the boundary, that A is an elliptic self-adjoint operator on ∂M, and P_+ denotes its spectral projection

$$P_+ = \frac{A + |A|}{2|A|},$$

which corresponds to the nonnegative part of the spectrum.

Further simplification of the boundary value problem (2.2) for general operators can be obtained if one assumes some additional properties on the projection P_+ that defines the subspace of boundary values. In the following section, we consider one such condition.

3 Boundary value problems in even subspaces

1. Let us consider spectral boundary value problems (2.2) with *even* spectral projections, i.e., the principal symbol of the projection is an even function with

$$\sigma\left(P\right)\left(x,-\xi\right) = \sigma\left(P\right)\left(x,\xi\right)$$

on the cosphere bundle.

It turns out that on an odd-dimensional closed manifold X (for boundary value problems $X = \partial M$), even projections admit a simple homotopy classification. In [SS99] the following isomorphism obtained:

$$\chi : S^{ev}\left(X\right) \otimes \mathbb{Z}\left[\frac{1}{2}\right] \xrightarrow{\simeq} K\left(X\right) \otimes \mathbb{Z}\left[\frac{1}{2}\right] \oplus \mathbb{Z}\left[\frac{1}{2}\right], \tag{3.1}$$

where $S^{ev}\left(X\right)$ is the group of stable homotopy classes of even projections. The first component of χ is determined by the principal symbol of the projection, while the second one is a functional on the set of projections. This dyadic functional, denoted by d, is characterized by the following properties:

1. additivity and homotopy invariance;
2. (normalization) $d\left(P_{C^\infty(X,E)}\right) = 0$;
3. (dimension) $d\left(P\right) = \operatorname{rank} P$ for a finite-rank projection P.

By $P_{C^\infty(X,E)}$ we denote the projection onto the space of vector bundle sections.

The isomorphism in (3.1) implies that for an even projection P there exists a natural number N and a homotopy of projections such that

$$2^N P \sim P_{C^\infty(X,E')} \oplus P', \tag{3.2}$$

where P' has finite rank.

2. The homotopy classification of projections (3.1) leads to the homotopy classification of corresponding spectral boundary value problems. Denote by $\mathrm{Ell}^{ev}\left(M\right)$ the group of stable homotopy classes of spectral boundary value problems for first order operators with even projections. The following isomorphism is valid

$$\chi : \mathrm{Ell}^{ev}\left(M\right) \otimes \mathbb{Z}\left[\frac{1}{2}\right] \xrightarrow{\simeq} K_c\left(T^*\left(M\backslash\partial M\right)\right) \otimes \mathbb{Z}\left[\frac{1}{2}\right] \oplus \mathbb{Z}\left[\frac{1}{2}\right]$$

where the first component of χ is determined by the principal symbol of the operator, while the second one is equal to the functional d evaluated on the corresponding spectral projection.

As a corollary to the homotopy classification, one obtains an index formula for the spectral boundary value problem (2.2)

$$\operatorname{ind}(D, P_+) = \frac{1}{2}\operatorname{ind}(D \cup \alpha^* D) - d(P_+),$$

here, $D \cup \alpha^* D$ denotes an elliptic operator on the double $M \cup_{\partial M} M$. Its principal symbol is defined by glueing the principal symbols $\sigma(D)$ and $\alpha^* \sigma(D)$, where α^* is induced by the antipodal involution $\alpha(x, \xi) = (x, -\xi)$ of the cotangent bundle.

3. In [SS00] the so-called *odd projections* were considered. These pseudodifferential projections satisfy the equation

$$\sigma(P)(x, -\xi) + \sigma(P)(x, \xi) = Id.$$

In that paper, the homotopy classification of projections was obtained. Without going into the details, let us only mention that odd projections are considered on even-dimensional manifolds and, for example, the following spectral boundary value problem index formula is valid

$$\operatorname{ind}(D, P_+) = \frac{1}{2}\operatorname{ind}\left(D \cup \alpha^* D^{-1}\right) - d(P_+);$$

here, the operator on the double is defined in terms of the principal symbols $\sigma(D)$ and $\alpha^* \sigma(D)^{-1}$.

Acknowledgements

The first and the third authors are supported by RFBR grants NN 99-01-01254, 99-01-01100, 00-01-00161 and by the Arbeitsgruppe Partielle Differentialgleichungen und Komplexe Analysis, Institut für Mathematik, Universität Potsdam.

A preliminary version of this paper was published in [SSS99].

References

[AB64]　M.F. Atiyah and R. Bott. The index problem for manifolds with boundary. In *Bombay Colloquium on Differential Analysis*, pages 175–186, Oxford, 1964. Oxford University Press.

[APS75]　M. Atiyah, V. Patodi, and I. Singer. Spectral asymmetry and Riemannian geometry I. *Math. Proc. Cambridge Philos. Soc.*, 77:43–69, 1975.

[BBW93]　B. Booß-Bavnbek and K. Wojciechowski. *Elliptic Boundary Problems for Dirac Operators*. Birkhäuser, Boston–Basel–Berlin, 1993.

[BdM71]　L. Boutet de Monvel. Boundary problems for pseudodifferential operators. *Acta Math.*, 126:11–51, 1971.

[Cal76]   A.P. Calderón. *Lecture Notes on Pseudo-Differential Operators and Elliptic Boundary Value Problems, I.* Consejo Nacional de Investigaticnes V Technicas. Instituto Argentino de Mathematica, Buenos AIRES, 1976.

[Hör85]   L. Hörmander. *The Analysis of Linear Partial Differenticl Operitors. III.* Springer–Verlag, Berlin Heidelberg New York Tokyo, 1985.

[SS99]   A.Yu. Savin and B.Yu. Sternin. Elliptic operators in even subspaces. *Matem. sbornik*, 190(8):125–160, 1999. English transl.: Sbornik: Mathematics **190**, N 8 (1999), p. 1195–1228; math.DG/9907027.

[SS00]   A.Yu. Savin and B.Yu. Sternin. Elliptic operators in odd suospaces. *Matem. sbornik*, 191(8):89–112, 2000. English transl.: Sbornik: Mathematics **191**, N 8 (2000), math.DG/9907039.

[Sch01]   B.-W. Schulze. An algebra of boundary value problems not requiring Shapiro–Lopatinskij conditions. J. Funct. Anal. **179** (2001), 374-408.

[SSS98]   B.-W. Schulze, E. Sternin, and V. Shatalov. On general bourdary value prcblems for elliptic equations. *Math. Sb.*, 189(10):145–160, 1998. English transl.: Sbornik: Mathematics **189**, N 10 (1998), p. 1573–1586.

[SSS99]   A. Savin, B.-W. Schulze, and B. Sternin. *The Homotopy Classification and the Index of Boundary Value Problems for General Elliptic Operators.* Univ. Potsdam, Institut für Mathematik, Potsdam, Oktober 1999. Preprint N 99/20.

Addresses

ANTON SAVIN, Department of Computational Mathematics ard Cybernetics, Moscow State University, Vorob'evy Gory 119899, Moscow. Russia

E-MAIL: antonsavin@mtu-net.ru

BERT-WOLFGANG SCHULZE, Institut für Mathematik, Universität Potsdam, Am Neuen Palais 10 , 14469 Potsdam

E-MAIL: schulze@math.uni-potsdam.de

BORIS STERNIN, Department of Computational Mathematics and Cybernetics, Moscow State University, Vorob'evy Gory 119899, Moscow, Russia

E-MAIL: sternine@mtu-net.ru

2000 Mathematics Subject Classification. Primary 58J32 ; Secondary 58J20, 19M05

Operator Theory:
Advances and Applications, Vol. 126
© 2001 Birkhäuser Verlag Basel/Switzerland

Bosons in a Trap: Asymptotic Exactness of the Gross-Pitaevskii Ground State Energy Formula

ROBERT SEIRINGER

Abstract. Recent experimental breakthroughs in the treatment of dilute Bose gases have renewed interest in their quantum mechanical description, respectively in approximations to it. The ground state properties of dilute Bose gases confined in external potentials and interacting via repulsive short range forces are usually described by means of the Gross-Pitaevskii energy functional. In joint work with Elliott H. Lieb and Jakob Yngvason its status as an approximation for the quantum mechanical many-body ground state problem has recently been rigorously clarified. We present a summary of this work, for both the two- and three-dimensional case.

1   Introduction

The Gross-Pitaevskii (GP) functional was introduced in the early sixties as a phenomenological description of the order parameter in superfluid He$_4$ [G61,P61,G63]. It has come into prominence again because of recent experiments on Bose-Einstein condensation of dilute gases in magnetic traps. The paper [DGPS99] brings an up to date review of these developments.

The present contribution is based on the joint work [LSY00a,LSY00b] with Elliott H. Lieb and Jakob Yngvason (see also [LSY00c]). The starting point of our investigation is the Hamiltonian for N identical bosons moving in $\mathbb{R}^D$, $D = 2$ or 3, that interact with each other via a radially symmetric pair-potential $v(|\boldsymbol{x}_i - \boldsymbol{x}_j|)$ and are confined by an external potential $V(\boldsymbol{x})$:

$$H = \sum_{i=1}^{N}\{-\Delta_i + V(\boldsymbol{x}_i)\} + \sum_{1 \leq i < j \leq N} v(|\boldsymbol{x}_i - \boldsymbol{x}_j|). \tag{1.1}$$

The Hamiltonian acts on *symmetric* wave functions in $\otimes^N L^2(\mathbb{R}^D, d\boldsymbol{x})$. The pair interaction v is assumed to be *nonnegative* and of short range, more precisely, we demand it to have a finite scattering length. (For a definition of the scattering length in arbitrary dimension see [LY00].) The potential V that represents the trap is locally bounded and $V(\boldsymbol{x}) \to \infty$ as $|\boldsymbol{x}| \to \infty$. By shifting the energy scale we can assume that $\min_{\boldsymbol{x}} V(\boldsymbol{x}) = 0$.

Units are chosen so that $\hbar = 2m = 1$, where m is the particle mass. A natural energy unit is given by the ground state energy $\hbar\omega$ of the one particle Hamiltonian

$-(\hbar^2/2m)\Delta + V$. The corresponding length unit, $\sqrt{\hbar/(m\omega)}$, measures the effective extension of the trap.

We are interested in the ground state energy $E^{\mathrm{QM}} = \inf \operatorname{spec} H$. Besides N it depends on the potentials V and v, but with V fixed and

$$v(r) = (a_1/a)^2 v_1(a_1 r/a), \tag{1.2}$$

where v_1 has scattering length a_1 and is regarded as *fixed*, E^{QM} is a function of N and a only. The corresponding ground state density is given by

$$\rho^{\mathrm{QM}}(\boldsymbol{x}) = N \int |\Psi_0(\boldsymbol{x}, \boldsymbol{x}_2, \ldots, \boldsymbol{x}_N)|^2 d\boldsymbol{x}_2 \ldots d\boldsymbol{x}_N, \tag{1.3}$$

where Ψ_0 is a ground state wave function of H.

Note that v given in (1.2) has scattering length a. Here a is dimensionless and really stands for $a\sqrt{m\omega/\hbar}$. Hence a scaling of v like (1.2) is equivalent to scaling the external potential V at fixed v. In particlar the limit $a \to 0$ with fixed V is equivalent to the limit $\omega \to 0$ with fixed v, if one introduces the scaling $V(\boldsymbol{x}) = \omega V_1(\omega^{1/2}\boldsymbol{x})$ for some fixed V_1.

Recent experiments on Bose-Einstein condensation are usually interpreted in terms of a function $\Phi^{\mathrm{GP}}(\boldsymbol{x})$ of $\boldsymbol{x} \in \mathbb{R}^D$, which minimizes the *Gross-Pitaevskii energy functional*

$$\mathcal{E}^{\mathrm{GP}}[\Phi] = \int_{\mathbb{R}^D} \left(|\nabla\Phi|^2 + V|\Phi|^2 + 4\pi g|\Phi|^4 \right) d\boldsymbol{x} \tag{1.4}$$

under the subsidiary condition $\int |\Phi|^2 = N$. The corresponding energy is

$$E^{\mathrm{GP}}(N, g) = \inf_{\int |\Phi|^2 = N} \mathcal{E}^{\mathrm{GP}}[\Phi] = \mathcal{E}^{\mathrm{GP}}[\Phi^{\mathrm{GP}}]. \tag{1.5}$$

The parameter g is different in dimensions 2 and 3. However, for any value of $g > 0$ and $N > 0$ it can be shown that a unique, strictly positive Φ^{GP} exists [LSY00a]. It depends on these parameters, of course, and when this is important we denote it by $\Phi^{\mathrm{GP}}_{N,g}$.

The motivation of the term $4\pi g|\Phi|^4$ in the GP functional comes from the ground state energy density, $\varepsilon_0(\rho)$, of a a dilute, thermodynamically infinite, homogeneous Bose gas of density ρ, interacting via a repulsive potential with scattering length a. The formulas for this quantity are older than the GP functional [B47,HY57,S71], at least for $D = 3$, but they have only very recently been derived rigorously for suitable interparticle potentials. See [LY98] and [LY00]. They are given by

$$\varepsilon_0(\rho) \approx 4\pi a\rho^2 \qquad \text{for } D = 3,$$
$$\varepsilon_0(\rho) \approx 4\pi\rho^2 |\ln(a^2\rho)|^{-1} \qquad \text{for } D = 2, \tag{1.6}$$

where $\approx$ means that the formulas are valid for *dilute* gases, where $c^D\rho \ll 1$. Hence the natural choice of the parameter g is

$$g = a \qquad \text{for } D = 3, \tag{1.7}$$
$$g = |\ln(a^2\bar\rho)|^{-1} \qquad \text{for } D = 2, \tag{1.8}$$

where $\bar\rho$ is the *mean GP density*

$$\bar\rho = \frac{1}{N}\int |\Phi^{\mathrm{GP}}(x)|^4 dx. \tag{1.9}$$

Note that Φ^{GP} depends on g, so (1.8) together with (1.9) are non-linear equations for g. Alternatively, one could define g using the minimizer for $g = 1$ in the definition of $\bar\rho$. Since $\bar\rho$ appears only under a logarithm, this would not effect our leading order calculations. For the same reason one could use the TF minimizer (see below) instead of the GP minimizer to define g. Note also that unlike in the three-dimensional case g depends on N in the two-dimensional case.

The idea is now that with this choice of g one should, for *dilute* gases, have that

$$E^{\mathrm{GP}} \approx E^{\mathrm{QM}} \quad \text{and} \quad \rho^{\mathrm{QM}}(x) \approx \left|\Phi^{\mathrm{GP}}(x)\right|^2 \equiv \rho^{\mathrm{GP}}(x). \tag{1.10}$$

This is made precise in the following theorems. Note that by scaling

$$E^{\mathrm{GP}}(N,g) = NE^{\mathrm{GP}}(1,Ng) \quad \text{and} \quad \Phi^{\mathrm{GP}}_{N,g}(x) = N^{1/2}\Phi^{\mathrm{GP}}_{1,Ng}(x). \tag{1.11}$$

Hence Ng is the natural parameter in GP theory. With this in mind we can state our first main result.

Theorem 1.1 (The GP limit of the QM ground state energy and density).
If $N \to \infty$ with Ng fixed, then

$$\lim_{N\to\infty} \frac{E^{\mathrm{QM}}(N,a)}{E^{\mathrm{GP}}(N,g)} = 1, \tag{1.12}$$

and

$$\lim_{N\to\infty} \frac{1}{N}\rho^{\mathrm{QM}}(x) = \left|\Phi^{\mathrm{GP}}_{1,Ng}(x)\right|^2 \tag{1.13}$$

in the weak L_1-sense.

Note that by hypothesis of the theorem above it really applies to dilute gases, since for fixed Ng (which we refer to as the GP case) the mean density $\bar\rho$ is of order N and

$$a^3\bar\rho \sim N^{-2} \quad \text{for} \quad D = 3, \quad a^2\bar\rho \sim \exp(-N) \quad \text{for} \quad D = 2. \tag{1.14}$$

Especially for $D = 2$ this is an unsatisfactory restriction, since a has to decrease exponentially with N. For a slower decrease Ng tends to infinity with N, and the same holds for $D = 3$ if a does not decrease at least as N^{-1}. In this case, the gradient term in the GP functional becomes negligible compared to the other terms and the so-called *Thomas-Fermi (TF) functional*

$$\mathcal{E}^{\mathrm{TF}}[\rho] = \int_{\mathbb{R}^D} \left(V\rho + 4\pi g\rho^2\right) d\boldsymbol{x} \tag{1.15}$$

arises. It is defined for nonnegative functions ρ on $\mathbb{R}^D$. Its ground state energy E^{TF} and density ρ^{TF} are defined analogously to the GP case. Our second main result is that minimization of (1.15) reproduces correctly the ground state energy and density of the many-body Hamiltonian in the limit when $N \to \infty$, $a^D\bar{\rho} \to 0$, but $Ng \to \infty$ (which we refer to as the TF case), provided the external potential is reasonably well behaved. We will assume that V is asymptotically equal to some function W that is homogeneous of some order $s > 0$ and locally Hölder continuous (see [LSY00b] for a precise definition). This condition can be relaxed, but it seems adequate for most practical applications and simplifies things considerably.

Theorem 1.2 (The TF limit of the QM ground state energy and density).
Assume that V satisfies the conditions stated above. If $\gamma \equiv Ng \to \infty$ as $N \to \infty$, but still $a^D\bar{\rho} \to 0$, then

$$\lim_{N\to\infty} \frac{E^{\mathrm{QM}}(N, a)}{E^{\mathrm{TF}}(N, g)} = 1, \tag{1.16}$$

and

$$\lim_{N\to\infty} \frac{\gamma^{D/(s+D)}}{N}\rho^{\mathrm{QM}}(\gamma^{1/(s+D)}\boldsymbol{x}) = \tilde{\rho}^{\mathrm{TF}}(\boldsymbol{x}) \tag{1.17}$$

in the weak L_1-sense, where $\tilde{\rho}^{\mathrm{TF}}$ is the minimizer of the TF functional under the condition $\int \rho = 1$, $g = 1$, and with V replaced by W.

Remark. The theorems are independent of the interaction potential v_1 in (1.2). This means that in the limit we consider only the scattering length effects the ground state properties, and not the details of the potential. Note also that the particular limit we consider is *not* a mean field limit, since the interaction potential is very hard in this limit; in fact the term $4\pi g|\Phi|^4$ is mostly kinetic energy.

In the following, we will present a brief sketch of the proof of Theorems 1.1 and 1.2. We will derive appropriate upper and lower bounds on the ground state energy E^{QM}. The convergence of the densities follows from the convergence of the energies in the usual way by variation with respect to the external potential. We refer to [LSY00a] and [LSY00b] for details.

2 Upper Bound to the QM Energy

To derive an upper bound on E^{QM} we use a generalization of a trial wave function of Dyson [D57], who used this function to give an upper bound on the ground state energy of the homogeneous hard core Bose gas. It is of the form

$$\Psi(\boldsymbol{x}_1,\ldots,\boldsymbol{x}_N) = \prod_{i=1}^{N} \Phi^{\mathrm{GP}}(\boldsymbol{x}_i) F(\boldsymbol{x}_1,\ldots,\boldsymbol{x}_N), \tag{2.1}$$

where F is constructed in the following way:

$$F(\boldsymbol{x}_1,\ldots,\boldsymbol{x}_N) = \prod_{i=1}^{N} f(t_i(\boldsymbol{x}_1,\ldots,\boldsymbol{x}_i)), \tag{2.2}$$

where $t_i = \min\{|\boldsymbol{x}_i - \boldsymbol{x}_j|, 1 \le j \le i-1\}$ is the distance of $\boldsymbol{x}_i$ to its *nearest neighbor* among the points $\boldsymbol{x}_1,\ldots,\boldsymbol{x}_{i-1}$, and f is a function of $r \ge 0$. We choose it to be

$$f(r) = \begin{cases} f_0(r)/f_0(b) & \text{for} \quad r < b \\ 1 & \text{for} \quad r \ge b, \end{cases} \tag{2.3}$$

where f_0 is the solution of the zero-energy scattering equation (see [LY00]) and b is some cut-off parameter of order $b \sim \bar{\rho}^{-1/D}$. The function (2.1) is not totally symmetric, but for an upper bound it is nevertheless an acceptable test wave function since the bosonic ground state energy is equal to the *absolute* ground state energy.

The result of a somewhat lengthy computation is the upper bound

$$E^{\mathrm{QM}}(N,a) \le E^{\mathrm{GP}}(N,g) \times \begin{cases} 1 + O(a\bar{\rho}^{1/3}) & \text{for} \quad D = 3 \\ 1 + O(|\ln(a^2\bar{\rho})|^{-p}) & \text{for} \quad D = 2, \end{cases} \tag{2.4}$$

with the power p equal to 1 in the GP case and $s/(s+2)$ in the TF case (where V is asymptotically homogeneous of order s).

3 Lower Bound to the QM Energy

To obtain a lower bound for the QM energy the strategy is to divide space into boxes and use the estimate on the homogeneous gas, given in [LY98] and [LY00], in each box with *Neumann* boundary conditions. One then minimizes over all possible divisions of the particles among the different boxes. This gives a lower bound to the energy because discontinuous wave functions for the quadratic form defined by the Hamiltonian are now allowed. We can neglect interactions among particles in different boxes because $v \ge 0$. Finally, one lets the box size tend to zero. However, it is not possible to simply approximate V by a constant potential in each box. To see this consider the case of noninteracting particles, i.e., $v = 0$ and hence $a = 0$. Here $E^{\mathrm{QM}} = N\hbar\omega$, but a 'naive' box method gives only 0 as lower bound, since it clearly pays to put all the particles with a constant wave function in the box with the lowest value of V.

For this reason we start by separating out the GP wave function in each variable and write a general wave function Ψ as

$$\Psi(\boldsymbol{x}_1,\ldots,\boldsymbol{x}_N) = \prod_{i=1}^{N} \Phi^{\mathrm{GP}}(\boldsymbol{x}_i) F(\boldsymbol{x}_1,\ldots,\boldsymbol{x}_N). \tag{3.1}$$

This defines F for a given Ψ because Φ^{GP} is everywhere strictly positive, being the ground state of the operator $-\Delta + V + 8\pi g |\Phi^{\mathrm{GP}}|^2$. We now compute the expectation value of H in the state Ψ. Using partial integration and the variational equation for Φ^{GP}, we see that for computing the ground state energy of H we have to minimize the normalized quadratic form

$$Q(F) = \sum_{i=1}^{N} \frac{\int \prod_{k=1}^{N} \rho^{\mathrm{GP}}(\boldsymbol{x}_k) \left(|\nabla_i F|^2 + \sum_{j=1}^{i-1} v(|\boldsymbol{x}_i - \boldsymbol{x}_j|)|F|^2 - 8\pi g \rho^{\mathrm{GP}}(\boldsymbol{x}_i)|F|^2 \right)}{\int \prod_{k=1}^{N} \rho^{\mathrm{GP}}(\boldsymbol{x}_k)|F|^2}. \tag{3.2}$$

Compared to the expression for the energy involving Ψ itself we have thus obtained the replacements

$$V(\boldsymbol{x}) \to -8\pi g \rho^{\mathrm{GP}}(\boldsymbol{x}) \quad \text{and} \quad \prod_{i=1}^{N} d\boldsymbol{x}_i \to \prod_{i=1}^{N} \rho^{\mathrm{GP}}(\boldsymbol{x}_i) d\boldsymbol{x}_i \tag{3.3}$$

(recall that $\rho^{\mathrm{GP}}(\boldsymbol{x}) = |\Phi^{\mathrm{GP}}(\boldsymbol{x})|^2$). We now use the box method on *this* problem. More precisely, labeling the boxes by an index α, we have

$$\inf_F Q(F) \geq \inf_{\{n_\alpha\}} \sum_\alpha \inf_{F_\alpha} Q_\alpha(F_\alpha), \tag{3.4}$$

where Q_α is defined by the same formula as Q but with the integrations limited to the box α, F_α is a wave function with particle number n_α, and the infimum is taken over all distributions of the particles with $\sum n_\alpha = N$. Approximating ρ^{GP} by a constant in each box, we can use the bound on the homogeneous case ([LY98] and [LY00]) in each box. To control the error terms, we need a lower bound on the ratio $\rho_{\alpha,\mathrm{min}}/\rho_{\alpha,\mathrm{max}}$ in each box, which stems from the measures in (3.2), where $\rho_{\alpha,\mathrm{max}}$ and $\rho_{\alpha,\mathrm{min}}$, respectively, denote the maximal and minimal values of ρ^{GP} in box α.

The problem is that for any fixed size of boxes $\rho_{\alpha,\mathrm{min}}/\rho_{\alpha,\mathrm{max}}$ tends rapidly to zero for boxes far from the origin. This problem can be solved by enclosing the whole system in a big box Λ_R of side length R, with Neumann conditions on the boundary. Replacing Φ^{GP} by Φ_R^{GP}, which is the minimizer of $\mathcal{E}^{\mathrm{GP}}$ restricted to Λ_R (and satisfies Neumann conditions), and restricting the integrations to Λ_R we can let the side lengths of the small boxes tend to zero and be sure that $\rho_{\alpha,\mathrm{min}}/\rho_{\alpha,\mathrm{max}} \to 1$ uniformly for all the boxes α. However, we must control the error made by enclosing the system in the big box. Let $E_R^{\mathrm{QM}}(N,a)$ denote the quantum mechanical ground state energy in the box Λ_R. The essential step is

Lemma 3.1. *There is an $R_0 < \infty$, depending only on Ng, such that*

$$E^{\mathrm{QM}}(N,a) \geq E_R^{\mathrm{QM}}(N,a) \tag{3.5}$$

for all $R \geq R_0$ and all N, a with Ng fixed.

This lemma follows from $V(x) \to \infty$ for $|x| \to \infty$, together with an estimate for the chemical potential

$$E_R^{\mathrm{QM}}(N+1,a) - E_R^{\mathrm{QM}}(N,a) \leq e(Ng)(1+o(1)) \tag{3.6}$$

where $e(Ng)$ depends only on Ng and is independent of R. The proof of (3.6) is similar to the proof of the upper bound (2.4).

Now $\rho_{\alpha,\mathrm{min}}/\rho_{\alpha,\mathrm{max}}$ is bounded below uniformly in each small box contained in Λ_R. We first let the size of the boxes tend to zero as $N \to \infty$, and finally take the limit $R \to \infty$. This proves the desired lower bound (for fixed Ng, i.e. the GP case).

If $Ng \to \infty$ as $N \to \infty$ the method above does not work since the R_0 of Lemma 3.1 tends to infinity with Ng. However, using the explicit form of the TF minimizer, namely

$$\rho_{N,g}^{\mathrm{TF}}(x) = \frac{1}{8\pi g}[\mu^{\mathrm{TF}} - V(x)]_+, \tag{3.7}$$

where $[t]_+ \equiv \max\{t,0\}$ and μ^{TF} is chosen so that the normalization condition $\int \rho_{N,g}^{\mathrm{TF}} = N$ holds, we can use $V(x) \geq \mu^{\mathrm{TF}} - 8\pi g\rho^{\mathrm{TF}}(x)$ to get an replacement as in (3.3) without changing the measure. Moreover, ρ^{TF} has compact support, so, applying again the box method described above, the boxes far out do not contribute to the energy. However, μ^{TF} (which depends only on the combination Ng) tends to infinity as $Ng \to \infty$. Therefore we need to control the asymptotic behaviour of the external potential, and this leads to the restrictions on V described in the paragraph preceding Theorem 1.2.

After controlling all error terms properly, we arrive at the result

$$\liminf_{N \to \infty} \frac{E^{\mathrm{QM}}(N,a)}{E^{\mathrm{TF}}(N,g)} \geq 1 \tag{3.8}$$

in the limit $N \to \infty$, $a^3\bar{\rho} \to 0$ and $Ng \to \infty$. Together with the upper bound (2.4) and the fact that $E^{\mathrm{GP}}(N,g)/E^{\mathrm{TF}}(N,g) = E^{\mathrm{GP}}(1,Ng)/E^{\mathrm{TF}}(1,Ng) \to 1$ as $Ng \to \infty$ this proves Theorem 1.2.

References

[B47]   N.N. Bogoliubov, J. Phys. (U.S.S.R.) **11**, 23 (1947); N.N. Bogoliubov and D.N. Zubarev, Sov. Phys.-JETP **1**, 83 (1955).

[DGPS99]   F. Dalfovo, S. Giorgini, L.P. Pitaevskii, and S. Stringari, *Theory of Bose-Einstein condensation in trapped gases*, Rev. Mod. Phys. **71**, 463–512 (1999).

[D57]   F.J. Dyson, *Ground-State Energy of a Hard-Sphere Gas*, Phys. Rev. **106**, 20–26 (1957).

[G61]   E.P. Gross, *Structure of a Quantized Vortex in Boson Systems*, Nuovo Cimento **20**, 454–466 (1961).

[G63]   E.P. Gross, *Hydrodynamics of a superfluid condensate*, J. Math. Phys. **4**, 195–207 (1963).

[HY57]   K. Huang, C.N. Yang, Phys. Rev. **105**, 767-775 (1957); T.D. Lee, K. Huang, and C.N. Yang, Phys. Rev. **106**, 1135-1145 (1957); K.A. Brueckner, K. Sawada, Phys. Rev. **106**, 1117-1127, 1128-1135 (1957); S.T. Beliaev, Sov. Phys.-JETP **7**, 299-307 (1958); T.T. Wu, Phys. Rev. **115**, 1390 (1959); N. Hugenholtz, D. Pines, Phys. Rev. **116**, 489 (1959); M. Girardeau, R. Arnowitt, Phys. Rev. **113**, 755 (1959); T.D. Lee, C.N. Yang, Phys. Rev. **117**, 12 (1960); E.H. Lieb, Phys. Rev. **130**, 2518–2528 (1963).

[LSY00a]   E.H. Lieb, R. Seiringer, and J. Yngvason, *Bosons in a trap: a rigorous derivation of the Gross-Pitaevskii energy functional*, Phys. Rev. A **61**, 043602-1-13 (2000).

[LSY00b]   E.H. Lieb, R. Seiringer, and J. Yngvason, *A rigorous derivation of the Gross-Pitaevskii energy functional for a two-dimensional Bose gas*, arXiv: cond-mat/0005026, to appear in Commun. Math. Phys.

[LSY00c]   E.H. Lieb, R. Seiringer, J. Yngvason, *The Ground State Energy and Density of Interacting Bosons in a Trap*, in *Quantum Theory and Symmetries*, Goslar, 1999, H.-D. Doebner, V.K. Dobrev, J.-D. Hennig and W. Luecke, eds., pp. 101-110, World Scientific (2000). arXiv math-ph/9911026, mp_arc 99-439.

[LY98]   E.H. Lieb and J. Yngvason, *Ground state energy of the low density Bose gas*, Phys. Rev. Lett. **80**, 2504–2507 (1998).

[LY00]   E.H. Lieb and J. Yngvason, *Ground State Energy of a Dilute Two-dimensional Bose Gas*, arXiv: math-ph/0002014, J. Stat. Phys. (in press)

[P61]   L.P. Pitaevskii, *Vortex lines in an imperfect Bose gas*, Sov. Phys. JETP, **13**, 451–454 (1961).

[S71]   M. Schick, *Two-Dimensional System of Hard Core Bosons*, Phys. Rev. A **3**, 1067–1073 (1971).

Address

ROBERT SEIRINGER, Institut für Theoretische Physik, Universität Wien, Boltzmanngasse 5, A-1090 Vienna, Austria

E-MAIL: rseiring@ap.univie.ac.at

2000 Mathematics Subject Classification. Primary 81V70; Secondary 35Q55, 46N50

Operator Theory:
Advances and Applications, Vol. 126
© 2001 Birkhäuser Verlag Basel/Switzerland

Eigenfunction Expansions Associated with Relativistic Schrödinger Operators

Tomio Umeda[*]

Abstract. We first examine action of $\sqrt{-\Delta}$ on distributions, so as to define $\sqrt{-\Delta}f$ for f in some weighted Sobolev space, which is useful in time-independent scattering theory. We then derive the eigenfunction expansion associated with the relativistic Schrödinger operator $\sqrt{-\Delta}+V(x)$ $(x \in \mathbf{R}^3)$. The potential $V(x)$ is assumed to be a real valued function satisfying $|V(x)| \leq C(1+|x|)^{-\sigma}$ for some $\sigma > 2$. Also, we show that the generalized eigenfunctions obtained here satisfy Lippmann-Schwinger type integral equations.

1 Introduction

The aim of this contribution is to derive the eigenfunction expansion associated with the relativistic Schrödinger operator

$$\sqrt{-\Delta} + V(x) \qquad (x \in \mathbf{R}^3),$$

where V is assumed to be a real-valued measurable function satisfying

$$|V(x)| \leq C(1+|x|)^{-\sigma} \qquad (\sigma > 2).$$

The starting point for the eigenfunction expansion is the fact that the function $e_0(x, k) = \exp(ix \cdot k)$ satisfies the equation

$$(1.1) \qquad \sqrt{-\Delta_x}e_0(x, k) = |k|e_0(x, k) \quad \text{in} \quad \mathcal{S}'(\mathbf{R}_x^n)$$

for every $k \in \mathbf{R}^n$, where $\mathcal{S}'(\mathbf{R}_x^n)$ is the space of tempered distributions. Here and in the sequel, the configuration space can have any dimension. In the equation (1.1), however, the meaning of $\sqrt{-\Delta_x}e_0(x, k)$ is not clear because the symbol $|\xi|$ is singular at the origin and $e_0(x, k)$ does not belong to the Sobolev space of any order as a function of x. Therefore, we must begin with a discussion on how to justify the equation (1.1); in other words, how to extend the action of $\sqrt{-\Delta_x}$ on distributions so that $\sqrt{-\Delta_x}e_0(x, k)$ has a clear meaning.

The difficulty in the definition

$$\langle \sqrt{-\Delta}f, \varphi \rangle := \langle f, \overline{\sqrt{-\Delta}\,\overline{\varphi}} \rangle \qquad (\varphi \in \mathcal{D}(\mathbf{R}^n))$$

---

[*] Dedicated to Professor Kiyoshi Mochizuki on his sixtieth birthday

for $f \in \mathcal{D}'(\mathbf{R}^n)$, the space of distributions, is that $\sqrt{-\Delta}\,\overline{\varphi}$ may neither belong to $\mathcal{D}(\mathbf{R}^n)$, the space of test functions, nor $\mathcal{S}(\mathbf{R}^n)$, the space of rapidly decreasing functions. Actually, in the 1-dimensional case, there exists a function $\varphi_0 \in \mathcal{S}(\mathbf{R})$ such that $\sqrt{-\Delta}\varphi_0$ belongs to $L^{2,\,s}(\mathbf{R})$ for all $s < 3/2$, but does not belong to $L^{2,\,s}(\mathbf{R})$ for all $s \geq 3/2$ (cf. [U00]). Here $L^{2,\,s}(\mathbf{R}^n)$ denotes the weighted L^2 space which is defined by

$$L^{2,\,s}(\mathbf{R}^n) = \{f \in \mathcal{S}'(\mathbf{R}^n) \mid \langle x \rangle^s f \in L^2(\mathbf{R}^n)\}.$$

Hence $\sqrt{-\Delta}\varphi_0$ does neither belong to $\mathcal{D}(\mathbf{R})$ nor $\mathcal{S}(\mathbf{R})$. What we can show in general is the following fact.

Lemma 1.1. *If $s < n/2+1$, then $\sqrt{-\Delta}$ maps $\mathcal{S}(\mathbf{R}^n)$ continuously into $L^{2,\,s}(\mathbf{R}^n)$.*

Note that $e_0(\cdot,\, k) \in L^{2,\,-s}(\mathbf{R}^n)$ for any $s > n/2$. In view of Lemma 1.1, it follows that $\sqrt{-\Delta_x}e_0(x,\, k)$ makes sense as a tempered distribution:

$$(1.2) \qquad \langle \sqrt{-\Delta}e_0(\cdot,\, k),\, \varphi \rangle = \langle e_0(\cdot,\, k),\, \overline{\sqrt{-\Delta}\,\overline{\varphi}} \rangle \qquad (\varphi \in \mathcal{S}(\mathbf{R}^n)).$$

The right hand side of (1.2) becomes

$$\int e^{ix \cdot k} \overline{\sqrt{-\Delta}\,\overline{\varphi}\,(x)}\, dx = (2\pi)^{n/2}\overline{\mathcal{F}[\sqrt{-\Delta}\overline{\varphi}](k)}$$

$$= (2\pi)^{n/2}|k|\,\overline{\mathcal{F}[\overline{\varphi}](k)} = \langle |k|e_0(\cdot,\, k),\, \varphi \rangle.$$

Here $\mathcal{F}[\varphi]$ denotes the Fourier transform of φ. Thus we have shown that (1.1) does make sense.

On the other hand, in spectral and scattering theory, weighted L^2 spaces and weighted Sobolev spaces, which are defined by

$$H^{m,\,s}(\mathbf{R}^n) = \{f \in \mathcal{S}'(\mathbf{R}^n) \mid \langle x \rangle^s \langle D \rangle^m f \in L^2(\mathbf{R}^n)\},$$

play important roles. Therefore, extending the action of $\sqrt{-\Delta}$ to some weighted Sobolev spaces is necessarily a basic problem. As to this problem, the following theorem gives an answer([U00]).

Theorem 1.1. *Let m be in $\mathbf{R}$.*

(a) *If $s \geq 0$ and $t < \min\{1,\, s - n/2\}$, then $\sqrt{-\Delta}$ can be extended to a bounded operator from $H^{m,\,s}(\mathbf{R}^n)$ to $H^{m-1,\,t}(\mathbf{R}^n)$.*

(b) *If $-n/2-1 < s < 0$ and $t < s-n/2$, then the same conclusion as in (a) holds.*

Thanks to Theorem 1.1, we can establish a uniqueness theorem for the equation

$$(\sqrt{-\Delta} - \lambda)u = f \quad \text{in} \quad \mathcal{S}'(\mathbf{R}^n)$$

with a radiation condition.

Theorem 1.2. *Let $n \geq 2$ and $1/2 < s < 1$. Suppose that $u \in L^{2,-s}(\mathbf{R}^n) \cap H^1_{loc}(\mathbf{R}^n)$ satisfies the equation*

$$(\sqrt{-\Delta} - \lambda)u = 0 \quad in \quad \mathcal{S}'(\mathbf{R}^n) \qquad (\lambda > 0),$$

and, in addition, that u satisfies either

$$\left(\frac{\partial}{\partial x_j} - i\lambda\frac{x_j}{|x|}\right)u \in L^{2,s-1}(\mathbf{R}^n), \qquad j = 1, \cdots, n$$

or

$$\left(\frac{\partial}{\partial x_j} + i\lambda\frac{x_j}{|x|}\right)u \in L^{2,s-1}(\mathbf{R}^n), \qquad j = 1, \cdots, n$$

Then u vanishes identically.

2   The Generalized Eigenfunctions

In this section and in the following section, we confine ourselves to the 3 dimensional case. Construction of the generalized eigenfunctions of $\sqrt{-\Delta} + V(x)$ is based on the limiting absorption principle, which was established by Ben-Artzi and Nemirovski [BN97] in a much more general setting. In what follows, the self-adjoint realization of $\sqrt{-\Delta} + V(x)$ in $L^2(\mathbf{R}^3)$ is denoted by H, and the resolvent is denoted by $R(z) = (H - z)^{-1}$. We write $\mathbf{C}^{\pm} = \{ z \mid \pm \Im z > 0 \}$.

Theorem.(Ben-Artzi and Nemirovski)

(a) *The continuous spectrum $\sigma_c(H) = [0, \infty)$ is absolutely continuous, except possibly for a discrete sequence Λ_i of imbedded eigenvalues, which can accumulate only at 0 or ∞.*

(b) *The resolvent $R(z)$, $z \in \mathbf{C}^+$, (respectively $z \in \mathbf{C}^-$) can be extended continuously, with respect to the operator norm topology of $\mathbf{B}(L^{2,s}, H^{1,-s})$, $s > 1/2$, to $\mathbf{C}^+ \cup \big((0, \infty) \setminus \Lambda_i\big)$ (respectively $\mathbf{C}^- \cup \big((0, \infty) \setminus \Lambda_i\big)$).*

The extensions of $R(z)$ to $\mathbf{C}^{\pm} \cup \big((0, \infty) \setminus \Lambda_i\big)$ will be denoted by $R^{\pm}(z)$, respectively. Thanks to Theorem 1.1 again, we are able to show that for $f \in L^{2,s}(\mathbf{R}^3)$

$$(\sqrt{-\Delta} + V(x) - \lambda)R^{\pm}(\lambda)f = f \quad in \quad \mathcal{S}'(\mathbf{R}^3) \qquad (\lambda \in (0, \infty) \setminus \Lambda_i).$$

This fact leads to the construction of the generalized eigenfunctions:

$$\varphi^{\pm}(x, k) := e_0(x, k) - R^{\mp}(|k|)[V(\cdot)e_0(\cdot, k)](x)$$

for k such that $|k| \in (0, \infty) \setminus \Lambda_i$. With this definition, one can prove

$$(\sqrt{-\Delta_x} + V(x))\varphi^\pm(x, k) = |k|\varphi^\pm(x, k) \quad \text{in} \quad \mathcal{S}'(\mathbf{R}_x^3)$$

when $|k| \in (0, \infty) \setminus \Lambda_i$. The generalized eigenfunctions constructed here are complete in the sense described by Properties (a) and (b) of the following theorem.

Theorem 2.1. *There exist partial isometries $\mathcal{F}_\pm$ with the following properties:*

(a) $(\mathcal{F}_\pm u)(k) = (2\pi)^{-3/2} \int u(x)\overline{\varphi^\pm(x, k)}\, dx$;

(b) *The initial sets and the final sets of $\mathcal{F}_\pm$ are $\mathcal{H}_c(H)$, the subspace of continuity with respect to H, and $L^2(\mathbf{R}^3)$ respectively;*

(c) *$\mathcal{F}_\pm$ diagonalize H: $\mathcal{F}_\pm H u = |k|\mathcal{F}_\pm u$ for $u \in \mathrm{Dom}(H)$.*

Sketch of the proof. (Following the idea of Kuroda [K79].) Define

$$\mathcal{F}_\pm = \mathcal{F} W_\pm^*,$$

where $W_\pm = \underset{t \to \pm\infty}{\text{s-}\lim}\ e^{itH} e^{-itH_0}$ and H_0 is the self-adjoint realization of $\sqrt{-\Delta}$ in $L^2(\mathbf{R}^3)$. It is well-known (cf. Simon [S79]) that $W_\pm$ are complete, i.e. $\mathrm{Ran}(H) = \mathcal{H}_c(H)$. On the other hand, one can derive the representation of the wave operators in terms of the extended resolvents:

$$W_\pm = \int_0^\infty \{1 - V R^\pm(\lambda)\}^* \frac{1}{2\pi i} \{R_0^+(\lambda) - R_0^-(\lambda)\}\, d\lambda,$$

where $R_0^\pm(\lambda) = (H_0 - \lambda \mp i0)^{-1}$. Combining this representation with the completeness of the wave operators, one can see that $\mathcal{F}_\pm$ possess the desired properties.

3　The Lippmann-Schwinger Type Integral Equations

Thanks to Theorem 1.1 once again and to Theorem 1.2, we can prove that $I - R^\pm(\lambda)V$, $\lambda \in (0, \infty) \setminus \Lambda_i$, is invertible in $L^{2, -s}$ when $3/2 < s < \sigma - 1/2$, and that

$$(I - R^\pm(\lambda)V)^{-1} = I + R_0^\pm(\lambda)V.$$

Using the formula (Lieb and Loss [LL97])

$$\exp(-tH_0)f(x) = \int_{\mathbf{R}^3} \frac{1}{\pi^2} \frac{t}{[t^2 + |x - y|^2]^2} f(y)\, dy,$$

one can compute explicitly the integral kernels of $R_0^\pm(\lambda)$. This enables us to show that the generalized eigenfunctions $\varphi^\pm(x, k)$ are characterized as the unique solutions to the Lippmann-Schwinger type integral equations

$$f(x) = e_0(x, k) - \int_{\mathbf{R}^3} \frac{1}{2\pi i}\Big\{ \frac{1}{|x - y|^2} - \frac{\lambda}{|x - y|} G_\lambda^\mp(x - y)\Big\} V(y)f(y)\, dy$$

where

$$G_\lambda^\pm(x) = \left\{ \mathrm{Ci}\,(\lambda|x|) \mp i\pi \right\} \sin(\lambda|x|) - \left\{ \mathrm{si}\,(\lambda|x|) + \pi \right\} \cos(\lambda|x|)$$

$$\mathrm{Ci}\,\rho = -\int_\rho^\infty \frac{\cos\tau}{\tau}\, d\tau, \quad \mathrm{Si}\,\rho = \int_0^\rho \frac{\sin\tau}{\tau}\, d\tau, \quad \mathrm{si}\,\rho = \mathrm{Si}\,\rho - \frac{\pi}{2}.$$

Acknowledgements

Research on this paper is supported by Grant-in-Aid for Scientific Research(Basic Research(C), No. 09640212), Japan Society for the Promotion of Science.

References

[BN97] Ben-Artzi, M., Nemirovski, J., Remarks on relativistic Schrödinger operators and their applications, *Ann. Inst. Henri Poincaré* , 67:29-39, 1997.

[K79]   Kuroda, S.T., Spectral theory II, *Iwanami Shoten, Tokyo*, 1979.(In Japanese)

[LL97] Lieb, E.H., Loss, M., Analysis, *American Mathematical Society, Providence*, 1997.

[S79]   Simon, B., Phase space analysis of simple scattering systems: Extensions of some work of Enss, *Duke Math. J.*, 46:119-168, 1979.

[U00]   Umeda, T., Action of $\sqrt{-\Delta}$ on distributions, *preprint* (mp_arc 00-263).

Address

Tomio Umeda, Department of Mathematics, Himeji Institute of Technology, Himeji 671-2201, Japan

E-mail: umeda@sci.himeji-tech.ac.jp

2000 Mathematics Subject Classification. Primary 35P10, 35P25 ; Secondary 47A10, 47A40

Operator Theory:
Advances and Applications, Vol. 126
© 2001 Birkhäuser Verlag Basel/Switzerland

The Time-Dependent Approach to Inverse Scattering

RICARDO WEDER

Abstract. In this talk we present a time-dependent approach for the solution of inverse scattering problems. With our approach we prove uniqueness and we give a method for the reconstruction of the potentials. This approach has been applied to many linear and nonlinear scattering problems. We review the literature and we discuss in particular the application of this method to the case of potential scattering for the linear Schrödinger equation and to the nonlinear Schrödinger equation with a potential. In the later case we uniquely reconstruct the potential and the nonlinearity.

To illustrate our method we first consider a simple inverse scattering problem. Namely, the case of the linear Schrödinger equation with a short-range potential. On the Hilbert space $L^2(\mathbb{R}^n), n \geq 2$, the free Hamiltonian is $H_0 := -\frac{1}{2m}\Delta$, where Δ is the self-adjoint realization of the Laplacian with domain equal to the Sobolev space $W_{2,2}$ and $m > 0$ is the mass of the particle. For simplicity of the presentation we discuss here the case of bounded continuous potentials. The class of short-range potentials consists of multiplication operators V from the set,

$$\mathcal{V}_{SR} := \left\{ V \in C(\mathbb{R}^n) : |\sup_{|x|\geq R} |V(x)| \in L^1([0,\infty), dR) \right\}. \tag{1.1}$$

The interacting Hamiltonian, $H := H_0 + V$, is self-adjoint on the domain of H_0. The linear Schrödinger equation is,

$$i\frac{\partial}{\partial t}u(t,x) = H_0 u(t,x) + V(x)u(t,x); u(0,x) = \phi(x), \tag{1.2}$$

where $t \in \mathbb{R}, x \in \mathbb{R}^n, n \geq 2$. As is well known the wave operators,

$$W_\pm := s - \lim_{t\to\pm\infty} e^{itH} e^{-itH_0}, \tag{1.3}$$

exist and are complete and the scattering operator,

$$S := W_+^* W_-, \tag{1.4}$$

is unitary on L^2. As a general reference for mathematical scattering theory see [RS79]. Let us consider states, ϕ, with compact momentum support on the ball of radius $m\hat{R}$ and center zero,

$$\hat{\phi} \in C_0^\infty \left(B_{m\hat{R}}(0) \right), \tag{1.5}$$

where $\hat{\phi}$ denotes the Fourier transform of ϕ. The boosted state

$$\phi_{\mathbf{v}} := e^{im\mathbf{v}\cdot x}\phi \ \leftrightarrow\ \hat{\phi}_{\mathbf{v}} \in C_0^\infty \left(B_{m\hat{R}}(m\mathbf{v})\right), \tag{1.6}$$

has velocity support of radius $\hat{R}$ around $\mathbf{v}$. In the theorem below we use the high-velocity limit in an arbitrary fixed direction $\hat{\mathbf{v}} := \frac{\mathbf{v}}{|\mathbf{v}|}, v := |\mathbf{v}| \to \infty$.

Theorem 1.1. *For $\phi_{\mathbf{v}}, \psi_{\mathbf{v}}$ as defined in (1.6) the following high-velocity limit exists for any $\hat{R}, \hat{\mathbf{v}}$,*

$$\lim_{v\to\infty} iv\left(\phi_{\mathbf{v}}, (S-I)\psi_{\mathbf{v}}\right) = \left(\phi, \int_{-\infty}^{\infty} d\tau V(x+\hat{\mathbf{v}}\tau)\psi\right). \tag{1.7}$$

Moreover, the scattering operator, S, determines uniquely the potential $V \in \mathcal{V}_{SR}$.

Theorem 1.1 is proven in [EW1-95]. The mathematical proof closely follows physical intuition. The key issue is that at high energies translation of the wave packets dominates over spreading during the interaction time. In fact, in the high-energy limit it is sufficient for the calculation of the scattering operator to consider translation of wave packets rather than their correct free evolution. Since on this limit spreading occurs only when and where the interaction is negligible, i.e. when the free and the interacting time evolutions are almost the same, the effect of spreading does not appear on the scattering operator. For this reason scattering simplifies on the high-energy limit and we can uniquely reconstruct the potential inverting the Radon (or X-ray) transform of the potential. The time-dependent approach also gives error bounds and is quite flexible. It has been applied to many inverse scattering problems. In [EW2-95], [EW3-95] to N-Body systems with singular and long-range potentials, in [W96] to the N-Body Stark effect, and in [EW96] to two-cluster scattering. In [W1-97] the case of N-Body systems with time-dependent potentials was treated. The case of magnetic fields was considered in [A97], [A1-98], and [A2-98]. The relativistic Schrödinger operator, and the Dirac and Klein-Gordon equations where studied in [J96] and [J97]. In [I98] the Dirac equation with time-dependent electromagnetic potentials was considered. In all these papers the direct scattering problem is linear. The study, with stationary methods, of multidimensional inverse scattering problems in the case where the direct problem is linear has a long and distinguished history going back to the works of Fadeev [F56] and Berezanskii [B58]. For further references on the stationary theory see [W91] and[EW3-95].

The time-dependent approach does not uses the linearity of the direct scattering problem in an essential way and, as a consequence, it has a natural extension to the case where the direct scattering problem is nonlinear. The goal in this case is to give a method to uniquely reconstruct the potential and the nonlinearity. The nonlinear Schrödinger equation was treated in [W2-97], [W1-00], [W2-00], [W1-01] and [W2-01] and the nonlinear Klein-Gordon equation in [W3-01] and [W4-01].

Below we discuss the case of the nonlinear Schrödinger equation on the line. For the multidimensional case see [W2-01].

We study the following nonlinear Schrödinger equation with a potential

$$i\frac{\partial}{\partial t}u(t,x) = -\frac{d^2}{dx^2}u(t,x) + V_0(x)u(t,x) + F(x,u), u(0,x) = \phi(x),\qquad (1.8)$$

where $t,x \in \mathbb{R}$, the potential, V_0, is a real-valued function and $F(x,u)$ is a complex-valued function. We first introduce some standard notations and definitions. We say that $F(x,u)$ is a C^k function of u in the real sense if for each $x \in \mathbb{R}$, $\Re F$ and $\Im F$ are C^k functions with respect to the real and imaginary parts of u. We will assume that F is C^2 in the real sense and that $\left(\frac{\partial}{\partial x}F\right)(x,u)$ is C^1 in the real sense. If $F = F_1 + iF_2$ with F_1, F_2 real-valued, and $u = r + is, r,s \in \mathbb{R}$ we denote,

$$F^{(2)}(x,u) := \sum_{j=1}^{2}\left[\left|\frac{\partial^2}{\partial r^2}F_j(x,u)\right| + \left|\frac{\partial^2}{\partial r\partial s}F_j(x,u)\right| + \left|\frac{\partial^2}{\partial s^2}F_j(x,u)\right|\right],\qquad (1.9)$$

$$\left(\frac{\partial}{\partial x}F\right)^{(1)}(x,u) := \sum_{j=1}^{2}\left[\left|\frac{\partial}{\partial r}\left(\frac{\partial}{\partial x}F_j\right)(x,u)\right| + \left|\frac{\partial}{\partial s}\left(\frac{\partial}{\partial x}F_j\right)(x,u)\right|\right].\qquad (1.10)$$

For any $\gamma \in \mathbb{R}$, L_γ^1 denotes the Banach space of all complex-valued measurable functions, ϕ, defined on $\mathbb{R}$ and such that

$$\|\phi\|_{L_\gamma^1} := \int |\phi(x)|\,(1+|x|)^\gamma\,dx < \infty.\qquad (1.11)$$

If $V_0 \in L_1^1$ the differential expression $\tau := -\frac{d^2}{dx^2} + V_0(x)$ is essentially self-adjoint on the domain

$$D(\tau) := \left\{\phi \in L_C^2 : \phi \text{ and } \frac{d}{dx}\phi \text{ are absolutely continuous and } \tau\phi \in L^2\right\},\qquad (1.12)$$

where L_C^2 denotes the set of all functions on L^2 that have compact support. We denote by H the unique self-adjoint realization of τ. It is known that H has a finite number of negative eigenvalues, that it has no positive or zero eigenvalues, that it has no singular-continuous spectrum and that the absolutely-continuous spectrum is $[0,\infty)$. By H_0 we denote the unique self-adjoint realization of $-\frac{d^2}{dx^2}$ with domain equal to the Sobolev space $W_{2,2}$. The wave operators are defined as in (1.3). It well known (see [S81]) that the $W_\pm$ exist and are complete. The scattering operator for the linear Schrödinger equation ((1.8) with $F = 0$) is defined as follows:

$$S_L := W_+^* W_-.\qquad (1.13)$$

For any pair u,v of solutions to the stationary Schrödinger equation:

$$-\frac{d^2}{dx^2}u + V_0 u = k^2 u,\ k \in C,\qquad (1.14)$$

let $[u, v]$ denotes the Wronskian of u, v:

$$[u, v] := \left(\frac{d}{dx} u \right) v - u \frac{d}{dx} v. \tag{1.15}$$

Let $f_j(x, k), j = 1, 2, \Im k \geq 0$, be the Jost solutions to (1.14) (see for example [DT79]). A potential V_0 is said to be *generic* if $[f_1(x, 0), f_2(x, 0)] \neq 0$ and V_0 is said to be *exceptional* if $[f_1(x, 0), f_2(x, 0)] = 0$. If V_0 is *exceptional* there is a bounded solution to (1.14) with $k^2 = 0$ (a half-bound state or a zero-energy resonance). The trivial potential, $V_0 = 0$, is *exceptional*. Let us designate,

$$M := \left\{ u \in C(\mathbb{R}, W_{1,p+1}) : \sup_{t \in \mathbb{R}} (1 + |t|)^d \|u\|_{W_{1,p+1}} < \infty \right\},$$

$$\text{with norm} : \|u\|_M := \sup_{t \in \mathbb{R}} (1 + |t|)^d \|u\|_{W_{1,p+1}}, \tag{1.16}$$

where $p \geq 1$, and $d := \frac{1}{2} \frac{p-1}{p+1}$. For functions $u(t, x)$ defined in $\mathbb{R}^2$ we denote $u(t)$, for $u(t, \cdot)$. The small-amplitude scattering operator is given in the following theorem.

Theorem 1.2. *Suppose that $V_0 \in L^1_\gamma$, where in the generic case $\gamma > 3/2$ and in the exceptional case $\gamma > 5/2$, that H has no negative eigenvalues, and that*

$$N(V_0) := \sup_{x \in \mathbb{R}} \int_x^{x+1} |V_0(y)|^2 \, dy < \infty. \tag{1.17}$$

Furthermore, assume that F is C^2 in the real sense, that $F(x, 0) = 0$, and that for each fixed $x \in \mathbb{R}$ all the first order derivatives, in the real sense, of F vanish at zero. Moreover, suppose that $\frac{\partial}{\partial x} F$ is $C^{(1)}$ in the real sense. We assume that the following estimates hold:

$$F^{(2)}(x, u) = O\left(|u|^{p-2}\right), \quad \left(\frac{\partial}{\partial x} F\right)^{(1)}(x, u) = O\left(|u|^{p-1}\right), \; u \to 0, \tag{1.18}$$

uniformly for $x \in \mathbb{R}$, for some $\rho < p < \infty$, and where ρ is the positive root of $\frac{1}{2} \frac{\rho-1}{\rho+1} = \frac{1}{\rho}$. Then, there is a $\delta > 0$ such that for all $\phi_- \in W_{2,2} \cap W_{1,1+\frac{1}{p}}$ with $\|\phi_-\|_{W_{2,2}} + \|\phi_-\|_{W_{1,1+\frac{1}{p}}} \leq \delta$ there is a unique solution, u, to (1.8) such that $u \in C(\mathbb{R}, W_{1,2}) \cap M$ and,

$$\lim_{t \to -\infty} \|u(t) - e^{-itH} \phi_-\|_{W_{1,2}} = 0. \tag{1.19}$$

Moreover, there is a unique $\phi_+ \in W_{1,2}$ such that

$$\lim_{t \to \infty} \|u(t) - e^{-itH} \phi_+\|_{W_{1,2}} = 0. \tag{1.20}$$

Furthermore, $e^{-itH}\phi_{\pm} \in M$ and

$$\left\| u - e^{-itH}\phi_{\pm} \right\|_M \leq C \left\| e^{-itH}\phi_{\pm} \right\|_M^p, \tag{1.21}$$

$$\left\| \phi_+ - \phi_- \right\|_{W_{1,2}} \leq C \left[\left\| \phi_- \right\|_{W_{2,2}} + \left\| \phi_- \right\|_{W_{1,1+\frac{1}{p}}} \right]^p. \tag{1.22}$$

The scattering operator, $S_{V_0} : \phi_- \hookrightarrow \phi_+$ is injective on $W_{1,1+\frac{1}{p}} \cap W_{2,2}$.

Note that we do not restrict F in such a way that energy is conserved. Moreover, $\rho \approx 3.56$.

To reconstruct the potential, V_0, we introduce below the scattering operator that relates asymptotic states that are solutions to the linear Schrödinger equation with potential zero:

$$S := W_+^* S_V W_-. \tag{1.23}$$

We have proven the following results.

Theorem 1.3. *Suppose that the assumptions of Theorem 1.2 are satisfied. Then for every $\phi \in W_{2,2} \cap W_{1,1+\frac{1}{p}}$*

$$\left. \frac{d}{d\epsilon} S(\epsilon\phi) \right|_{\epsilon=0} = S_L\phi, \tag{1.24}$$

where the derivative in the left-hand side of (1.24) exists in the strong convergence sense in $W_{1,2}$.

Corollary 1.1. *Under the conditions of Theorem 1.2 the scattering operator, S, determines uniquely the potential V_0.*

In the case where $F(x,u) = \sum_{j=1}^{\infty} V_j(x)|u|^{2(j_0+j)}u$ we can also reconstruct the $V_j, j = 1, 2, \cdots$.

Lemma 1.1. *Suppose that the conditions of Theorem 1.2 are satisfied, and more-over, that $F(x,u) = \sum_{j=1}^{\infty} V_j(x)|u|^{2(j_0+j)}u$, where j_0 is an integer such that, $j_0 \geq (p-3)/2$, for $|u| \leq \eta$, for some $\eta > 0$, and where $V_j \in W_{1,\infty}$ with $\|V_j\|_{W_{1,\infty}} \leq C^j, j = 1, 2, \cdots$, for some constant C. Then, for any $\phi \in W_{2,2} \cap W_{1,1+\frac{1}{p}}$ there is an $\epsilon_0 > 0$ such that for all $0 < \epsilon < \epsilon_0$:*

$$i\left((S_{V_0} - I)(\epsilon\phi), \phi\right)_{L^2} = \sum_{j=1}^{\infty} \epsilon^{2(j_0+j)+1} \left[\int\int dt\, dx\, V_j(x) \left| e^{-itH}\phi \right|^{2(j_0+j+1)} + Q_j \right], \tag{1.25}$$

where $Q_1 = 0$ and $Q_j, j > 1$, depends only on ϕ and on V_k with $k < j$. Moreover, for any $x' \in \mathbb{R}$, and any $\lambda > 0$, we denote, $\phi_\lambda(x) := \phi(\lambda(x - x'))$. Then, if $\phi \neq 0$:

$$V_j(\acute{x}) = \frac{\lim_{\lambda\to\infty} \lambda^3 \int\int dt\, dx\, V_j(x) \left| e^{-itH}\phi_\lambda \right|^{2(j_0+j+1)}}{\int\int dt\, dx\, \left| e^{-itH_0}\phi \right|^{2(j_0+j+1)}}. \tag{1.26}$$

Corollary 1.2. *Under the conditions of Lemma 1.1 the scattering operator, S, determines uniquely the potentials $V_j, j = 0, 1, \cdots$.*

The method to reconstruct the potentials $V_j, j = 0, 1, \cdots$, is as follows. First we obtain S_L from S using (1.24). By any standard method for inverse scattering for the linear Schrödinger equation on the line we reconstruct V_0. We then reconstruct S_{V_0} from S using (1.23). Finally (1.25) and (1.26) give us, recursively, $V_j, j = 1, 2, \cdots$. Theorems 1.2, 1,3, Lemma 1.1 and Corollaries 1.1, 1.2 are proven in [W1-01] where also a discussion of the literature is given. The key issue for the proof is the following time-dependent $L^p - L^{p'}$ estimate that we proved in [W1-00],

$$\left\| e^{-itH} P_c \right\|_{B\left(L^p, L^{p'}\right)} \leq \frac{C}{t^{\left(\frac{1}{p} - \frac{1}{2}\right)}}, t > 0, \tag{1.27}$$

for some constant $C, 1 \leq p \leq 2$, and $\frac{1}{p} + \frac{1}{p'} = 1$ and where P_c denotes the projector onto the space of continuity of H. The $L^p - L^{p'}$ estimate (1.27) expresses the dispersive nature of the solutions to the linear Schrödinger equation with initial data on the continuous subspace of H. It gives a quantitative meaning to the *spreading of the wave packets*. In the typical applications the nonlinearity, F, is a high-enough power of u. This type of nonlinearities make the solutions to (1.8) bigger where they are already big. On the other hand, the *spreading* of the associated linear equation prevents the solution from becoming to big; provided that the initial data was small enough to start with. It is the balance from these two phenomena that is at the heart of small-amplitude scattering theory. Eventually, the *spreading* prevents the solution from blowing up in a finite time, and for large times the evolution is dominated by the linear part in the sense that the solution is asymptotic to a solution of the linearized equation . This is the physical content of Theorem 1.2. By the same argument, on the small amplitude limit the nonlinear effects become negligible and scattering is dominated by the linear term. This fact is expressed in a quantitave way by Theorem 1.3 that allows us to reconstruct the linear scattering operator from the derivative at zero of the nonlinear scattering operator. It is interesting to observe that the *spreading of the wave packets*, that is irrelevant on the high-energy limit in the linear case (Theorem 1.1), is actually essential on the low-energy (small amplitude) limit in the nonlinear case.

Acknowledgements

Fellow Sistema Nacional de Investigadores. Research partially supported by Proyecto PAPIIT-DGAPA IN105799.

References

[A97]   Arians, S., Geometric Approach to inverse scattering for the Schrödinger equation with magnetic and electric potentials. *J. Math. Phys.* 38:2731 - 2773, 1997.

[A1-98]   Arians, S., Geometric approach to inverse scattering for hydrogen like systems in a homogeneous magnetic field. *J. Math. Phys.* 39:1730 - 1743, 1998.

[A2-98]   Arians, S., Inverse Streutheorie für die Schrödingergleichung mit Magnetfeld. *Dissertation RWTH Aachen ; Logos-Verlag*, Berlin, 1998.

[B58]   Berezanskii, Yu. M.,The uniqueness theorem in the inverse problem of spectral analysis for the Schrödinger equation. *Trudy Moscow Mat. Obshch.* 7:3 - 62, 1958; English transl. in Amer. Math. Soc. Transl. Ser.,2, 35: 167 - 235, 1964.

[DT79]   Deift, P. and Trubowitz, E., Inverse scattering on the line. *Commun. Pure Appl. Math.*: XXXII:121 - 251, 1979.

[EW1-95]   Enss, V., Weder, R., Inverse potential scattering: A geometrical approach. In: *Mathematical Quantum Theory II: Schrödinger Operators, CRM Proc. Lecture Notes* vol 3, pp. 151 - 162, *American Mathematical Society*, Providence, 1995 (Proceedings Vancouver 1993).

[EW2-95]   Enss, V., Weder, R., Uniqueness and reconstruction formulae for inverse N-particle scattering . In: *Differential Equations and Mathematical Physics*, pp. 55-66, *International Press*, Boston, 1995 (Proceedings Birmingham AL 1994).

[EW3-95]   Enss, V., Weder, R., The geometrical approach to multidimensional inverse scattering. *J. Math. Phys.*: 36, 3902 - 3921, 1995.

[EW96]   Enss, V., Weder, R., Inverse two-cluster scattering. *Inverse Problems*:12, 409 - 418, 1996.

[F56]   Faddeev, L.D., Uniqueness of the solution of the inverse scattering problem. *Vestnik Leningrad Univ.*, 11: 126 - 130, 1956.

[I98]   Ito, H., T., An inverse scattering problem for the Dirac equation with time-dependent electromagnetic fields. *Pub. Res. Inst. Math. Sci.* 34:355 - 381, 1998.

[J96]   Jung, W., Der geometrische Ansatz zur inversen Streutheorie bei der Dirac-Gleichung, *Diplomarbeit RWTH Aachen*, 1996.

[J97]   Jung, W., Geometrical approach to inverse scattering for the Dirac equation. *J. Math. Phys.*38: 39 - 48, 1997.

[RS79]   Reed, M., Simon, B., Methods of Modern Mathematical Physics III. Scattering Theory, *Academic Press*, New York, 1979.

[S81]   Schechter, M., Operator Methods in Quantum Mechanics, *North Holland*, New York, 1981.

[W91]   Weder, R., Characterization of the scattering data in multidimensional inverse scattering theory. *Inverse Problems* 7:461 - 489, 1991.

[W96]   Weder, R., Multidimensional inverse scattering in an electric field. *J. Funct. Anal.* 139:441 - 465, 1996.

[W1-97]   Weder, R., Inverse scattering for N-body systems with time-dependent potentials. In:*Inverse Problems of Wave Propagation and Diffraction*, pp. 27 - 46, *Lecture Notes in Physics, Springer-Verlag*, Berlin, 1997 (Proceedings Aix les Bains 1996).

[W2-97]   Weder, R., Inverse scattering for the nonlinear Schrödinger equation. *Comm. Partial Differential Equations* 22:2089 - 2103, 1997.

[W1-00] Weder, R., $L^p - L^{p'}$ estimates for the Schrödinger equation on the line and inverse scattering for the nonlinear Schrödinger equation with a potential. *J. Funct. Anal.* 170:37 - 68, 2000.

[W2-00] Weder, R., Uniqueness of inverse scattering for the nonlinear Schrödinger equation and reconstruction of the potential and the nolinearity. In: *Proceedings of the Fifth International Conference on Mathematical and Numerical Aspects of Wave Propagation*, pp. 631 - 634, *SIAM Proceedings Series, Society for Industrial and Applied Mathematics*, Philadelphia, 2000 (Proceedings Santiago de Compostela 2000).

[W3-00] Weder, R., $L^p - L^{p'}$ estimates for the Schrödinger equation and inverse scattering. In:*Differential Equations and Mathematical Physics*, pp. 435 - 448, *American Mathematical Society and International Press*, Providence, 2000 (Proceedings Birmingham AL 1999).

[W1-01] Weder, R., Inverse scattering for the nonlinear Schrödinger equation. Reconstruction of the potential and the nonlinearity. *Preprint 1999*, to appear in *Mathematical Methods in the Applied Sciences*.

[W2-01] Weder, R., Inverse scattering for the nonlinear Schrödinger equation II. Reconstruction of the potential and the nolinearity in the multidimensional case. *Preprint 2000*, to appear in Proceedings of the American Mathematical Society.

[W3-01] Weder, R., Inverse Scattering on the line for the nonlinear Klein-Gordon equation with a potential. *J. Math. Anal. Appl.* 252:102 - 123, 2000.

[W4-01] Weder, R., Multidimensional inverse scatering for the nonlinear Klein-Gordon equation with a potential. *Preprint 2000*.

Address

RICARDO WEDER, Instituto de Investigaciones en Matemáticas Aplicadas y en Sistemas. Universidad Nacional Autónoma de México. Apartado Postal 20-726. México D.F. 01000. México.

E-MAIL: weder@servidor.unam.mx

2000 Mathematics Subject Classification. Primary 35R30, 35Q40, 35Q55 ; Secondary 35P25, 81U40.

Operator Theory:
Advances and Applications, Vol. 126
© 2001 Birkhäuser Verlag Basel/Switzerland

Cone Conormal Asymptotics

INGO WITT

Abstract. Asymptotic types for cone conormal asymptotics are constructed refining Schulze's notion of discrete asymptotic type. This extends previous joint work with Liu Xiaochun (Wuhan University) for the case of Fuchsian differential operators to general cone pseudodifferential operators.

1 The Main Result

Let X be a C^∞ manifold with boundary, ∂X. Further let $\mathcal{C}^0(X, \delta) = \mathcal{C}^0(X; (\delta, \delta, (-\infty, 0]))$ be Schulze's cone calculus on $X^\circ = X \setminus \partial X$ with respect to some fixed conormal order δ and asymptotic information carried on the weight strip $\{z \in \mathbb{C}; \Re z < \dim X/2 - \delta\}$ of infinite width. The typical operator $A \in \mathcal{C}^0(X, \delta)$ appears in (3.1) below. For more details, see [Sch98, Ch. 2]. Note that X becomes a manifold with conical point after ∂X is shrunk to a point.

Let $C^{\infty, \delta}_{\mathrm{as}}(X)$ be the space of all C^∞ functions $u(x)$ on X° admitting a conormal asymptotic expansion as $x \to \partial X$ of conormal order at least δ. The latter means that, in a collar neighbourhood $\mathcal{U}$ of ∂X, $\mathcal{U}$ is diffeomorphic to $[0, 1) \times Y$ via $x \mapsto (t, y)$,

$$u(x) \sim \sum_{j=1}^{N} \sum_{k=0}^{m_j - 1} t^{-p_j} \log^k t \, c_{jk}(y) \quad \text{as } t \to +0, \tag{1.1}$$

where $N \in \mathbb{N} \cup \{\infty\}$, $p_j \in \mathbb{C}$, $\Re p_j < \dim X/2 - \delta$, $\Re p_j \to -\infty$ as $j \to \infty$ when $N = \infty$, $m_j \in \mathbb{N}$, $m_j \geq 1$, and $c_{jk} \in C^\infty(Y)$. Asymptotics are understood in an increasing order of flatness. They are uniquely determined provided that they exist and $c_{jm_j-1} \neq 0$ holds for each j. The space $C^{\infty, \delta}_{\mathrm{as}}(X)$ only depends on δ, but not on the chosen splitting of coordinates $x \mapsto (t, y)$. Recall that

$$A\colon C^{\infty, \delta}_{\mathrm{as}}(X) \to C^{\infty, \delta}_{\mathrm{as}}(X)$$

for all $A \in \mathcal{C}^0(X, \delta)$.

With asymptotic expansions of the form (1.1), Schulze has assigned discrete asymptotic types, P, which are sequences $\{(p_j, m_j, L_j)\}_{j=1}^{N}$, where p_j, m_j are as above and $L_j \subset C^\infty(Y)$ are finite-dimensional linear subspaces. (In case $\dim X = 1$, the spaces $L_j = \mathbb{C}$ disappear from the discussion.) For $u \in C^{\infty, \delta}_{\mathrm{as}}(X)$ to have conormal asymptotic expansion of type P, it is then additionally required that the coefficients $c_{jk}(y)$ for $0 \leq k \leq m_j - 1$ belong to L_j. See [RS89], [Sch91].

Let $C^\infty_{\mathcal{O}}(X)$ be the space of all C^∞ functions on X° vanishing to the infinite order on ∂X. The subscript anticipates the empty asymptotic type, $\mathcal{O}$. In general, the space $C^\infty_{\mathcal{O}}(X)$ is not left invariant by operators belonging to $\mathcal{C}^0(X,\delta)$. This observation leads to the refined asymptotic types.

Let $\mathrm{Lat}(C^{\infty,\delta}_{\mathrm{as}}(X)/C^\infty_{\mathcal{O}}(X))$ be the complete lattice of all linear subspaces of the quotient space $C^{\infty,\delta}_{\mathrm{as}}(X)/C^\infty_{\mathcal{O}}(X)$. By a Borel summation argument, the space $C^{\infty,\delta}_{\mathrm{as}}(X)/C^\infty_{\mathcal{O}}(X)$ is the space of all *formal asymptotic expansions* of the form (1.1). Below we shall introduce the set $\underline{\mathrm{As}}^\delta(X)$ of all asymptotic types on X as a certain sublattice of $\mathrm{Lat}(C^{\infty,\delta}_{\mathrm{as}}(X)/C^\infty_{\mathcal{O}}(X))$.

One of the quantities we are interested in is

$$L_A = \big(A^{-1}(C^\infty_{\mathcal{O}}(X)) + C^\infty_{\mathcal{O}}(X)\big)/C^\infty_{\mathcal{O}}(X) \tag{1.2}$$

characterizing the amount of asymptotics annihilated by $A \in \mathcal{C}^0(X,\delta)$. Then it turns out that

$$\underline{\mathrm{As}}^\delta(X) = \big\{L_A;\ A \in \mathcal{C}^0(X,\delta) \text{ admits a distinguished parametrix}\big\},$$

see Definition 2.4. Note that operators A appearing as indices are elliptic. Furthermore, for all $P \in \underline{\mathrm{As}}^\delta(X)$, $A \in \mathcal{C}^0(X,\delta)$, the push-forward of P under A,

$$P^A = \big(A\pi^{-1}(P) + C^\infty_{\mathcal{O}}(X)\big)/C^\infty_{\mathcal{O}}(X), \tag{1.3}$$

where $\pi\colon C^{\infty,\delta}_{\mathrm{as}}(X) \to C^{\infty,\delta}_{\mathrm{as}}(X)/C^\infty_{\mathcal{O}}(X)$ is the canonical projection, is dominated by an asymptotic type, again. This property expresses a certain local finite-dimensionality involved in the definition of asymptotic type.

Now we can state our main result.

Theorem 1.1. *Let P_0, Q_0, P_1, $Q_1 \in \underline{As}^\delta(X)$, where we additionally assume that $P_0 \wedge Q_0 = P_1 \wedge Q_1 = \mathcal{O}$ ($=$ the greatest lower bound) if $\dim X = 1$. Then there exists an elliptic $A \in \mathcal{C}^0(X,\delta)$ admitting a distinguished parametrix, B, such that*

$$L_A = P_0,\ L_B = Q_0,\ L_{A^*} = Q_1,\ and\ L_{B^*} = P_1.$$

Here adjoints are taken with respect to the scalar product in $\mathcal{H}^{0,\delta}(X)$.

Here $\mathcal{H}^{0,\delta}(X)$ is the Hilbert space of all square integrable functions on X° with weight δ. Recall that if $A \in \mathcal{C}^0(X,\delta)$ is elliptic, then there exists a parametrix, B, to A, i.e., an operator $B \in \mathcal{C}^0(X,\delta)$ such that

$$AB - 1,\ BA - 1 \in C_G(X,\delta),$$

where $C_G(X,\delta) = C_G(X;(\delta,\delta,(-\infty,0]))$ is the class of Green operator. For details, see [Sch98, Ch. 2].

2 Asymptotic Types

There is a general concept for introducing asymptotic types if a unital algebra acting on some linear space, with its distiguished linear subspace, is given, like the algebra $\mathcal{C}^0(X,\delta)$, the space $C_{as}^{\infty,\delta}(X)$, and its subspace $C_{\mathcal{O}}^{\infty}(X)$. More details on that can be found in [Wit00].

2.1 Basic definitions

First we provide another representation of the quotient space $C_{as}^{\infty,\delta}(X)/C_{\mathcal{O}}^{\infty}(X)$, in fixed normal coordinates $x \mapsto (t,y)$ close to ∂X. See (2.1) below.

A discrete subset $V \subset \mathbb{C}$ is said to be a *carrier for asymptotics* if its intersection with all strips $\{z \in \mathbb{C};\ \alpha_0 < \Re z < \alpha_1\}$, where $\alpha_0, \alpha_1 \in \mathbb{R}$, $\alpha_0 < \alpha_1$, is finite. V is associated with the conormal order δ if $V \subset \{z \in \mathbb{C};\ \Re z < \dim X/2 - \delta\}$. The set of all these V is denoted by $\mathcal{C}^\delta$.

Let $[C^\infty(Y)]^\infty = \bigcup_{m\in\mathbb{N}}[C^\infty(Y)]^m$ denote the space of all finite sequences in $C^\infty(Y)$, where $[C^\infty(Y)]^m = \underbrace{C^\infty(Y) \times \ldots \times C^\infty(Y)}_{m \text{ times}}$ is the m-fold product and $[C^\infty(Y)]^m$ is identified with a subspace of $[C^\infty(Y)]^{m+1}$ via $(\phi_0, \phi_1, \ldots, \phi_{m-1}) \mapsto (0, \phi_0, \ldots, \phi_{m-1})$, i.e., by adding a leading zero. The *right shift operator* T on $[C^\infty(Y)]^\infty$ acts by the rule $(\phi_0, \ldots, \phi_{m-2}, \phi_{m-1}) \mapsto (0, \phi_0, \ldots, \phi_{m-2})$.

For $V \in \mathcal{C}^\delta$, we set $\mathcal{E}_V^\delta(Y) = \prod_{p\in V}[C^\infty(Y)]_p^\infty$, where $[C^\infty(Y)]_p^\infty$ is an isomorphic copy of $[C^\infty(Y)]^\infty$, and define $\mathcal{E}^\delta(Y)$ to be the space of all families $\Phi \in \mathcal{E}_V^\delta(Y)$ for some $V \in \mathcal{C}^\delta$ depending on Φ. Thereby, $\Phi \in \mathcal{E}_V^\delta(Y)$, $\Phi' \in \mathcal{E}_{V'}^\delta(Y)$ for possibly different $V, V' \in \mathcal{C}^\delta$ are identified if $\Phi(p) = \Phi'(p)$ for all $p \in V \cap V'$, while $\Phi(p) = 0$ for $p \in V\setminus V'$, $\Phi'(p) = 0$ for $p \in V'\setminus V$. The right shift operator T acts component-wise on $\mathcal{E}^\delta(Y)$, i.e., $T\Phi(p) = T(\Phi(p))$ for $\Phi \in \mathcal{E}_V^\delta(Y)$, $p \in V$.

Now we give a *non-canonical* isomorphism

$$C_{as}^{\infty,\delta}(X)/C_{\mathcal{O}}^{\infty}(X) \to \mathcal{E}^\delta(Y); \tag{2.1}$$

namely, with a vector $\Phi \in \mathcal{E}_V^\delta(Y)$, $\Phi(p) = (\phi_0^{(p)}, \ldots, \phi_{m_p-1}^{(p)})$ for $p \in V$, we associate the formal conormal asymptotic expansion

$$u(x) \sim \sum_{p\in V} \sum_{k+l=m_p-1} \frac{(-1)^k}{k!}\, t^{-p} \log^k t\, \phi_l^{(p)}(y) \quad \text{as } t \to +0. \tag{2.2}$$

For $\Phi \in \mathcal{E}^\delta(Y)$, we further define $\text{c-ord}(\Phi) = \dim X/2 - \max\{\Re p;\ \Phi(p) \neq 0\}$. Note that $\text{c-ord}(\Phi) > \delta$ if $\Phi \in \mathcal{E}^\delta(Y)$. Note also that, for $\Phi_i \in \mathcal{E}^\delta(Y)$, $\alpha_i \in \mathbb{C}$ for $i = 1, 2, \ldots$ satisfying $\text{c-ord}(\Phi_i) \to \infty$ as $i \to \infty$, the series $\sum_{i=1}^{\infty} \alpha_i\Phi_i$ is explained in $\mathcal{E}^\delta(Y)$ in a natural fashion.

2.2 Properties of asymptotic types

We are now going to consider linear subspaces $J \subset \mathcal{E}_V^\delta(Y)$, for some $V \in \mathcal{C}^\delta$, satisfying the following two conditions:

(a) $TJ \subseteq J$;

(b) $\dim J^{\delta+j} < \infty$ for all $j \in \mathbb{N}$.

Here $J^{\delta+j} = J / (J \cap \mathcal{E}^{\delta+j}(Y))$. Note that T induces a linear operator on $J^{\delta+j}$, since $TJ \subseteq J$ and $T(J \cap \mathcal{E}^{\delta+j}(Y)) \subseteq J \cap \mathcal{E}^{\delta+j}(Y)$. This operator, again denoted by T, is nilpotent, since $\dim J^{\delta+j} < \infty$. Let $\Pi_j \colon J \to J^{\delta+j}$ be the canonical projection.

A sequence $\{\Phi_i\}_{i=1}^e \subset \mathcal{E}_V^\delta(Y)$, where $e \in \mathbb{N} \cup \{\infty\}$, is called a *characteristic basis* of J if c-ord$(\Phi_i) \to \infty$ as $i \to \infty$ when $e = \infty$ and there are numbers $m_i \in (\mathbb{N} \setminus \{0\}) \cup \{\infty\}$ such that each $\Phi \in J$ can be written as

$$\Phi = \sum_{i=1}^e \sum_{k=0}^{m_i-1} \alpha_{ik} T^k \Phi_i \tag{2.3}$$

with certain *uniquely determined* $\alpha_{ik} \in \mathbb{C}$. The latter condition means that, if $m_{ij} \in \mathbb{N}$ are chosen so that c-ord$(T^{m_{ij}-1}\Phi_i) \leq \delta + j$, while c-ord$(T^{m_{ij}}\Phi_i) > \delta + j$ ($m_{ij} = 0$ if c-ord$(\Phi_i) > \delta + j$), then $m_{ij} \geq 1$ for only finitely many i and $\Pi_j\Phi_i, T\Pi_j\Phi_i, \ldots, T^{m_{ij}-1}\Pi_j\Phi_i$, where i runs through all these i, forms a Jordan basis for the nilpotent operator T on $J^{\delta+j}$. Note that $m_{i1} \leq m_{i2} \leq \ldots$ and $m_i = \sup_j m_{ij}$.

Note that a characteristic basis need not exist under the mere conditions (a), (b).

We introduce one further notion: $\Phi \in \mathcal{E}_V^\delta(Y)$ is called a *special vector* if there is a $p \in \mathbb{C}$, $\Re p < \dim X/2 - \delta$, such that $\Phi(p') = 0$ for all $p' \in V$, $p' \notin p - \mathbb{N}$. Obviously, if $\Phi \neq 0$, then p is uniquely determined by Φ, by the additional requirement that $\Phi(p) \neq 0$.

Definition 2.1. An *asymptotic type*, P, of conormal order δ is represented by a linear subspace $J \subset \mathcal{E}_V^\delta(Y)$ for some $V \in \mathcal{C}^\delta$ satisfying conditions (a), (b) above and admitting a characteristic basis consisting completely of special vectors. The set of all these asymptotic types is denoted by $\underline{\mathrm{As}}^\delta(X)$.

By representation we mean that P, which is a linear subspace of $C_{as}^{\infty,\delta}(X)/C_O^\infty(X)$, is mapped onto J under the identification (2.1). Recall that this mapping depends on the chosen splitting of coordinates $x \mapsto (t,y)$ close to ∂X.

Proposition 2.2. *The notion of asymptotic type as introduced above is invariant under coordinate changes.*

We shall say that a function $u \in C_{as}^{\infty,\delta}(X)$ possesses asymptotics of type P if (2.2) holds with some $\Phi \in J$, in the fixed splitting of coordinates $x \mapsto (t\ y)$ close to ∂X. The space of all these u is denoted by $C_P^\infty(X)$.

The set $\underline{As}^\delta(X)$ of asymptotic types is partially ordered by inclusion. We shall write, however, $P \preccurlyeq P'$ instead of $P \subseteq P'$. Likewise, $\vee$ and $\wedge$ for the least upper and the greatest lower bound, respectively.

Proposition 2.3. *The set $\underline{As}^\delta(X)$ is a lattice containing the empty asymptotic type, $\mathcal{O}$, and possessing the property that every non-empty subset, $\mathcal{S}$, admits a greatest lower bound, $\bigwedge \mathcal{S}$.*

Note that the empty asymptotic type, $\mathcal{O}$, is represented by the trivial subspace $\{0\} \subset \mathcal{E}^\delta(Y)$. The second part of the proposition implies that every bounded subset of $\underline{As}^\delta(X)$ admits a least upper bound. This, in fact, constitutes one of the fundamental principles in constructing asymptotic types with certain prescribed properties.

2.3   Distinguished parametrices

Generalizing (1.3), for $Q \in \mathrm{Lat}(C_{as}^{\infty,\delta}(X)/C_{\mathcal{O}}^\infty(X))$, $A \in \mathcal{C}^0(X,\delta)$, we define the push-forward of Q under A by $Q^A = \left(A\pi^{-1}(Q) + C_{\mathcal{O}}^\infty(X)\right)/C_{\mathcal{O}}^\infty(X)$. Accordingly, the spaces $C_Q^\infty(X)$, $C_{Q^A}^\infty(X)$ are introduced, and $A: C_Q^\infty(X) \to C_{Q^A}^\infty(X)$.

Definition 2.4. Let $A \in \mathcal{C}^0(X,\delta)$ be elliptic. A parametrix, B, to A is called distinguished if

$$AB - 1: C_{as}^{\infty,\delta}(X) \to C_{\mathcal{O}^A}^\infty(X), \qquad BA - 1: C_{as}^{\infty,\delta}(X) \to C_{\mathcal{O}^E}^\infty(X),$$
$$A^*B^* - 1: C_{as}^{\infty,\delta}(X) \to C_{\mathcal{O}^{A^*}}^\infty(X), \quad B^*A^* - 1: C_{as}^{\infty,\delta}(X) \to C_{\mathcal{O}^{E^*}}^\infty(X).$$

Note that every elliptic $A \in \mathcal{C}^0(X,\delta)$ can be written in the form $A = A_0 + G$, where $A_0 \in \mathcal{C}^0(X,\delta)$ admits a distinguished parametrix and $G \in \mathcal{C}_G(X,\delta)$. Furthermore, each Fuchsian differential operator that is elliptic with respect to the weight δ admits a distinguished parametrix.

Proposition 2.5. *Let A admit a distinguished parametrix, B. Then:*

(a) $P^A \in \underline{As}^\delta(X)$ *for any $P \in \underline{As}^\delta(X)$;*

(b) $L_A = \mathcal{O}^B$, $L_B = \mathcal{O}^A$, $L_{A^*} = \mathcal{O}^{B^*}$, *and* $L_{B^*} = \mathcal{O}^{A^*}$;

(c) *There is an order-preserving bijection*

$$\{P \in \underline{As}^\delta(X);\ P \succcurlyeq L_A\} \to \{Q \in \underline{As}^\delta(X);\ Q \succcurlyeq L_B\}, \quad P \to P^A$$

with its inverse given by $Q \mapsto Q^B$.

Proposition 2.5 (c) especially improves the fine control of the asymptotics of solutions to elliptic equations. See [LW00].

Remark 2.6. There is an ongoing program in constructing restricted calculi of cone pseudodifferential operators, with a better control on the way in which asymptotics are mapped. In particular, each elliptic operator in such a calculus admits a distinguished parametrix. See [Wit99a], [Liu00].

3   Sketch of Proof

We proceed as in [LW00]. First we assign to $A \in \mathcal{C}^0(X, \delta)$ its complete sequence $\{\sigma_c^{-j}(A)(z)\}_{j=0}^\infty \subset \mathcal{M}_{as}^0(Y)$ of conormal symbols. For the notation, see [Sch98]. Up to a remainder in the Green class $C_G(X, \delta)$, the sequence $\{\sigma_c^{-j}(A)(z)\}_{j=0}^\infty$ determines completely the manner in which asymptotics are mapped by A. The observation used in [LW00] was that, for Fuchsian differential operators, this holds even exactly, i.e., without any remainder term. The key observation now is that this continues to hold if A admits a distinguished parametrix.

Note that the behaviour of the complete conormal symbol under compositions is regulated by the Mellin translation product, i.e.,

$$\sigma_c^{-l}(AB)(z) = \sum_{j+k=l} \sigma_c^{-j}(A)(z-k)\sigma_c^{-k}(B)(z), \quad l = 0, 1, 2, \ldots$$

With respect to the Mellin translation product, the space of all infinite sequences in $\mathcal{M}_{as}^0(Y)$ becomes a unital algebra, where invertibility within this algebra is determined by the parameter-dependent ellipticity of the zeroth term of the sequence under consideration. Furthermore, the elements of this algebra act in a canonical way on the space $\mathcal{E}^\delta(Y)$.

Thus, having given asymptotic types $P_0, Q_0, P_1, Q_1 \in \underline{\mathrm{As}}^\delta(X)$ represented by $J_0, J_1, K_0, K_1 \subset \mathcal{E}_V^\delta(Y)$, respectively, for some $V \in \mathcal{C}^\delta$, and such that $P_0 \wedge Q_0 = P_1 \wedge Q_1 = \mathcal{O}$ if $\dim X = 1$, we first choose an invertible sequence $\mathfrak{s} = \{\mathfrak{s}_j(z)\}_{j=0}^\infty \subset \mathcal{M}_{as}^0(Y)$, with $\mathfrak{t} = \{\mathfrak{t}_k(z)\}_{k=0}^\infty \subset \mathcal{M}_{as}^0(Y)$ being its inverse under the Mellin translation product, such that $L_\mathfrak{s} = J_0$, $L_\mathfrak{t} = K_0$, $L_{\mathfrak{s}^*} = K_1$, and $L_{\mathfrak{t}^*} = J_1$, where $\mathfrak{s}^* = \{\mathfrak{s}_j(1 - j - 2\delta - \bar{z})^*\}_{j=0}^\infty$, $\mathfrak{t}^* = \{\mathfrak{t}_k(1 - j - 2\delta - \bar{z})^*\}_{k=0}^\infty$. Here $L_\mathfrak{s}$ etc. has the same meaning as L_A in (1.2). See [LW00]. The construction of the sequence $\mathfrak{s}$ relies on the factorization result of [Wit99b].

We decompose $\mathfrak{s}_j(z) = \mathfrak{s}_j^0(z) + \mathfrak{s}_j^1(z)$ with $\mathfrak{s}_j^0(z) \in \mathcal{M}_{\mathcal{O}}^0(Y)$ and $\mathfrak{s}_j^1(z) \in \mathcal{M}_{as}^{-\infty}(Y)$. Further we choose an $h(t, z) \in C^\infty(\overline{\mathbb{R}}_+; \mathcal{M}_{\mathcal{O}}^\mu(Y))$ such that $(1/j!)\, \partial^j h(0, z)/\partial t^j = \mathfrak{s}_j^0(z)$ for each j. Then we set

$$A = \omega(t)\mathrm{op}_M^{(n+1)/2-\delta}(h(t,z))\omega_0(t) + (1-\omega(t))A_\psi(1-\omega_1(t))$$

$$+ \sum_{j=0}^{\infty} \omega(c_j t)t^j \mathrm{op}_M^{(n+1)/2-\delta+\rho_j}(\mathfrak{s}_j^1(z))\omega_0(c_j t) + G. \tag{3.1}$$

Here $\dim X = n+1$, $\omega, \omega_0, \omega_1$ are cut-off functions supported on $[0,1)$, $\omega\omega_0 = \omega$, $\omega\omega_1 = \omega_1$, $0 \le \rho_j \le j$ for each j, where $\min\{\rho_j, j-\rho_j\} \to \infty$ as $j \to \infty$, $c_j \to \infty$ sufficiently fast such that the series $\sum_{j=0}^{\infty} \omega(c_j t)t^j \mathrm{op}_M^{(n+1)/2-\delta+\rho_j}(s_j^1(z))\omega_0(c_j t)$ converge in $C_{M+G}(X,(\delta,\delta,(-\infty,0]))$, and the pseudodifferential operator $A_\psi \in \Psi_{cl}^0(X^\circ)$ is chosen to make A an elliptic operator.

Then the crucial point is to choose the Green operator $G = \sum_{j=0}^{\infty} \omega(c_j t)G_j\omega_0(c_j t)$ in a way that the finite-rank operators $G_j \in C_G(X,\delta)$ compensate for additional asymptotics arising when $\rho_j > 0$ passes over poles of $\mathfrak{s}_j^1(z)$, but not producing new asymptotics, and the same for the adjoints. $\square$

Acknowledgements

I acknowledge discussion with Liu Xiaochun (Wuhan University) on this subject.

References

[Liu00]   X. Liu, A cone pseudo-differential calculus on the half-line with respect to conormal asymptotics of a given type. *Ph.D. thesis, Wuhan University, Wuhan,* 2000.

[LW00]   X. Liu and I. Witt, Asymptotic expansions for bounded solutions to semilinear Fuchsian equations. *Preprint 2001/01 Institute of Mathematics, University of Potsdam, Potsdam,* 2001.

[RS89]   S. Rempel and B.-W. Schulze, Asymptotics for Elliptic Mixed Boundary Problems. *Math. Research, Vol. 50, Akademie-Verlag,* Berlin, 1989.

[Sch91]   B.-W. Schulze, Pseudo-Differential Operators on Manifolds with Singularities. *Stud. Math. Appl., Vol. 24, North-Holland,* Amsterdam, 1991.

[Sch98]   B.-W. Schulze, Boundary Value Problems and Singular Pseudo-differential Operators. *Wiley Ser. Pure Appl. Math., J. Wiley,* Chichester, 1998.

[Wit99a]   I. Witt, Explicit algebras with the Leibniz-Mellin translation product. *Preprint 99/2, Institute of Mathematics, University of Potsdam, Potsdam,* 1999.

[Wit99b]   I. Witt, On the factorization of meromorphic Mellin symbols. To appear in: *Advances in Partial Differential Equations.*

[Wit00]   I. Witt, Asymptotic algebras. Submitted to: *Proceedings of the Workshop "Microlocal Analysis and Asymptotic Analysis of PDE," RIMS Kyoto,* Oct. 16–20, 2000.

Address

INGO WITT, University of Potsdam, Institute of Mathematics, PF 60 15 53,
D-14415 Potsdam, Germany

E-MAIL: ingo@math.uni-potsdam.de

2000 Mathematics Subject Classification. Primary 35S05; Secondary 35B40, 35J70

List of Participants

Agranovich, Mikhail
Moscow State Institute of Electronics and Mathematics (MGIEM)
Moscow 109028, Russia
msa.funcan@mtu-net.ru

Arai, Asao
Department of Mathematics, Hokkaido University
Sapporo 060-0810, Japan
arai@math.sci.hokudai.ac.jp

Baro, Michael
Institut für Mathematik, TU Clausthal
Erzstraße 1, 38678 Clausthal-Zellerfeld, Germany
mamib@math.tu-clausthal.de

Behnke, Henning
Institut für Mathematik, TU Clausthal
Erzstraße 1, 38678 Clausthal-Zellerfeld, Germany
mahb@math.tu-clausthal.de

Bellissard, Jean
Laboratoire de Physique Quantique, Université Paul Sabatier
118 Route de Narbonne, 31062 Toulouse Cedex, France
jeanbel@irsamc2.ups-tlse.fr

Ben-Artzi, Matania
Hebrew University, Givat Ram Campus
91904 Jerusalem, Israel
mbartzi@math.huji.ac.il

Berger, Günter
Fakultät für Mathematik/Informatik, Universität Leipzig
Augustusplatz 10/11, 04109 Leipzig, Germany
berger@mathematik.uni-leipzig.de

van den Berg, Michiel
School of Mathematics, University Walk
Bristol BS8 1TW, England
M.vandenBerg@bristol.ac.uk

Bolthausen, Erwin
Institute for Mathematics, University of Zürich
CH-8057 Zürich, Switzerland
eb@amath.unizh.ch

Brüning, Erwin
Department of Mathematics and Applied Mathematics, University of Durban
Private Bag X54001, Durban 4000, South Africa
ebruning@pixie.udw.ac.za

Bruneau, Vincent
Dep. de Math., Univ. Bordeaux I
351, Cours de la Libération, F-33405 Talence, France
vbruneau@math.u-bordeaux.fr

Buslaev, Vladimir
Department of Math. Physics, NIIF, St. Petersburg University
Ulyanovskaya 1, 198904 St. Petersburg-Petrodworetz, Russia
buslaev@mph.phys.spbu.ru

Chen, Hua
School of Mathematical Science, Wuhan University
Wuhan 430072, China
chenhua@whu.edu.cn

Chen, Shuxing
Institute of Mathematics, Fudan University
Shanghai 200433, China
sxchen@fudan.ac.cn

Coriasco, Sandro
Dipartimento di Matematica, Université di Torino
Via Carlo Alberto, 10, I - 10123 Torino, Italy
coriasco@dm.unito.it

Demuth, Michael
Institut für Mathematik, TU Clausthal
Erzstraße 1, 38678 Clausthal-Zellerfeld, Germany
mamd@math.tu-clausthal.de

Denk, Robert
NWF I - Mathematik, Universität Regensburg
93040 Regensburg, Germany
robert.denk@mathematik.uni-regensburg.de

Djawadi, Schahin
Institut für Mathematik, TU Clausthal
Erzstraße 1, 38678 Clausthal-Zellerfeld, Germany
insd@math.tu-clausthal.de

Dobrev, Vladimir
Institute of Nuclear Research and Nuclear Energy, Bulgarian Academy of Sciences
72 Tsarigradsko Chaussee, 1784 Sofia, Bulgaria
dobrev@inrne.bas.bg

Donig, Jörg
FB 11, Gerhard-Mercator Universität
Lotharstr. 65, 47048 Duisburg, Germany
donig@unidui.uni-duisburg.de

Dreher, Michael
Institute of Mathematics, University of Tsukuba
Tennodai 1-1-1, Tsukuba-shi, Ibasaki 305-8571, Japan
dreher@math.tsukuba.ac.jp

Duclos, Pierre
Université de Toulon et du Var et Centre de Physique Théorique de Marseille
Luminy-Case 907, F-13288 Marseille Cedex 9, France
duclos@cpt.univ-mrs.fr

Exner, Pavel
Department of Theoretical Physics, NPI, Academy of Sciences
CZ - 25068 RCE - Prague, Czechia
exner@ujf.cas.cz

Faierman, Melvin
Department of Mathematics, University of Witwatersrand
Johannesburg, WITS 2C50, South Africa
036mef@cosmos.wits.ac.za

Garello, Gianluca
Dipartimento di Matematica, Université di Torino
Via Carlo Alberto, 10, I - 10123 Torino, Italy
garello@dm.unito.it

Georgescu, Vladimir
Dep. de Mathématiques, Univ. de Cergy-Pontoise
2 avenue Adolphe Chauvin, 95302 Cergy-Pontoise Cedex, France
vlad@oara.pst.u-cergy.fr

Giere, Eckhard
Institut für Mathematik, TU Clausthal
Erzstraße 1, 38678 Clausthal-Zellerfeld, Germany
maeg@math.tu-clausthal.de

Gil, Juan
Department of Mathematics, Temple University
Philadelphia, PA 19122, USA
gil@math.temple.edu

Gramsch, Bernhard
FB Mathematik, Johannes Gutenberg-Universität
Saarstr. 21, 55099 Mainz, Germany
gramsch@mathematik.uni-mainz.de

Hagedorn, George
Department of Mathematics, Virginia TECH Blacksburg
VI 24061 Blacksburg, USA
hagedorn@math.vt.edu

Hempel, Rainer
Institut für Mathematik, TU Braunschweig,
Pockelsstr. 14, 38106 Braunschweig, Germany
r.hempel@tu-bs.de

Ichinose, Takashi
Department of Mathematics, Faculty of Science, Kanazawa University
Kanazawa 920 - 1192, Japan
ichinose@kappa.s.kanazawa-u.ac.jp

Iftimovici, Andrei
Dep. de Mathématiques, Univ. de Cergy-Pontoise
2 avenue Adolphe Chauvin, 95302 Cergy-Pontoise Cedex, France
andrei.iftimovici@math.u-cergy.fr

Ivanovic, Decan
Faculty of Mechanical Engineering, Unviversity of Montenegro
81000 Podgorica, Yugoslavia
decan@1.cis.cg.ac.yu

Kapanadze, David
A. Razmadze Mathematical Institute, Academy of Sciences of Georgia
1 M. Alexidze Str., Tbilisi 93, Georgia
daka@rmi.acnet.ge

Karp, Lavi
Department of Applied Mathematics, Ort Braude College
P.O. Box 78, 21982 Karmiel, Israel
karp@techunix.technion.ac.il

Klassert, Steffen
Fachbereich Mathematik, J.W.Goethe-Universität
Robert Mayer Str. 6, 60325 Frankfurt Main, Germany
klassert@math.uni-frankfurt.de

Klein, Markus
Institut für Mathematik, Universität Potsdam
Postfach 601553, 14415 Potsdam, Germany
mklein@math.uni-potsdam.de

Kochubei, Anatoly
Institute of Mathematics, National Academy of Sciences of Ukraine
Tereshchenkivska 3, Kiev 01601, Ukraine
ank@ank.kiev.ua

Korey, Michael
Institut für Mathematik, Universität Potsdam
Postfach 601553, 14415 Potsdam, Germany
mike@math.uni-potsdam.de

Korotyaev, Evgeni
Institut für Mathematik, Universität Potsdam
Postfach 601553, 14415 Potsdam, Germany
evgeni@math.uni-potsdam.de

Koshmanenko, Volodymyr
Institute of Mathematics
Tereshchenkivska 3, Kyiv-4, 252601 GSP, Ukraine
kosh@imath.kiev.ua

Kounchev, Ognyan
Institute of Mathematics, Bulgarian Academy of Sciences
Acad. G. Bonchev Str. 8, 1113 Sofia, Bulgaria
kounchev@cblink.net

Kozhevnikov, Alexander
Department of Mathematics, University of Haifa
Haifa 31905, Israel
kogevn@mathcs2.haifa.ac.il

Krainer, Thomas
Universität Potsdam, Institut für Mathematik
Postfach 601553, 14415 Potsdam, Germany
krainer@math.uni-potsdam.de

Kravchenko, Vladislav
Depto. de Telecomunicaciones, SEPI-ESIME Zac, Instituto Politécnico Nacional
Edificio 1,2-do piso, C.P. 07738 Mexico, Mexico
vkravche@maya.esimez.ipn.mx

Kurasov, Pavel
Department of Mathematics, Stockholm University
10691 Stockholm, Sweden
kuraspbs@rz.ruhr-uni-bochum.de

Kuzmin, Alexander
Institut for Mathematics and Mechanics, St. Petersburg University Petrodvorets
St. Petersburg 198504, Russia
Alexander.Kuzmin@pobox.spbu.ru

Lafleur, Paul
Instiute for Theor. Physik
Toernooivelt, NL-6525 ED, Nymegen, Netherlands

Laptev, Ari
Department of Mathematics, KTH
10044 Stockholm, Sweden
laptev@math.kth.se

Lenz, Daniel
FB Mathematik, J.W. Goethe-Universität
D-60054 Frankfurt/Main, Germany
dlenz@math.uni-frankfurt.de

Li, Yongsheng
Department of Mathematics, Huazhong University of Science and Technology
Wuhan 430074 Hubei, China
liys@public.wh.hb.cn

Lim, Swee Cheng
Department of Physics, Universiti Kebangsaan
43600 Bangi, Selandor, Malasia
sclim@pkrisc.cc.ukm.my

Loss, Michael
Department of Mathematics, Georgia Institute of Technology
Atlanta, GA 30332, USA
loss@math.gatech.edu

Lumer, Günther
Inst. de Math. et d'Informatique, Univ. de Mons-Hainaut
Avenue Maistriau 15, 7000 Mons, Belgium
dufour@umh.ac.be

Ma, Li
Department of Mathematical Sciences, Tsinghua University
Beijing 100084, China
lma@math.tsinghua.edu.cn

Mazýa, Vladimir
Department of Mathematics, Linköping University
S-58183 Linköping, Sweden
vlmaz@mai.liu.se

Möller, Manfred
Department of Mathematics, University of the Witwatersrand
Johannesburg , 2050 WITS, South Africa
036man@cosmos.wits.ac.za

Nazaikinskii, Vladimir
Institut for Problems in Mechanics, Russian Academy of Sciences
Vernadskogo 101-1, Moscow 117526, Russia
nazaikinskii@mtu-net.ru

Neubrander, Frank
Department of Mathematics, L.S.U.
Baton Rouge, LA 70803, USA
neubrand@bellsouth.net

Noll, André
AG Angewandte Analysis, Institut für Mathematik, TU Darmstadt
Schloßgartenstraße 7, 64289 Darmstadt, Germany
noll@mathematik.tu-darmstadt.de

Novikov, Roman
Dept. de Math., Univ. de Nantes
2, rue de la Houssinière, BP 92208, F-44322 Nantes Cedex 03, France
novikov@math.univ-nantes.fr

Okaji, Takashi
Department of Mathematics, Kyoto University
Kyoto 606-8502, Japan
okaji@kusm.kyoto-u.ac.jp

Oviedo, Hector
Ingenierós Civiles 22, Col. Jardines de Ch., Mexico D.F.Z.P. 09410, Mexico
oviemac@mail.internet.com.mx

Paneah, Boris
Department of Mathematics, Technion Haifa
32000 Haifa, Israel
peter@techunix.technion.ac.il

Popivanov, Petar
Institute of Mathematics, Bulgarian Academy of Sciences
Acad. G. Bonchev Str. 8, 1113 Sofia, Bulgaria
popivano@banmatpc.math.bas.bg

Posilicano, Andrea
Dipartimento di Sci., Uni. dell'Insubria
Via Lucini 3, I-22100 Como, Italia
posilicano@mat.unimi.it

Post, Olaf
Institut für Reine und Angewandte Mathematik, Rheinisch-Westfälische Technische Hochschule Aachen
Templergraben 55, 52062 Aachen, Germany
post@iram.rwth-aachen.de

Prellberg, Thomas
Theoretical Physics, TU-Clausthal
Arnold-Sommerfeld-Str. 6, 38678 Clausthal-Zellerfeld, Germany
thomas.prellberg@tu-clausthal.de

Purice, Radu
Institute of Mathematics of the Romanian Academy
P.O. Box 1-764, 70700 Bucharest, Romania
`radu.purice@imar.ro`

Qi, Minyou
Institut of Mathematics, Wuhan University
Wuhan 430072, China
`qimy@wuhan.cngb.com`

Rabinovich, Vladimir
Depto. de Telecomunicaciones
ESIME-Zacatenco del IPN, Edif. 1, 2-do piso, Av IPN S/N, C.P. 07738 Mexico, D.F., Mexico
`rabinov@maya.esimez.ipn.mx`

Raikov, Georgi
Institute of Mathematics, Bulgarian Academy of Sciences
Acad. G. Bonchev Str. 8, 1113 Sofia, Bulgaria
`gdraikov@omega.bg`

Ramm, Alexander
Mathematics Department, Kansas State University
Manhattan, KS 66506-2602, USA
`ramm@math.ksu.edu`

Rehberg, Joachim
Weierstraß-Institut für Angewandte Analysis und Stochastik
Mohrenstraße 39, 10117 Berlin, Germany
`rehberg@wias-berlin.de`

Remling, Christian
FB Mathematik, Universität Osnabrück
49069 Osnabrück, Germany
`cremling@mathematik.uni-osnabrueck.de`

Render, Hermann
Gerhard-Mercator Universität, FB 11
Lotharstr. 65, 47048 Duisburg, Germany
`render@math.uni-duisburg.de`

Renger, Walter
Institut für Mathematik, TU Clausthal
Erzstraße 1, 38678 Clausthal-Zellerfeld, Germany
`mawr@math.tu-clausthal.de`

Rodino, Luigi
Dipartimento di Mathematica, Universitá di Torino
Via Carlo Alberto 10, 10123 Torino, Italy
rodino@dm.unito.it

Rosenberger, Elke
Institut für Mathematik, Universität Potsdam
Postfach 601553, 14415 Potsdam, Germany
erosen@rz.uni-potsdam.de

Savin, Anton Yu.
Dept. of Comp. Math. and Cybernetic, Moscow State University
119899 Moscow, Russia
antonsavin@mtu-net.ru

Schmidt, Andreas
Fachbereich Mathematik, J.W. Goethe-Universität
Robert Mayer Str. 6, 60325 Frankfurt Main, Germany
aschmidt@math.uni-frankfurt.de

Schulze, Bert-Wolfgang
Institut für Mathematik, Universität Potsdam
Postfach 601553, 14415 Potsdam, Germany
schulze@math.uni-potsdam.de

Seiler, Jörg
Institut für Mathematik, Universität Potsdam
Postfach 601553, 14415 Potsdam, Germany
seiler@math.uni-potsdam.de

Seiringer, Robert
Institute for Math. Physics, Universität Wien
Pasteurgass 6/6, A-1090 Wien, Austria
rseiring@ap.univie.ac.at

Shaposhnikova, Tatyana
Department of Mathematics, Linköping University
S-58183 Linköping, Sweden
tasha@mai.liu.se

Sheftel, Zinovi
Backhausstraße 11, 37081 Göttingen, Germany

Shimada, Shin-ichi
Department of Mathematics and Physik, Setsunan University
Ikeda-Nakamachi 17-8, Neyagawa 572 - 8508, Japan
shimada@mpg.setsunan.ac.jp

Sobol, Zeev
Department of Mathematics, University of Bristol
Bristol BS8 1TW, United Kingdom
z.sobol@bris.ac.uk

Sternin, Boris
Dept. of Comput. Math. and Cybernetic, Moscow State University
119899 Moscow, Russia
sternine@.mtu.net.ru

Stollmann, Peter
Fakultät für Mathematik, TU Chemnitz
Reichenhainer Str. 41/711, 09107 Chemnitz
P.Stollmann@mathematik.tu-chemnitz.de

Szafraniec, Francizek H.
Instytut Matematyki, Uniwersytet Jagiellónski
ul. Reymonta 4, 30059 Kraków, Poland
fhszafra@im.uj.edu.pl

Tarkhanov, Nikolai
Institut für Mathematik, Universität Potsdam
Postfach 601553, 14415 Potsdam, Germany
tarkhan@math.uni-potsdam.de

Umeda, Tomio
Department of Mathematics, Himeji Institut of Technology
Himeji 671-2201, Japan
umeda@sci.himeji-tech.ac.jp

Us, Oleksiy
Department of Mathematics, University of Bristol
Bristol BS8 1TW, United Kingdom
a.us@bris.ac.uk

vanCasteren, Jan
Department of Mathematics and Computer Science, University of Antwerp (UIA)
Universitätspleint 1, 2610 Antwerpen, Belgium
vcaster@uia.ua.ac.be

Volevich, Leonid
Keldysh Inst. of Applied Mathematics, Russia Academy of Science
Miusskaya sqz 4, 125047 Moscow, Russia
volevich@spp.keldysh.ru

Weder, Ricardo Instituto de Investigaciones en Matemáticas Aplicadas y en
Sistemas, Universidad Nacional Autónoma de México
Apartado Postal 20-726, México D.F. 01000, México
weder@servidor.unam.mx

Weikard, Rudi
Department of Mathematics, University of Alabama at Birmingham
Birmingham, AL 35294-1170, USA
`rudi@math.uab.edu`

Witt, Ingo
Institut für Mathematik, Universität Potsdam
Postfach 601553, 14415 Potsdam, Germany
`ingo@math.uni-potsdam.de`

Yamada, Osanobu
Department of Mathematics, Ritsumeikan University
Noji, Kusatsu, Shiga 525-8577, Japan
`yamadaos@se.ritsumei.ac.jp`

List of Talks

Agranovich, Mikhail: "On some boundary value problems and spectral boundary value problems in Lipschitz domains"

Arai, Asao: "Some aspects of infinite dimensional Dirac operators on Boson-Fermion Fock spaces"

Bellissard, Jean: "Coherent and incoherent quantum transport"

Ben-Artzi, Matania: "Global properties of nonlinear parabolic equations"

van den Berg, Michiel: "Area versus capacity in the crushed ice problem"

Bolthausen, Erwin: "On the effective mass of the polaron"

Brüning, Erwin: "On non convex minimazation"

Bruneau, Vincent: "Semiclassical resolvent estimates and spectral asymptotics for trapping perturbations"

Buslaev, Volodya: "Semiclassical operators with double discontinuous symbols and their applications"

Chen, Hua: "Some new results on the nonlinear singular partial differential equations"

Chen, Shuxing: "Solution to supersonic flow past a pointed body"

Coriasco, Sandro: "Fourier integral operators defined by means of global weighted symbols and their applications"

Denk, Robert: "Weakly parameter-elliptic boundary value problems"

Dobrev, Vladimir: "Quantum group deformations of invariant differential equations"

Dreher, Michael: "Wedge Sobolev spaces and weakly hyperbolic equations"

Duclos, Pierre: "On the one-dimensional behaviour of the atoms in intense magnetic fields"

Exner, Pavel: "Magnetoresonances in quantum-waveguide resonators"

Faierman, Melvin: "Eigenvalue asymptotics for an elliptic boundary problem involving an indefinite weight."

Garello, Gianluca: "Pseudodifferential operators with non-regular symbols"

Georgescu, Vladimir: "C*-algebras of Hamiltonians on a Fock space and spectral analysis of quantum field models"

Giere, Eckhard: "Absence of the singular continuous spectrum for perturbations of the polyharmonic operator"

Gil, Juan: "On the heat trace for operators of Fuchs type with vanishing defect indices"

Gramsch, Bernhard: "Analytic Fredholm families and stochastic partial differential equations"

Hagedorn, George: "An exponentially accurate time-dependent Born-Oppenheimer approximation"

Hempel, Rainer: "Eigenvalues in gaps for magnetic Schrödinger operators"

Ichinose, Takashi: "On the norm convergence of the Trotter-Kato product formula"

Iftimovici, Andrei: "A nonperturbative method of computation of the essential spectrum and of proving the Mourre estimate"

Ivanovic, Decan: "Partial differential equation in generalized similarity form of unsteady boundary layer on porous contour"

Kapanadze, David: "Pseudo-differential crack theory"

Karp, Lavi: "A priori estimates for the Laplace operator in bounded and in unbounded domains"

Klein, Markus: "Spectral theory and Markov chains of disordered mean-field-models"

Kochubei, Anatoly: "Heat equation in the non-Archimedean infinite-dimensional analysis"

Korotyaev, Evgeni: "Inverse scattering, resonances and conformal mapping for Schrödinger operator"

Koshmanenko, Volodymyr: "To spectral theory of Schrödinger operator with fractal potential"

Kounchev, Ognyan: "Wavelet analysis through solutions of elliptic equations"

Kozhevnikov, Alexander: "Complete scale of isomorphisms for elliptic pseudodifferential boundary problems"

Krainer, Thomas: "Long-time asymptotics of solutions to time-dependent parabolic pseudodifferential operators"

Kravchenko, Vladislav: "On a class of quarternionic solutions of Maxwell's and Diracs equations"

Kurasov, Pavel: "Supersingular perturbations of self-adjoint operators in perturbation theory"

Kuzmin, Alexander: "Equations of mixed elliptic-hyperbolic type: Boundary value problems and applications"

Laptev, Ari: "Trace formula for matrix-valued Schrödinger operators"

Lenz, Daniel: "Hirachies and uniform spectral properties of one-dimensional quasicrystals"

Li, Yongsheng: "Global attractor for the derivative 2D Ginzburg-Landau equation"

Lim, Swee Cheng: "Some properties of multifractional Brownian motion"

Loss, Michael: "On the self energy of the electron in QED"

Lumer, Günther: "Asymptotic methods and generalized Laplace transforms applied to singular parabolic problems and dynamical systems"

Ma, Li: "Positive solutions of an elliptic PDE"

Mazýa, Vladimir: "The Schrödinger operator on the energy space: Boundedness and compactness criteria"

Möller, Manfred: "On the essential spectrum of operators in mathematical physics"

Nazaikinskii, Vladimir: "Surgery and the relative index in elliptic theory"

Neubrander, Frank: "Remarks on von Neumann stable and/or stabilizable approximation methods "

Noll, André: "R-boundedness and Mikhlin's theorem for operator-valued Fourier multipliers"

Novikov, Roman: "On determination of a gauge field on $\mathbf{R}^d$ from its non-abelian Radon transform along oriented straight lines"

Okaji, Takashi: "Propagation of wave packets and its applications"

Oviedo, Hector: "Some applications of quaternionic analysis to the solution of the Maxwell equations"

Paneah, Boris: "On a maximum principle and on sharp and coercive estimates in nonelliptic boundary problems"

Popivanov, Petar: "Gevrey and analytic properties of the solutions of several classes of partial differential equations"

Posilicano, Andrea: "Singular perturbations of self-adjoint operators"

Post, Olaf: "Periodic manifolds, spectral gaps and eigenvalues in gaps"

Purice, Radu: "Hardy type inequalities, Mourre estimate and a-priori decay for eigenfunctions"

Qi, Minyou: "A unified approach to the theory of fundamental solutions"

Rabinovich, Vladimir: "Finite section method for boundary value problems in unbounded domains"

Raikov, Georgi: "Asymptotic properties of the IDOS for a random magnetic Schrödinger operator"

Ramm, Alexander: "Property C for ODE and applications."

Rehberg, Joachim: "Estimates for differential operators including general boundary conditions"

Remling, Christian: "Schrödinger operators and deBranges spaces"

Render, Hermann: "An interpolation problem for cardinal polysplines"

Rodino, Luigi: "Partial differential operators with multiple symplectic characteristics"

Savin, Anton Yu.: "Homotopy invariants of boundary value problems"

Schulze, Bert-Wolfgang: "Ellipticity and parametrices on manifolds with boundary under global projection conditions"

Seiringer, Robert: "A rigorous derivation of the Gross-Pitaevskii energy functional for two-dimensional Bose gas"

Shaposhnikova, Tatyana: "Pointwise interpolation inequalities for derivatives and potentials"

Sheftel, Zinovi: "General elliptic problems with a shift for elliptic systems"

Shimada, Shin-ichi: "The phase shift formula for Aharonov-Bohm Hamiltonians"

Sobol, Zeev: "Positive semigroups generated by non-sectorial forms"

Sternin, Boris: "The eta invariant and elliptic operators in subspaces"

Stollmann, Peter: "Multi-scale analysis implies strong dynamical localization"

Szafraniec, Francizek H.: "Subnormality in the quantum harmonic oscillator"

Tarkhanov, Nicolai: "Pseudodifferential analysis on manifolds with cupsidal corners"

Umeda, Tomio: "Eigenfunction expansions associated with the relativistic Schrödinger operators"

Us, Oleksiy: "Strong uniqueness for Dirichlet operators"

vanCasteren, Jan: "Some problems in stochastic analysis and semigroup theory"

Volevich, Leonid: "Parameter-elliptic boundary value problems connected with the Newton polygon"

Weder, Ricardo: "The time dependent approach to inverse scattering theory"

Weikard, Rudi: "Commuting differential operators and integrable systems"

Witt, Ingo: "Asymptotic expansions for bounded solutions to semilinear Fuchsian equations"

Yamada, Osanobu: "Essential self-adjointness of Dirac operators with a variable mass term"

MIX
Papier aus verantwortungsvollen Quellen
Paper from responsible sources
FSC® C105338

If you have any concerns about our products,
you can contact us on
ProductSafety@springernature.com

In case Publisher is established outside the EU,
the EU authorized representative is:
**Springer Nature Customer Service Center GmbH
Europaplatz 3, 69115 Heidelberg, Germany**

Printed by Libri Plureos GmbH
in Hamburg, Germany